STUDY GUIDE FOR

EARTH REVEALED

INTRODUCTORY GEOLOGY

by

Ruth Y. Lebow

KENDALL/HUNT PUBLISHING COMPANY
4050 Westmark Drive   Dubuque, Iowa 52002

Earth Revealed is a televison-based course produced by
the Southern California Consortium

Copyright ©1992 by The Corporation for Community College Television and
The Corporation for Public Broadcasting

ISBN 0-8403-6368-0

Printed in the United States of America

10 9 8 7 6 5 4 3 2

CONTENTS

PREFACE

Earth, the third planet from the sun, is unique in that life evolved here and not on the other planets, and is still changing and evolving. Geology, the study of Earth, covers a time span of about 4.5 billion years, and is a fascinating story, still full of mystery and uncertainty but with scenes of intense drama and suspense. The visual richness and intellectual excitement of geology are intriguing for both life-long learners and undergraduate students. Few of us can remain impassive when viewing the fiery eruptions of volcanoes, or seeing the effects of destructive earthquakes. Understanding the causes of such geologic hazards and how to predict or mitigate their effects is important, no matter where we live.

The goals of this course are closely related to the new discoveries about how Earth works: about its hot churning interior that shifts great crustal plates, builds mountains, opens and closes ocean basins, and continually reshapes continental margins. The theory of plate tectonics has changed the way we regard our planet and ourselves. We are learning how life and Earth are interactive and interrelated. With this knowledge, we have come to realize that we must become the guardians of all life on the planet and of the life-supporting environment. The problems of exploitation of natural resources, the probable warming of the atmosphere, and the existence of the ozone hole over the South Pole must be addressed now, because waiting may prove the effects to be irreversible.

Geology is a many-faceted subject, and this study guide has been created through the efforts of many experts in their own field. The idea and planning for this course originated with the Instructional Telecommunications Consortium. Sally Beaty, of the Southern California Consortium, supplied the direction, organization, and constant encouragement in the preparation of the study guide. The academic advisors added their own expertise as they reviewed every lesson for scientific accuracy and instructional quality and offered thoughtful and invaluable comments. Special acknowledgement and thanks go to Richard Hazlett, Pomona College; Thomas Hartnett, Rancho Santiago College; J. Lawford Anderson, University of Southern California; Gary Houlette, Oklahoma City Community College; and Robert M. Norris, University of California, Santa Barbara. Consultation with Glenn Kammen, Senior Producer;

Robert Lattanzio, Producer; and others at the Consortium's production office were crucial in keeping the study guide coordinated with the video component.

A special thanks to David Lane and Evelyn Brzezinksi, Mary Lewis, and Kathryn Opsahl, at Interwest Applied Research. They provided the instructional design for the study guide, many hours of consultation, editing, encouragement, and general impetus to keep the project going. In addition, Hy Field early in the project, and then Hilda Moskowitz, both on behalf of the Annenberg/CPB Project, contributed their experience in creating successful telecourse projects and study guides.

This is truly the Golden Age of Geology, with its exciting new concepts and discoveries, and with opportunities to make a lasting effect on the quality of life. This is the best possible time to explore the wonders and beauty of Earth, our home planet.

– Ruth Lebow

MODULE I

INTRODUCTION

1

DOWN TO EARTH

This is the first of 26 lessons that will open your eyes – in the broadest sense of the phrase – to the geologic wonders of our beautiful Earth. As you know, each lesson primarily consists of three parts: readings from your textbook, a video program, and this study guide. In addition, you may have an assigned laboratory activity to go along with some of the lessons. While you've read many textbooks and seen many television programs, you may never have worked with a document such as the study guide. So before you read the content for this first chapter, we'll give you a little information about the purpose of each section that will appear in each of the 26 study guide chapters. Knowing how the guide is organized will help you throughout your "revealing" study of the planet Earth.

GOAL: This section presents the overall purpose of the lesson. It helps you keep in mind the overriding principle guiding the lesson.

LEARNING OBJECTIVES: The specific objectives that will be the instructional focus of this lesson are presented in this section. Learning these objectives will enable you learn the material and reach the Goal of the lesson.

INTRODUCING THE LESSON: This section is designed as an advance organizer – a motivator – for what the lesson covers. Organizers such as this can help make learning more meaningful, ensure better retention, and pique your interest.

LESSON ASSIGNMENT: Here, you will be given a concise, easy-to-locate place in each study guide chapter where the steps, assignments, and activities are listed that you should follow to master the Objectives and achieve the lesson's Goal. The word "step" is used to reinforce the idea that learning is a process.

BEFORE YOU BEGIN LESSON 1

TEXT ASSIGNMENT: This section lists the specific parts of the text which need to be read, reviewed, and studied. This section expands Step 1 in the Lesson Assignment by emphasizing the role of specific parts of the text in the course.

KEY TERMS AND CONCEPTS: The terms and concepts that need to be mastered will be noted here. If terms for a particular lesson are adequately defined in the text, you will simply be referred to the "Terms to Remember" section at the end of each textbook chapter. Definitions, then, will be given in the study guide for terms that are not defined in the text or need further elaboration and explanation.

VIEWING GUIDE: The Viewing Guide presents questions to be answered both before and after viewing the video portion of the lesson. This section is designed to guide your viewing, to help you "see what you should see" in the video.

PUTTING IT ALL TOGETHER: This section integrates the text and video, focusing learning in relation to the Lesson Objectives. Through summaries, case studies, questions, you are provided with information in different formats to help solidify your learning. Activities are provided for you to conduct on your own. They might be used as part of the course requirements given by your instructor or they can be done for personal interest.

SELF-TEST: Multiple choice items are presented to help you review material and prepare for the typical examinations used in distance education. Answers to the multiple choice questions will be provided in the back of the study guide.

IN CONCLUSION

Keeping in mind the purpose of each study guide section will support your learning as you proceed through the 26 lessons. And along with the videos and your text readings, using this study guide effectively will reinforce the themes for this course:

- An appreciation of the unifying model of plate tectonics.

- An understanding of the scientific process as it applies to geology.

- A recognition of Earth as a dynamic, constantly changing planet.

- An appreciation of the relationship and interaction of humans with the changing physical environment.

With that as an introduction, let's begin to explore the geology of our planet.

GOAL

This lesson will help you appreciate the wonders of Earth.

After reading the textbook assignments, completing the exercises in this study guide, and viewing the lesson's video portion, you will be able to:

1. Describe the field of geology.

2. Summarize the tasks and responsibilities of the geologist.

3. Describe the hazards and resources of Earth.

4. Give examples of the uniqueness of Earth.

5. Discuss why the resources of Earth are essential for life.

6. Discuss why the intelligent preservation of our environment depends on the knowledge and application of geology.

LEARNING OBJECTIVES

INTRODUCING THE LESSON

What a glorious sight from space, this blue planet! Photographs of Earth taken by astronauts on the moon fill us with awe. This glimpse from space, however, gives only the faintest clue to the dynamic and evolving planet beneath the swirling cloud cover. The infinitely varied landscape and the powerful geologic forces at work must be experienced and understood to truly appreciate the wonders of our planet home.

Geology is the study of Earth. The fact that Earth is where we live is reason enough to learn about it. In addition, this is the best of all possible times to study geology, in the midst of a revolution in the earth sciences. *Plate tectonics*, the framework of modern geology, is the new concept that in a few decades has radically altered our view of *terra firma* (solid Earth). No longer can we cling to the comforting idea of an unchanging Earth with ancient ocean basins and stable continents, but we must accept the fact that we live on a dynamic planet with a churning interior and a surface in constant and sometimes violent motion. We learned that Earth's *crust* and upper *mantle* consist of a series of rigid plates slipping about on a partially molten layer in the mantle, building mountains, opening and closing ocean basins, and shifting continents. With this knowledge came a better understanding of the causes of earthquakes, volcanoes, landslides, and other "natural disasters." This theory – literally Earth-shaking – is as important as any other great scientific idea about this planet and its forms of life.

From the many voyages of space craft and from our growing knowledge of other planets, we have learned that Earth is unique in many ways. Over 70 percent of Earth's surface is under water – a rare commodity in the solar family. Earth seems to be just the right distance from the sun to provide temperatures that allow water to exist as a solid, a liquid, and a gas simultaneously on Earth's surface. The inner planets are too hot and the outer planets are too cold for water to exist in a liquid state. Earth is also of sufficient size to generate enough gravity to hold an atmosphere and retain the precious light-weight molecules of water vapor and other gases essential to living things. Life, however, is the special property of Earth that is lacking on other planets in the solar system. As life forms changed and evolved over the eons of geologic time, so did the physical environment, partly because of geologic processes and partly because of life itself.

Geology, the study of how Earth works, is a challenging and exciting science. The geologist's workplace is the landscape, and geophysical data has expanded this workplace to include Earth's whole interior. Geologists do far more than collect fossils and analyze rocks. One of the major responsibilities of geologists is to search for and recover natural resources. When human populations were small, natural resources seemed limitless. In recent years, however, it has become apparent that some resources are becoming scarce, or are non-renewable or inaccessible or, even worse, are polluting our waters and our atmosphere through improper use. Since every aspect of life, from the smallest amoeba to the most complex industrial society, depends on bounty from Earth, we look to geologists to supply us with necessities and teach us to live with our planet home without damaging it.

Prevention of loss of life from geologic hazards is another responsibility of the geologist. Studies of belts of active volcanoes and earthquake-prone zones, using past historical records and instrumentation at the cutting edge of science, have improved our ability to detect precursors of possibly dangerous eruptions and Earth movements. Warnings given by the United States Geological Survey before the massive explosion of Mount St. Helens saved many lives, certainly hundreds and perhaps thousands. Prediction of earthquakes, however, is a much more complex problem that may one day yield to intensive study, but isn't yet at the level of volcanic prediction.

One of the great contributions of geology to human thought is the concept of *geologic time*. Ancient philosophers who conceived of an Earth no more than ten thousand years old would be reluctant to accept a planet over 4.5 *billion* years old, even though the scientific data are incontrovertible. There must be vast spans of geologic time to explain the enormous changes on Earth, to account for the life of today, and to understand the populations of living organisms and their relationship to some processes that act so slowly as to be almost imperceptible.

As we study the long history and ever-changing landscape of Earth and life, we must not forget the scientific creativity of generations of geologists as they struggled and worked to unravel the mysteries of this beautiful, complex planet.

LESSON ASSIGNMENT

Completing the following seven steps will help you master the lesson objectives and achieve the goal for this lesson:

Step 1: Read the TEXT ASSIGNMENT: Chapter 1, "Introduction to Physical Geology."

Step 2: Study the KEY TERMS AND CONCEPTS as noted in the study guide.

Step 3: Watch the VIDEO portion using the VIEWING GUIDE in the study guide.

Step 4: Read the study guide's PUTTING IT ALL TOGETHER section, which will help you summarize and integrate all of the information in this lesson.

Step 5: Complete any assigned lab exercises.

Step 6: Complete any assigned ACTIVITIES found in the study guide.

Step 7: Review the material in this lesson and complete the SELF-TEST found in the study guide.

Step 8: Go back to the LEARNING OBJECTIVES and make sure you have learned each one.

1. Read Chapter 1, pages 3 to 23. Be sure to include the "Introduction" and "Summary" in your reading.

2. Note the beautiful photograph of Mount Robson facing page 3. Also, look at the photographs in the "Table of Contents," pages v-xii, to give you a sense of what you will study in this course.

3. Using the physiographic map on pages 8 and 9, find where you live. Learn to use this map as the various regions are discussed in each lesson.

4. Study all the photographs and diagrams carefully. This chapter will lay the groundwork for much of the course.

5. Review Box 1.2, "Plate Tectonics and the Scientific Method," pages 19-20. These concepts will help you understand how scientists arrive at their theories.

6. Now is a good time to learn the eras of geologic time, Table 1.1, page 22.

TEXT ASSIGNMENT

The Key Terms listed under "Terms to Remember" that follow every chapter in the text will aid your understanding of the lesson's material. Make sure you look up the definitions in the glossary, text chapter, or study guide. This is part of the basic vocabulary in geology, and becoming familiar with the terms will be useful in this and succeeding lessons.

Consider starting your own "Geologic Dictionary" in which you list the Key Terms and their definitions, adding to the list as you study each lesson. It may be advantageous to rearrange the words in "Terms to Remember" according to category rather than trying to learn them in alphabetical order or some other arbitrary order. For example: *crust, lithosphere, asthenosphere, mantle,* and *core* all refer to the internal structure of Earth. *Igneous rock, sedimentary rock,* and *metamorphic rock*

KEY TERMS AND CONCEPTS

should be studied together. *Plate tectonics* should include *continental drift, converging boundary, hypothesis, mid-oceanic ridge, scientific method, spreading center, subduction zone, tectonic forces, theory,* and *transform boundary. Convection* and *heat engine* might also be added to this category. It is much easier to remember terms that are related to each other or to a specific theme.

VIEWING GUIDE

Once you have read the text material for the lesson and studied the terms, you are ready to view the lesson's video portion. Review the questions that appear in this section to help you watch for important points and to prepare you for what you will see in the video.

Before viewing, answer the following questions:

1. Explain why water can exist on Earth as a liquid, solid, and gas at the same time.

2. What are the three major zones of the interior of Earth? Briefly describe each one.

3. "The present is the key to the past" is an important concept in geology. What does it mean?

After viewing, answer the following questions:

1. What does a geologist do? How is the role of the geologist changing?

2. In what ways is Earth unique in the solar system? What conditions exist on Earth that make it a suitable habitat for life?

3. Why is Earth regarded as a dynamic planet? Describe the two *heat engines.*

4. How have our perceptions of Earth changed? How are the applications of geology changing?

5. Briefly describe some of the modern techniques used in the exploration for petroleum.

6. Groundwater is one of our most important resources, but there are severe problems with our groundwater supply. What are some of these problems and what is being done in Orange County, California, for example, to preserve good quality groundwater?

7. What suggestions were made to improve soil conservation, and what government agency is involved in saving this vital resource?

8. What are some of the signs that tell geologists there is a strong probability of an imminent volcanic eruption?

9. According to Dr. Evelyn Roeloffs, what are the three goals of the Parkfield experiment?

10. In what ways will this course be useful to you?

You have read the text assignment and seen the video portion of the lesson. This section, made up of a SUMMARY, QUESTIONS, suggestions for ACTIVITIES, and a SELF-TEST, will help you pull the information together.

The field of geology is the study of Earth, from its earliest inception as part of a cloud of dust, gases, and particles surrounding an infant sun, to all of the processes and forces that led to the dynamic landscape we see today, about 4.5 billion years later. Earth is unique in the solar system in its ability to support life, from the microscopic simple cells of three billion years ago, to the diverse, complex fauna and flora that have inhabited the planet for the last half billion years. As life evolved, so did the planet. The arrangement of ocean and continents, mountains and valleys is specific to this period and was different in the past and will be different in the future. The changes are wrought by the interaction between the great *internal heat engine*, powered by heat from decay of radioactive minerals, and the *external heat engine*, which draws energy from the sun and controls the ocean currents, the atmosphere, the water cycle, and weather and climate. The great unifying theory of how Earth works is called *plate tectonics*, which provides the framework of this course as you explore *Earth Revealed*.

What Does a Geologist Do?

Look around you and consider every item you see. Is there anything that did not begin as an Earth resource? Even materials such as wood, paper, and all of our fruit, vegetable, and grain-based food grew in soil, a product of the weathering of rock. Petroleum, an organic product formed on the sea bed from the bodies of microscopic marine organisms, is a principal source of energy. Coal, another energy source, is derived from trees and other plants of ancient swamps and forests that captured and stored solar energy millions of years ago. Resources of all kinds, from those that provide energy to ores of metals and nonmetals, have been deposited and concentrated by geologic processes acting over the last 4.5 billion years. Huge volumes of our resources are being extracted, however, far faster than they can be replenished by natural processes.

Geologists are explorers often working under difficult circumstances to understand the landforms and processes that shape Earth. Some geologists are soil scientists; others explore mountains and specialize in geologic structures. Petroleum geologists constantly look for new oil fields to slake our great thirst for energy. Exploration geologists continue to search for new mineral deposits rich in copper, gold, chromium, iron, and other important commodities upon which our daily lives depend.

The prediction, preparation for, and mitigation of geologic hazards are tasks of geologists who work with volcanic eruptions, landslides, earthquakes, floods, even coastal erosion. Geochemists work with radioactive minerals to determine absolute ages of minerals and rocks; paleontologists study life of the past, both to date their enclosing rock formations and to study relationships between ancient extinct life forms and their living descendants. (The current theories that account for the rather sudden extinction of the dinosaurs has captivated the scientific community and general public alike. See Box 8.1, page 183). Geophysicists work far from their site of interest, Earth's interior, but through sophisticated instrumentation they can visualize the layers and structures that lie deep beneath our feet (Figure 1.9, page 14).

Look through the "Table of Contents" of your text. All the information about each topic listed has been gathered by geologists who are experts in each phase of geology, including astrogeology, the study of the planets. We look to geologists to help solve the problems of the *greenhouse effect* and the heating of the atmosphere. Geochemists are studying the *hole in the ozone layer*. Working with atmospheric scientists, they have discovered the chemicals that are disrupting Earth's protective *ozone shield*. And finally, geologists strive to unravel Earth's inner workings and figure out what combination of forces and factors actually drive this incredibly dynamic and restless planet. With exciting new theories and geologic activity spreading into many areas, this is truly the Golden Age of Geology.

What are Geologic Hazards?

An important part of the study of geology is an effort to understand and mitigate the effects of geologic hazards, the natural phenomena that threaten human lives and property. The first to come to mind are earthquakes – sudden violent shakings of Earth in response to movement of the great crustal plates (Figure 1.1, 1.2, and 1.3, pages 4 and 5). Californians will not soon forget the October 17, 1989, disaster that struck San Francisco, Oakland, and the Loma Prieta district. (See the photograph facing page 147). In fact, according to one set of statistics, more than 52,000 people died in earthquakes around the world in 1990 alone – nearly as many as the 57,500 who died through *all* of the 1980s. Planning for earthquakes has been an on-going project for decades in many earthquake-prone areas, and newer buildings constructed to meet quake-resistant standards generally seem to suffer less damage than older buildings. Earthquakes are not limited to California, but have occurred in many of the 50 states and are especially strong in Alaska (See Figure 7.10, page 155, and locate your home state). In addition, there are active earthquakes areas in other parts of the world such as in the countries surrounding the Pacific basin and in the East-West belt that extends from the Mediterranean Sea through China (Figure 7.20, page 163).

Erupting in 1980, Mount St. Helens was an unforgettable sight (Box 11.1, Figures 1 though 5, pages 240-243). Geologists had been studying the volcano for years and, based on past records and indications of an imminent eruption, issued a timely warning that saved hundreds, possibly thousands, of lives. Active volcanoes produce a series of tremors, ash and cinder eruptions, tilting summits, and bulging flanks that warn of an impending eruption. Earthquake prediction is much more difficult, however; and although the precursor events may be taking place, the exact time, place, and magnitude of the quake cannot, at this time, be accurately predicted.

Landslides and floods, while not as dramatic, perhaps, as earthquakes and volcanoes, have been responsible for more deaths and property damage – at least in the United States and Canada – than earthquakes or volcanoes (Figures 1.7 A and B, pages 12-13, and photograph facing page 285). Other geologic hazards that geologists investigate include wave erosion at coastlines and collapsing ground surfaces where the bed rock is soluble limestone or where mining has taken place.

As our cities grow and our population increases, many people are moving into or already living in geologically unstable areas. There is pressure on geologists to improve earthquake prediction as well as to identify areas where it is safe to build and live. Sources of soil and groundwater pollution must be identified as health hazards and the appropriate technology devised to clean them up. Modern society is becoming more and more dependent on the geologic sciences to provide safety from geologic hazards and to teach us how and where we can live to our best advantage.

What is it about Earth that Makes it Unique?

Earth is unique in that only on this single planet in the solar system has life developed. From what we know at present, life does not exist elsewhere in the Universe. Circling the sun at just the right distance, Earth has the proper temperatures for water to exist in all three states, solid (ice), liquid (water), and gas (water vapor). The fact that there is such an abundance of water on Earth and so little elsewhere in the solar system presents an interesting question. It is the liquid state that is necessary for life, since most life processes take place within the watery portions of cells. About 70 percent of Earth's surface is under water, a habitat that supports a rich and varied assortment of plants and animals. The source of the water is from within Earth, emitted from volcanoes and vents of the early, still hot, and steaming planet.

Earth is of a sufficient size (or mass) to generate enough gravity to hold water, water vapor, and other light active gases such as oxygen, carbon dioxide, and nitrogen in an atmospheric envelope. Gases such as ozone (a 3-atom form of oxygen) in the upper atmosphere form a *global shield* that prevents ultraviolet rays from the sun from destroying living cells on Earth's surface. Mars may have temperatures that life could endure but is so small that it lacks an adequate atmosphere. Venus, on the other hand, has such a thick atmosphere of carbon dioxide that its surface has never been seen but must be explored by instruments from satellites (Figures 2 and 3 in "Astrogeology" Box 10.2, page 232). The carbon dioxide is responsible for an intense greenhouse effect that has raised the temperature on Venus to a sizzling 465° C. The outer planets, such as Jupiter and Saturn, also have thick atmospheres mostly of hydrogen, helium, and toxic ammonia and methane, but are very cold because of their great distance from the sun.

The remarkable diversity of living things is a testament to the variety of physical conditions on Earth and to the ability of life to adapt to a broad range of environments, many of which are undergoing change. The role of the geologist is to question the nature of early Earth, why life could flourish here, and why Earth has changed so drastically during its 4.5 billion years. This is the story of Geology.

Questions

1. What is meant by Earth's internal heat engine?

A heat engine is a device that converts heat energy into mechanical energy or causes motion to take place (Figure 1.4, page 6). The internal heat of Earth results from radioactive decay of certain elements such as uranium. Hot material deep within Earth moves slowly upwards toward the surface, where it is cooled, then descends to be reheated and continue the cycle. This is called *convection*, a condition seen in the atmosphere as well as in a pot of heated water (Figure 1.8, page 13). Most substances, even such solids as rock, become less dense and will buoyantly rise when heated, while cooled material, becoming more dense, will sink under the force of gravity. These convection currents occur within Earth, moving very slowly but enough to spread sea floors, cause continental collisions, raise lofty mountains, and alter the face of Earth (Figure 1.10, page 15).

2. Is there also an external heat engine?

The external heat engine is essentially solar powered. Heat from the sun provides the energy that makes winds blow, circulates the waters of the seas, and causes evaporation to take place that moves water and water vapor through the hydrologic cycle (Figure 1.5, page 6). The moisture in the air causes rock to weather and streams to flow, seemingly quiet events that eventually wear away even the loftiest mountains created by the internal heat engine.

3. If one task of the geologist is to find mineral resources such as petroleum, how is the geologist also involved in protecting the environment?

After the Industrial Revolution, the need for energy and other resources led geologists to find and extract minerals without giving too much thought to possible adverse effects upon the environment. As populations grew, the use of petroleum and metal deposits increased at an unprecedented rate; they became harder to find, and it was obvious that they were non-renewable resources. In addition, acids from coal and metal mines affected water supplies; burning of fossil fuels caused increased emission of carbon dioxide, a *greenhouse* gas, that may lead to global atmospheric warming.

An understanding of these problems has come primarily from government agencies that employ geologists to assess the damage done to the environment from oil spills, toxic emissions, and disposal of toxic wastes. Public awareness has resulted in enactment of restrictions, some international, that may in time alleviate some of the problems (Box 1.1, page 10).

4. What is the significance of the new theory of plate tectonics?

Geologists have pondered some very basic questions about Earth, such as how do mountains rise, only to be worn down and then rise again to new, perhaps greater heights? Why are the rocks of the sea floor so different from those of the continents? In fact, why is there so much water on this planet and so little or none on the other planets? Why do volcanoes and earthquakes occur where they do? Why do they seem to be in belts and not spread evenly over Earth? How and when did life appear? Will we run out of oil and gas or other crucial resources? And how can we solve problems of depletion of resources and pollution of our water supplies and atmosphere? After centuries of neglect, can we renovate our depleted soils?

These are difficult problems, but we have gained considerable insight to how Earth works as a result of the new theory of plate tectonics. The theory is basically simple: it states that the crust and upper mantle of Earth are in constant motion, slipping on a partially molten layer called the *asthenosphere*. (See Figure 1.9, page 14.) The energy to move the massive plates comes from convection currents within Earth's mantle. It is at the boundaries between the plates that most of the geologic activity, especially earthquakes and volcanoes, is taking place. Mountains are thrust upward where plates collide, at what geologists call convergent boundaries. The Atlantic is getting wider as sea-floor spreading is taking place at a *divergent boundary* in the middle of the ocean. In California, the famous San Andreas fault is the *transform boundary* where the Pacific plate is grinding northwestward against the western edge of the North American plate. (See Figures 1.11, 1.12, 1.13 1.14, and 1.15, pages 16-18, to understand plates and plate boundaries.) Plate tectonics explains the great vertical as well as horizontal movements of Earth's crust and enables geologists to interpret events happening today as well as those of the distant past. (Read Box 1.2, "Plate Tectonics and the Scientific Method," page 19. The theory will be studied intensively in later lessons, but this introduction to the scientific method offers a clear explanation of why it is accepted in the scientific community after almost half a century of doubt, questioning, and controversy.)

Activities

1. This chapter presents an overview of much of the whole field of geology. There is a wealth of material to be learned, but be assured that all the concepts and vocabulary will be presented again. Nevertheless, several readings and thoughtful study of the photographs and diagrams will give you a head start on this fascinating topic. So take time to re-read the assignment for this lesson.

2. Look over the "List of Boxes" on page xiii. The boxes have very well-written articles that are current and relevant to the field of earth science. Pick one or two and read them for general interest and entertainment.

3. Start to think geologically. Look around your home and surroundings and think about geologic events or resources that contribute to your personal well-being. Think about trips you have taken away from home, possibly to the mountains, the beach, the desert, or perhaps to the Grand Canyon or Hawaii, and consider what questions you would like to ask a geologist. Travel is even more exciting when you have a geologic background.

4. In your text, look over the "Questions for Review" in your text and try to answer them. The "Questions for Thought" are just that and are also more difficult than the review questions. Later in the course, when you return to the questions in Chapter 1, you will find them very simple.

Self-Test

1. The energy released by earthquakes and the forces that move crustal plates are derived from

   a. the hydrologic cycle.
   b. uniformitarianism.
   c. the internal heat engine.
   d. erosion.

2. Earth is unique in that liquid water can exist in abundance on the surface. This is related primarily to the fact that

   a. the orbit is at the proper distance from the sun.
   b. other planets never had water in their history.
   c. all the other planets are too cold.
   d. the atmosphere of early Earth originally contained abundant water.

3. The ozone layer is important to life on Earth because

   a. plants use the gases for photosynthesis.
   b. animals use the gas for respiration.
   c. ozone is part of the hydrologic cycle.
   d. ozone shields living things from ultraviolet radiation from the sun.

4. The constant change and state of flux within Earth is believed to be caused by

   a. the hydrologic cycle and gravity working together.
   b. slow moving convection currents in the mantle.
   c. changing atmospheric pressures and temperatures on Earth.
   d. the moving plates of the surface.

5. The internal heat of Earth is considered a result of

   a. the influx of strong solar energy.
   b. friction between moving plates.
   c. heat from radioactive decay of certain elements.
   d. rapidly moving convection currents.

6. An example of a non-renewable energy resource is

   a. coal.
   b. wind power.
   c. hydrologic cycle.
   d. gravity.

7. Earth is able to hold the atmosphere, crucial to life, primarily because it

   a. is covered by oceans that supply water vapor.
   b. has a restricted temperature range.
   c. allows heat and light to enter through it.
   d. is of sufficient mass to generate enough gravity to hold gases.

8. The winds, ocean currents, and hydrologic cycle result from

   a. energy from internal convection currents.
   b. the size of the planet, as compared to the outer planets.
   c. solar energy and the external heat engine.
   d. heating by radioactive decay of such minerals as uranium.

9. The geologic events that cause the greatest loss of life and property are

   a. earthquakes.
   b. volcanoes.
   c. landslides and floods.
   d. wave erosion.

10. The theory of plate tectonics is considered a revolution in science because it

    a. proves the planet is very old.
    b. shows the planet is undergoing constant change.
    c. demonstrates the stability of mountains and ocean basins.
    d. allows geologists to predict earthquakes accurately.

11. One task of geologists is to study and predict natural disasters. But so far, which of the following have geologists been able to do?

    a. Issues warnings about the impending Loma Prieta earthquake in time for people to evacuate.
    b. Evacuate most of the village of Armero before the great mudflow.
    c. Eliminate most problems with landslides.
    d. Save many lives by predicting the eruption of Mount St. Helens.

12. Starting from the surface, the major divisions of Earth are, in order,

    a. crust, upper mantle, asthenosphere, mantle, outer core, inner core.
    b. crust, asthenosphere, lower mantle, outer core, inner core.
    c. asthenosphere, crust, upper mantle, outer core, inner core.
    d. upper mantle, crust, lower mantle, asthenosphere, outer core, inner core.

13. One of the great contributions to human understanding made by geology is the

    a. use of the scientific method.
    b. discovery of oil for energy.
    c. concept of the vastness of geologic time.
    d. importance of solar energy in the water cycle.

14. Did cave dwellers ever see a dinosaur?

    a. Yes; they lived in the same region as dinosaurs.
    b. No; the dinosaurs lived near the sea, and the humans lived inland.
    c. No; the dinosaurs were extinct before humans appeared.
    d. No; very early humans hunted the last dinosaurs to extinction before modern humans arrived.

15. The scientific method of solving problems starts with _____ and finally arrives at _____.

    a. a theory; a hypothesis
    b. data gathering; tentative solutions
    c. a hypothesis; data for confirmation
    d. data gathering; a theory

2

THE RESTLESS PLANET

This lesson will help you to understand the beginnings of our solar system and evolving Earth.

GOAL

After reading the textbook assignments, completing the exercises in this study guide, and viewing the lesson's video portion, you will be able to:

LEARNING OBJECTIVES

1. Summarize the various theories for the origin of our solar system.

2. Describe how Earth changed from a homogeneous body to a differentiated planet.

3. State the reasons for Earth's rising temperature as it formed.

4. Discuss how the production of free oxygen began.

5. Compare and contrast current theories on the origins of life.

6. Recognize that Earth continues to change because of internal and surficial processes.

7. Briefly describe the internal structure of Earth and how Earth's internal heat engine works.

8. Relate the beginnings of the theory of plate tectonics to the scientific method.

Imagine space without the sun or Earth. To better understand the workings of this unique planet, we must go back to our beginnings – to events only dimly seen, hidden by vast reaches of time. In this lesson, we will probe the origin of the solar system and the planets, the atmosphere of Earth, and Earth's internal and external processes. We will also examine some of the research into the origins of life.

The origin of the solar system is a problem for which there is no completely satisfactory answer. Scientists in the major disciplines –

INTRODUCING THE LESSON

astronomy, geology, chemistry, geophysics, biology, meteorology – have all contributed research and ideas as they continue to seek the most reasonable and scientifically acceptable theory for this intriguing puzzle.

Over the centuries, theories have been proposed, only to be discarded as new discoveries were made. In this lesson, you will learn about the current theories for the origin of the sun and its family of planets, moons, comets, asteroids, and meteorites. Today, most experts favor some variation on the *solar nebula theory*, which holds that the solar system was formed by the contraction of a vast rotating cloud of gas composed mostly of hydrogen, helium, and dust. The planets grew over millions of years by condensation of the gases and accretion of the particles to form planet-sized bodies.

At some time after accretion (defined in "Key Terms and Concepts"), Earth changed from a homogeneous composition to a layered body – consisting of a crust, a mantle, a non-solid outer core, and a solid inner core – in which the layers differed in chemical composition and physical properties. This process of *differentiation* was a major event in the history of the planet. It contributed to the formation of the atmosphere, the oceans, volcanoes, and even Earth's magnetic field. You will learn about this crucial change in the section presented later in this chapter called the *Iron Catastrophe*.

When did all this happen? From the evidence found on Earth alone, it is impossible to determine precisely the date of origin of the sun and planets. Continuous alteration by erosion and by igneous, metamorphic, and tectonic processes have obliterated rocks of the original crust. We must look to space for answers. Fortunately, we have actual samples of rock to help our quest for beginnings: *meteorites*, the extraterrestrial leftovers from the primitive solar system, and *moon rocks* brought back by our far-traveling astronauts. Using radiometric dating techniques on scores of meteorites, scientists have determined that the dates when the meteorites crystallized seem to cluster around 4.6 billion years ago. It was gratifying to find that the oldest moon rocks were also approximately 4.6 billion years old. This suggests that we can assign this same figure for the birth of the solar system and for the age of our planet, even though rocks of this great antiquity have never been found anywhere on Earth (Figures 8.17, 8.18, and 8.19, pages 186-187).

Another puzzle – how did life first evolve on primitive Earth and get to the stage of blue-green algae, the earliest life form for which we have abundant fossil evidence? Throughout much of human history, the subject of the origin of life was a theological and mythological concern. It is still a provocative question to which definitive answers are not yet available. Of all the planets in the solar system, only Earth is well suited for life as we know it. Earth is sufficiently large to hold by gravity the light nitrogen, oxygen, water vapor, and other gas molecules of the atmosphere. In addition, Earth's temperatures are such that most of its free water is liquid, the form that is essential to life. It is unlikely, however, that we will ever possess clear fossil evidence of the origin and early evolution of life on Earth, since cells break down easily. And even chemical components that might tell of past events can quickly deteriorate beyond recognition.

Certain indirect evidence does suggest that life is at least as old as the oldest rocks known. Graphite, a mineral consisting of pure carbon, is present in the oldest sedimentary rocks and is thought to be a result of carbon concentrations left by organisms. We know that the banded iron oxide formations – some of the oldest rocks known – were discovered in southern Greenland and were deposited as water-laid sediments (sediments deposited in an ocean or lake) at least 3.8 billion years ago. These very ancient formations indicate that liquid water was present and that oxygen in the water most likely was produced by photosynthetic organisms. Most of the banded iron formations accumulated between about 2.5 and 1.8 billion years ago, but none is younger.

Just as life has been evolving during the eons of Earth history, so the planet itself has changed and is still evolving. In this and succeeding lessons, you will learn about the theory of *plate tectonics* which is a revolution in earth science – on a par with the "Heliocentric (sun-centered) Theory" of Copernicus, Einstein's "Theory of Relativity," and Darwin's "Origin of the Species."

The main concept of plate tectonics theory is rather simple. Earth's crust and upper mantle are broken into about a dozen large, relatively thin rigid plates and many micro-plates. These plates slip around in relation to each other on a partially molten layer in the upper mantle. Most of the more dramatic geologic events – such as earthquakes, volcanoes, and mountain building – take place at the boundaries between the moving plates. The theory accounts for continental growth and uplift, and it explains ocean basins' openings and closings. It tells why the oceans have not filled up after receiving sediments from erosion of the continents since geologic time began. Plate tectonics is the great unifying theory that accounts for these seemingly unrelated geologic phenomena.

LESSON ASSIGNMENT

Completing the following eight steps will help you master the lesson objectives and achieve the goal for this lesson:

Step 1: Read the TEXT ASSIGNMENT: Review Chapter 1, "Introduction to Physical Geology," parts of Chapter 8, "Time and Geology," and parts of Chapter 2, "Earth's Interior." (In this lesson, you will find the readings located in different parts of your text rather than in a single chapter.)

Step 2: Study the KEY TERMS AND CONCEPTS as noted in the study guide.

Step 3: Watch the VIDEO portion of the lesson using the VIEWING GUIDE in the study guide.

Step 4: Read the PUTTING IT ALL TOGETHER section in the study guide, which will help you summarize and integrate all of the information in this lesson.

Step 5: Complete any assigned lab exercises.

Step 6: Complete any assigned ACTIVITIES found in the study guide.

Step 7: Review the material for the lesson and complete the SELF-TEST found in the study guide.

Step 8: Go back to the LEARNING OBJECTIVES and make sure you have learned each one.

TEXT ASSIGNMENT

1. Re-read and review Chapter 1 in your text, noting concepts that refer to material in the study guide for Lesson 2. Review the learning objectives printed at the beginning of this study guide lesson to direct your reading for this lesson.

2. Read Chapter 1's "Summary."

3. In Chapter 8, read the section on "Absolute Age," pages 184-187. This will be expanded in a later lesson but is relevant to techniques for dating the birth of Earth.

4. In Chapter 2, read "The Earth's Internal Structure," pages 27-32, as an introduction to the subject and to prepare you for Lesson 3. Be sure to examine the diagram showing the internal structure of Earth.

KEY TERMS AND CONCEPTS

Key terms will aid your understanding of the material in this course. "Terms to Remember," at the end of Chapter 1, page 22, should be more familiar to you now and will be used again in future lessons. It is important to add them to your geologic vocabulary now. Use your text and the glossary to learn these terms. Here are some other key terms:

Accretion: The gradual process of growth in size of the planets due to adding or accumulation of particles.

Nebula: A dense cloud of gas or dust, sometimes defined as the remains of the explosion of a supernova star. From the Latin *nebula* for mist or cloud.

Photosynthesis: The process whereby green plants containing chlorophyll use carbon dioxide and water, in the presence of sunlight, to make carbohydrates for their own nourishment and give off oxygen as a by-product of the process. The great abundance of microscopic green plants in the sea is considered the principal source of free oxygen in the atmosphere.

Once you have read the text material for the lesson and studied the terms, you are ready to view the lesson's video portion. Review the questions that appear in this section to help you watch for important points and to prepare you for what you will see in the video.

Before viewing, answer the following questions:

1. What is the evidence for the age of Earth?

2. What is the origin of the oxygen in Earth's present atmosphere?

3. List the main layers of the interior of Earth. Indicate which are considered non-solid.

After viewing, answer the following questions:

1. What are the two most abundant gasses in stars and nebulae? What was the origin of the dust particles and heavier elements found in Earth and other planets according to current scientific opinion?

2. By what processes did the planets form from the clouds of gas and dust? What are some of the main differences between the Earth-like planets and the giant outer planets such as Jupiter and Saturn?

3. What events led to heating and differentiation of Earth?

4. What natural events observed today probably resulted from differentiation?

5. How do scientists account for Earth's internal heat?

6. What was the source of Earth's atmosphere? Give two theories of origin of the water on this planet. How did Earth's atmosphere eventually change from toxic to life-supporting?

You have read the text assignment and seen the video portion of the lesson. This section, made up of a SUMMARY, a CASE STUDY, QUESTIONS, suggestions for ACTIVITIES, and a SELF-TEST, will help you pull the information together.

PUTTING IT ALL TOGETHER

Summary

Since humans discovered Earth was part of a family of planets, they have wondered about these wandering bodies, gleaming in the night sky while they move against the background of fixed stars. It has only been in the last decade that our space-travelling vehicles have been able to send back phenomenal photographs of our sister planets with their enigmatic moons and glowing rings. Astrogeology took a giant leap forward. From this newly acquired knowledge, scientists have gained increased insight into the Solar System and our home planet as well. In this lesson, we will speak of the origins of the sun and the planets and how Earth evolved over the eons, slowly changing into a planetary body suitable for life. You will learn that Earth is still in a state of transition, responding to both the internal heat engine that is slowly stirring the interior of the planet and external processes that gather energy from the sun and shape the geologic patterns of the surface.

The Solar System

Although we usually call our star *the sun*, the Latin name is *Sol*, and Earth and other bodies that travel with Sol are Sol's family or the *solar system*. The solar system consists of one star – shining by its own light (because of nuclear fusion of hydrogen into helium) – and nine planets shining by reflected light from the sun. The solar system includes the four inner or *terrestrial* planets (Mercury, Venus, Earth, and Mars), which are small, dense, rocky, metallic, and warm. Our solar system also includes the cold outer or giant planets (Jupiter, Saturn, Uranus, and Neptune), and small, rocky Pluto, which orbits at the edges of the system. The large outer planets are believed to contain very small rocky inner cores and vast layers of frozen gases, such as hydrogen and helium, ammonia and methane. These planets have such a low density that if there were a bathtub large enough, Saturn would float! A belt of asteroids lies between Jupiter and Mars, and most of the planets are circled by moons.

Theories for the Origin of the Solar System

Scientists routinely train their telescopes on multitudes of large stars in various stages of development, but unfortunately these stars are too far away to allow us to observe how planets may be forming around them. Most experts favor some variation on the *solar nebula theory*, which states that the planets and the sun were formed at about the same time from a rotating cloud or nebula made up mostly of hydrogen and helium gases and a small percentage of dust. This is consistent with observations that stars generally form by gravitational collapse of dense clouds of gas comprised mostly of hydrogen which is the lightest and most abundant element in the universe.

In the case of our solar system, because of the presence of heavy elements in the planets, it is postulated that the sun might have been a remnant of a much older, larger star that imploded or collapsed and created the heavy elements in its dense seething interior. Another variation on this theme is that a giant star exploded in a nova and shed some of its heavy matter into the solar nebula. (The present sun is neither large enough nor hot enough to create in its core – by nuclear synthesis – the required heavy elements within the planets. The total amount of heavy elements in all the planets, however, amounts to only a small percentage of the mass of the whole solar system.) At some point, gravitational attraction became a major factor and caused contraction of the cloud of gas and dust. As the cloud collapsed inward, most of the matter coalesced into a central mass that would become the sun. The speed of rotation increased, and the remaining gases tended to flatten into a disk.

It is significant that all the planets revolve in the same direction in nearly circular orbits and lie in nearly the same plane close to the solar equator, evidence that the planets formed simultaneously from a rapidly rotating disk-shaped dust cloud. Further collapse caused the sun's internal temperature to rise to about 1,000,000° C at which point thermonuclear fusion started . . . and the sun began to shine.

The Planets: Condensation

The release of immense energy from the sun, plus the increased rate of rotation, flung some gases and particles outward toward the cooler reaches of the solar system. Nearest the sun where temperatures were highest, only the heavier elements with high melting points were able to condense or solidify. Mercury, for example, the planet closest to the sun, is the densest and richest in iron. Gases and small chunks of minerals and rock circled outward, settling into orbits dictated by their temperatures of condensation. The rock-forming elements of silicon, magnesium, iron, oxygen, and other related solid materials collected in the orbits of the terrestrial (Earth-like) rocky planets. The lighter gaseous materials – such as hydrogen, helium, water, methane, and ammonia – were too volatile to remain on the inner planets where the temperatures were high. But these gases continued to spread outward, where they would condense into ices in the cold distant regions of the solar system. The ices collected to become the frozen giant planets of Jupiter, Saturn, Uranus, and Neptune. Thus each planet differs from the other planets in composition, mostly because of its location within the nebula. The distance from the sun affected the temperature, which in turn determined which substances would condense and be accreted.

The Planets: Impact and Accretion

As the grains and chunks of rock or *planetesimals* circled the new sun, their gravitational attraction toward each other resulted in countless collisions of particles large and small. The larger bodies pulled other materials into their orbit and continued to grow by impact and accretion to become the planets. The smaller bodies became the moons. Uncollected bits and pieces are the meteorites we see as glowing "shooting stars" that are being drawn into the planets to this day. The heavily cratered landscapes of Mercury and Earth's moon are vivid evidence of impacts during the long period of accretion and growth. In fact, recent discoveries indicate that cratering has been the dominant geologic process in the solar system and may have been the dominant process on Earth during the first half-billion years of its history. However, impact craters are rare on Earth. Can you think of a good reason why this is so?

Earth: The Iron Catastrophe

As you have learned, the planets grew by random accretion of the cold solid planetesimals. Earth was an unsorted conglomeration of silicon compounds, metallic oxides, and smaller amounts of all the natural chemical elements. How did Earth change from this essentially homogeneous body to a *differentiated* planet – one in which the interior is divided into layers with distinct chemical and physical properties?

Although the original planetesimals were cold, various processes began to operate that would heat up the growing planet. *Intense bombardment* by the accreting planetesimals created heat upon impact. *Compression* played a part as the inner parts of the planet were squeezed under the growing weight of the accumulating outer portions. A third factor was *radioactive elements* within Earth that emitted atomic particles whose energy of motion was also transferred into heat. Radioactive elements were much more abundant in the early accreting Earth because they decayed into their stable daughter elements over the next 4.6 billion years.

The combined effects caused a rise in temperature – not enough to initiate nuclear reactions as in the sun – but to a point where rock became molten. There is doubt that the whole planet became a molten mass, but at least partial melting certainly occurred. As heating increased, differentiation occurred, separating the heavier from the lighter elements. Any heavy metallic iron and nickel scattered within the planet sank toward the center and formed Earth's core. The elements with the lower melting points and lowest density became molten and floated upward towards the surface to form Earth's mantle and crust. Some experts believe this developmental stage was reached a few hundred million years after Earth was formed. This is based on the discovery in very ancient rocks – about 3.8 billion years old – that early Earth had a magnetic field which is a property of a molten core.

This differentiation was the *Iron Catastrophe*, so called because it was a geologically rapid event that produced the layered internal structure known today. Most likely all the terrestrial planets have undergone some differentiation, but they followed different evolutionary paths. None of the planets has developed a dynamic tectonic system like Earth's. Our moon and Mercury, for example, evolved rapidly in the first one or two billion years, then became inactive and geologically dead – no mountain building, no volcanism, and no earthquakes. In addition, without water there was no erosion. As a result, the innumerable scars of meteorite impacts are still preserved on those bodies' ancient pitted surfaces.

Earth: History of the Atmosphere

Just as Earth evolved, so did its atmosphere. Recent studies of the giant outer planets have shown that their frozen atmospheres consist primarily of the noxious gases ammonia and methane plus hydrogen and helium – probably representing part of the original nebula. Some planetary scientists believe that as the sun formed, it went through a stage of emitting powerful turbulent winds of charged atomic particles that swept by the planets, literally blowing away the thin gases surrounding the small inner planets. This is based on the discovery that the heavier inert gases such as neon, argon, and xenon are less abundant on Earth than in the cosmos, an indication that Earth lost its early atmosphere to space. The large outer planets captured the light gases, or retained their original atmospheres. It is also possible that as Earth was heating because of impacts and compression, the first gases "boiled" off into space.

It was during a period of intense volcanic activity, coincidental with or shortly after differentiation, that the next stage in the formation of the atmosphere occurred – *degassing* or loss of gases from Earth's interior. From an analysis of gases emitted during current volcanism, we get a clue to the gases that entered the atmosphere at that time: water vapor, carbon dioxide, carbon monoxide, nitrogen, sulfur dioxide, hydrogen chloride, and small amounts of other substances. Initially, the emitted gases were so hot they stayed in the gaseous state. Only after Earth cooled to surface temperatures of less than 100° C (the boiling point of water) did the water vapor in the atmosphere condense, fall as rain, which remain on Earth as oceans, lakes, and streams. The nitrogen (which is inert or chemically non-reactive) accumulated through geologic time to become 78 percent of the present atmosphere. Other components such as carbon dioxide dissolved in the growing oceans or were used much later during biologic activities.

Free oxygen, however, was lacking in the atmosphere of the young planet. The principal source of free oxygen then, as now, is photosynthesis by green plants. Oxygen was probably being released by photosynthesis from primitive marine green algae perhaps as early as 3.8 billion years ago and certainly as early as 3.2 to 3.5 billion years ago. But it stayed in the ocean, and not much of it reached the atmosphere. Oxygen from biologic sources did not begin to accumulate in the atmosphere until the sea water was saturated with oxygen and the iron that was dissolved in the sea was *oxidized* (combined with the oxygen to form new insoluble compounds such as iron oxide). Evidence of the oxidation of iron in the sea is found in the colorful red and brown deposits of iron oxide called *banded iron formations* that are found on land in Canada, South Africa, and other places around the world. After about 1.8 billion years, the additional oxygen could escape from the sea and start to accumulate in the atmosphere. Today, oxygen makes up about 21 percent of the atmosphere, primarily a result of photosynthetic activities of marine plant plankton over millions of years.

Dynamic Earth

During the time of the Iron Catastrophe, convection currents within the partially molten planet transferred much of the internal heat to the surface. The heat dissipated "rapidly" (in geologic time, that is), and the planet cooled. Most of the mantle beneath the thin solid crust also solidified except for a layer within the upper mantle called the *asthe-*

nosphere – about 100 kilometers (60 miles) below the surface – which has remained at least partially molten. The *outer core* of Earth is apparently completely molten even after 4 billion years.

The state of matter of rock, whether solid or molten, depends in part on pressure and temperature. Under high enough pressure and even at extremely high temperatures, rock will remain solid. Within the asthenosphere, the temperature rose faster than the confining pressure, and the rock is partially molten. The lower mantle is hot but remains solid because of the increased pressure of the overlying layers. At the molten outer core, the temperature again is higher than the pressure needed to keep it solid. Strangely enough, the inner core – still even hotter – is solid because the pressures are so very high at the center of the planet.

Dynamic Earth: Internal and External Heat Engines

The *internal heat engine* is responsible for slow convection that churns the insides of the planet and causes the tectonic plates to slip around on the asthenosphere. The plate motions, as we have mentioned, have opened and closed ocean basins, created the continents and sometimes rifted them apart, built mountains, and set off earthquakes and volcanoes. But the surface too is evolving, as the *external heat engine* resulting from the great influx of solar energy moves the waters through the hydrologic cycle, causes the winds to blow, and makes the seas turbulent. This causes erosion, which carves the landscape into the familiar but striking landforms characteristic of our planet home.

CASE STUDY:
Some Ideas about
the Origin of Life

Life is a very special phenomenon. As one scientist remarked, we can observe it, dissect it, and analyze it. We can discuss the attributes of life and can even state with a good deal of authority how living systems work. Yet our comprehension of life remains incomplete until we understand something of its origins. In spite of many years of thought and research, the origin of living organisms remains a matter of great speculation.

One of the first scientists to seriously consider the problem of the origin of life was A. I. Oparin, a Russian biochemist. He speculated in the 1920s and 1930s that early life forms developed under favorable conditions from common non-living substances that existed in the early oceans of Earth. Oparin postulated an oxygen-free environment containing the elements needed to form *amino acids* – the building-blocks of living things. He suggested an atmosphere containing ammonia, methane, hydrogen, and water vapor. At that time, this was the composition thought to have existed on early Earth. The oxygen-free environment was required because oxygen would have destroyed the amino acids by oxidation. In 1953, Stanley Miller exposed this mixture of gases to an electrical spark representing lightning. This experiment yielded a number of amino acids as well as hydrogen cyanide (HCN) and formaldehyde (a simple carbohydrate H_2CO); his experiment suggested that complex organic compounds essential to life can be synthesized from inorganic substances.

This experiment was repeated in 1957 by Philip Abelson, who used mainly the carbon monoxide, carbon dioxide, and nitrogen thought to be closer to the composition of the early atmosphere. As a result of his experiment, he obtained most of the amino acids necessary for life. It was later suggested that sources of energy besides lightning could have been ultraviolet radiation, radioactive decay, shock waves from meteor impacts, and heat from molten rock. The organic compounds thus formed would have accumulated in estuaries, river bottoms, lakes, or oceans to form what has been described as a "thin soup." Higher concentrations could have been achieved by evaporation of water in tide pools or mud flats and by adsorption on clay minerals or other mineral surfaces.

This concept of chemical evolution of life was only tentatively accepted by scientists because they could make no observations in the natural world to substantiate it. But two recent discoveries seem to validate the concept. The first occurred in 1969, when a *carbonaceous* meteorite fell at Murchison, Australia. This meteorite was dated as 4.5 billion years old and contained amino acids and other organic compounds — some of which have never been found on Earth! Since meteorites had the same origin and are the same age as Earth, it seems likely that carbohydrates and amino acids probably existed on the early planet as well. Since that time, astronomers have detected other carbon compounds in the spaces between the stars.

The second discovery was the presence of a form of cyanide and of amino acids in a hot spring on the sea floor. The only natural environment on Earth today containing methane and ammonia not produced by decaying organic material is the system of hot water springs called *hydrothermal vents* associated with the mid-ocean mountains. These mountains are found in every ocean basin where the sea floor is pulling apart and new crust is generated.

At these sites, sea water enters the faulted and fractured sea floor, emerges at high temperatures, and carries with it many dissolved minerals from within the crust. All of the compounds required for the synthesis of organic molecules — including those essential to cell division and the inheritance of traits — seem to be available at these vents. A remarkable group of animals thrives today in this strange habitat, but can we be certain the hot springs existed 4 billion years ago? It seems likely, for tectonic movements were probably initiated by then, and we have strong evidence that ocean water was present.

In the last decade or two, stimulated by the space program and the search for life on other planets, the study of the origin of life has become an active science linking biology with geology and astronomy. While we do not yet know all the steps that lead to life, we seem to be well on our way. It is exciting to discover how the origins of Earth, of the atmosphere and the oceans, and of life itself are intertwined in these complex interacting systems.

Questions

1. What is the evidence for the age of Earth and the solar system?

Both meteorites and moon rocks have been dated using radioactive elements, and both give maximum ages of about 4.6 billion years. From our present knowledge, we believe that the planets, the moons, and the meteorites all formed at about the same time and that this figure is valid for the age of Earth. However, recent computations based on the structure of the sun and the rate of energy production seem to indicate that the age of the sun is about 6 billion years and that Earth underwent differentiation about 4.5 billion years ago. As we said, these events are only poorly understood, and it is quite a challenge to determine exact dates. (See Chapter 8: "Absolute Age," pages 184-187.)

2. Why haven't rocks of the original crust been found on Earth? Why isn't the surface of Earth cratered like the moon's or that of Mercury?

The original crust has been obliterated by melting from impacts and later differentiation. The first continents to form were small, scattered, and very different from the great land masses of today. After a water-bearing atmosphere formed, weathering and the agents of erosion such as streams, glaciers, waves, and wind altered rocks of the continents and have continued to do so to this day. The most ancient rocks ever found, however, have been on the continents, such as the 3.9 billion year old discovery at Isua, West Greenland, and a 4 billion year old discovery in the Northwest Territories of Canada.

As a result of tectonic movements, the early oceanic crust was subducted down deep sea trenches, where it was recycled in the intense heat of Earth's interior. The oldest rocks on the sea floor today are a mere 200 million years old. If we are to find the oldest formations, we must continue to seek them on the continents – a worthy goal since we are still missing over one-half billion years of Earth history!

3. What was the composition of the solar nebula? Why do astronomers believe an older star imploded (collapsed) or exploded nearby?

The original nebula is believed to have been about 99 percent hydrogen and helium gas and about 1 percent cosmic dust. Hydrogen is the lightest and most abundant gas in the universe. All heavier elements form from hydrogen by a series of reactions called nuclear synthesis which takes place under very high temperatures and pressures such as those found in the centers of the more massive stars. Heavy elements occur on Earth and on the other terrestrial planets. As the sun is neither large enough nor hot enough to create heavy elements in its core, it is hypothesized that a much larger exploding star may have injected some of these elements into the solar nebula. Possibly the heavy metals we so take for granted are relics of some great exploding nova whose brilliant display lit up the sky eons before our star was born. It is true that in terms of the total mass of the solar system, the heavy elements are a very small percentage. But as components of the inner terrestrial planets, they are of great importance – perhaps to life itself.

4. What is the evidence for the differentiation of Earth (the Iron Catastrophe)?

From astronomic observations it seems that the planets grew by haphazard accretion of cold particles within the rotating disk of the nebula. The interiors of the planets were more or less a homogeneous mixture of the elements that condensed in each part of the nebula and would not reveal the layering discovered through studies of Earth's seismic waves.

The evidence that differentiation took place is indirect and is based on studies of Earth's interior made by seismologists and geophysicists. Seismic (earthquake) waves from a large earthquake may pass entirely through Earth (see Chapter 2, noting the diagrams). From the behavior of these waves, scientists are able to plot three main zones of Earth's interior: the crust, which forms a relatively thin skin on the surface; the mantle, a thick shell of rock that lies below the crust and makes up most of the mass of Earth; and the *core* or central part of Earth. The core is probably metallic (iron and nickel), molten in the outer portion, and the source of Earth's magnetic field (Figure 1.9, page 14). In other words, Earth is differentiated into layers having different physical properties such as temperature, density, and solid or molten state. The different *chemical properties* reflect the differences in mineral composition of the layers. This information has been verified in the field, from seismic studies, and from laboratory work on the properties of minerals at various temperatures, pressures, and combinations of elements.

5. What caused sufficient heating of the accreting planet to allow differentiation to occur?

Some scientists think all of the terrestrial planets had a molten outer layer due to the early violent bombardment by accreting particles. On Earth, all evidence of that ancient molten rock (magma) crust – if it ever existed – has been destroyed. The interiors of those planets were also heated by the collisions, but there is some question as to whether this caused melting all through the planet. More likely, it was the radioactive elements trapped within the planets that caused internal heating. As these elements decayed, they gave off heat from the kinetic energy of the emitted atomic particles. The inner rocks became compressed through gravity; the compression also contributed to the increasing temperature.

As time passed, the temperatures rose to a point at which magma could form. Gravitational collapse occurred and moved the metallic iron and nickel toward the center while minerals with the lowest melting temperature and density rose upward to form the crust. The planet probably never became a totally molten body, but pockets of molten rock might have been scattered throughout. Controversy about this aspect of Earth history continues.

6. What is the evidence that the early atmosphere did not contain free oxygen?

The original solar nebula – judging by the gases identified in present nebulae and star masses – did not contain free oxygen. The next stage in the evolution of the atmosphere was the cycle of violent volcanic activity and degassing of Earth's interior. Volcanoes today emit water vapor, carbon dioxide, carbon monoxide, nitrogen, sulfur, and other gases but no free oxygen. If there had been any free oxygen in the first volcanic emissions, it would have combined quickly with carbon monoxide to form carbon dioxide, effectively removing the oxygen from the atmosphere. Uranium and iron minerals occur as oxides in rocks younger than about 2 billion years old, but both minerals are found on five continents in older buried non-marine and shallow marine deposits in their unoxidized state.

The banded iron formations mentioned previously are of great interest. Not only are they among the oldest known rocks on Earth, they also account for most of the iron ore that is mined in the world today. Their origin is problematical. They are marine and sometimes occur near volcanic pillow lavas erupted under the sea (see Figure 11.26). Since the iron is oxidized, it suggests that oxygen was present in the oceans but not in the atmosphere until about 2 billion years ago.

7. What is the importance of the experiments by Oparin and Miller?

These experiments were important because they showed that organic compounds – such as amino acids, the building blocks of life – could be formed from the inorganic compounds that comprised the primitive atmosphere of Earth. The compounds the scientists used were ammonia (NH_3), methane (CH_4), hydrogen (H_2), and water vapor (H_2O); all were present in the early atmosphere and contained all the elements needed for life. The energy in the experiments was supplied by an electric spark, substituting for lightning. Later evidence suggests ultraviolet radiation from the sun might have provided the necessary energy.

8. What is the evidence for the internal heat engine? Why is it important to geologists?

Evidence for internal heat comes from many sources. We have direct evidence from volcanic eruptions bringing up molten lava at very high temperatures from chambers within the lower crust or upper mantle. Hot springs from deep circulating water and the rare but dramatic geysers are further evidence of the hot interior. We can measure the increase in heat as we descend into deep mines or in the hot oil that we pump from deep wells. Ocean-going geologists have heat probes that they lower into the sediment to measure the heat flow through the sea floor. Indirect evidence comes from seismic waves that are generated by earthquakes and slowed down or delayed in their arrival when they pass through molten portions of Earth. The partially molten asthenosphere and the molten outer core were discovered from seismic data.

When referring to the *internal heat engine*, we are discussing the processes whereby the materials within Earth are moved by heat flowing from the hot interior of Earth toward the cooler exterior. The term *convection* refers to the upward movement of hot low-density material and downward movement of cool high-density material that causes a *convection current* (Figure 1.10, page 15). Geologists believe that very slow convection currents occur in Earth's solid mantle, as well as in the partially molten asthenosphere. These currents are believed to be the source of energy that activates tectonic movements of Earth's great crustal plates.

9. What powers the external heat engine? What are some of the visible effects of the external processes?

The *external heat engine* is essentially solar powered. Because Earth's axis is tilted, the solar radiation strikes Earth with different intensities and causes differential heating. The Equator, for example, receives the greatest amount of solar radiation. This radiation heats Earth's surface which in turn heats the atmosphere. The hot air rises, but in polar regions or high latitudes the air is cold and descends. This leads to variations in atmospheric pressure, and these variations cause the winds to blow. The winds move the oceans' surface currents which can affect daily and seasonal weather. The warmed water evaporates and rises to form clouds; when the water condenses, it falls as rain.

The movement of water and water vapor from sea to air to land and back to the sea are elements of the hydrologic cycle (Figure 1.5, page 6). From the first drop that fell on the barren surface of the young planet, water leads to weathering and disintegration of the rocky surface. Streams, winds, glaciers, and waves are the forces of erosion that carry away the rocky debris of weathering; all are powered by solar energy and gravity. Even the greatest mountains originally raised by Earth's internal forces will be worn away by erosional processes driven by the external heat engine (Figure 1.16, page 21). The beautiful and varied landforms, the turbulent seas, the clouds of the atmosphere, and even life itself are all visible manifestations of both the internal and external forces that sustain Earth.

Activities

1. If there is a planetarium near your home, now is the time to go. The study of the planets has attracted much attention since the various space vehicles sent back the beautiful but enigmatic pictures of the planets and their moons. At the planetarium, you will find lectures, slide shows, exhibits, demonstrations, and hands-on experiments that illustrate features not only of the planets but of the sun, the moon, and Earth as seen from space.

2. Some observatories in local colleges and universities will bring out their telescopes on certain nights for the public to see the rings of Saturn or the moons of Jupiter. You might also see the great Nebula of Orion where young stars are being born, or you might possibly view a wandering glowing comet. Look up an amateur astronomers club and attend club meetings or see if the members have public viewings with their telescopes. The recent discoveries in space are thrilling and well worth your time to explore.

3. If you have a freshwater pond nearby and can borrow a low-power microscope or a high-power hand lens, place some algae or weeds in a little flat dish with pond water, and observe the tiny plants and animals that live in this environment. You might be able to see some of the algae giving off oxygen bubbles as they photosynthesize in the light. With a high-power microscope, you might see some *diatoms* – microscopic green plants that build tiny glassy shells of silicon dioxide. These elegant little plants are the pasturage of the sea, and their vast numbers form the base of the marine food chain. All through time, they have been the principal producers of oxygen in sea water and in the atmosphere. Take a deep breath and thank the diatoms.

Self-Test

1. The picture in the video of the spinning skater was used to show

   a. the formation of the planets by accretion.
   b. the effects of contraction on the rotating nebula.
   c. how rotation increased the heating of the growing planet.
   d. how heavy metals were added to the solar nebula.

2. The most abundant gas in the original nebula was most likely

   a. vaporized heavy metal.
   b. the heaviest inert gas.
   c. the lightest metal oxide.
   d. the lightest known element.

3. According to currently accepted theories, the first internal structure of the inner planets consisted of a

   a. homogeneous mixture of rocks, minerals, and elements.
   b. liquid interior with a solid crust.
   c. solid interior covered by a sea of molten magma.
   d. layered interior with a molten inner core.

4. The concept of a large exploding star or nova near the solar nebula is based upon the evidence that

    a. the heat in the nebula was not sufficient to bring about the Iron Catastrophe.
    b. the nebula alone did not contain enough water to fill our oceans.
    c. our sun is not large enough to generate the heavy metals in the inner planets.
    d. the larger star was needed to provide free oxygen for our atmosphere.

5. The evidence of life on this planet

    a. appeared shortly before the appearance of the dinosaurs.
    b. occurs in some of the oldest rocks found on Earth.
    c. occurs in the original crust of solidified molten magma.
    d. appears no more than 200 million years ago.

6. The source of water on Earth is related to

    a. rainfall during accretion of particles.
    b. volcanic degassing of Earth's interior.
    c. water vapor in the original atmosphere.
    d. photosynthetic activities by early life forms.

7. Of the following events, which is the least likely to have led to differentiation of Earth?

    a. Bombardment and impacts by planetesimals
    b. Compaction by the weight of overlying layers
    c. Turbulent winds of atomic particles from the new sun
    d. Radioactive elements in the planet's interior

8. The evidence for the Iron Catastrophe is based on the discovery of

    a. water-laid sediments in very ancient rocks.
    b. free oxygen in the atmosphere.
    c. the banded iron formations.
    d. the layered structure of Earth's interior.

9. One reason Earth is considered suitable for life as we know it is

    a. the temperature permits liquid water to exist on the surface.
    b. the atmosphere today contains enough ammonia and methane for amino acids to form.
    c. free oxygen in the original atmosphere permitted animal life to evolve.
    d. Earth's crust has stabilized and is no longer undergoing violent changes.

10. The importance of the Oparin-Miller experiments is that they

    a. produced living plants and animals from naturally occurring compounds.
    b. produced organic compounds from inorganic materials.

   c. demonstrated that life could only have started on Earth as no other planets had the necessary raw materials.

   d. proved that organic compounds must have come to Earth from some extraterrestrial source.

11. The external heat engine is powered by

   a. the molten core.
   b. the hydrologic cycle.
   c. agents of erosion.
   d. solar energy.

12. The movement of hot material within Earth upward toward the cool surface is considered responsible for all of the following except

   a. convection currents within Earth.
   b. movement of the tectonic plates.
   c. modification of Earth's weather.
   d. mountain building.

13. Most of the mass of Earth is contained within the

   a. crust.
   b. mantle.
   c. asthenosphere.
   d. inner core.

14. The carbonaceous meteorite that fell in Australia in 1969 was a momentous event because it

   a. was the first meteorite found that was the same age as Earth.
   b. was the largest meteorite ever recovered.
   c. contained amino acids and other organic compounds.
   d. was the oldest object ever found that contained living bacteria and green algae.

15. The age of Earth – about 4.6 billion years – is determined from

   a. analysis of meteorites and moon rocks.
   b. dating of the original crustal rocks found in Greenland.
   c. dating of rocks of the ocean floor.
   d. study of the banded iron formations of Canada and elsewhere.

MODULE II

PLATE TECTONICS:
THE UNIFYING MODEL

3

EARTH'S INTERIOR

This lesson will help you appreciate what geophysicists have learned about the inside of Earth using indirect tools of study – seismic waves and the measurement of gravity, heat flow, and Earth magnetism.

GOAL

After reading the textbook assignments, completing the exercises in this study guide, and viewing the lesson's video portion, you will be able to:

LEARNING OBJECTIVES

1. Describe how seismic exploration gives evidence of Earth's interior.

2. Name and briefly describe the three main zones of Earth's interior.

3. Compare and contrast the features of the crust and mantle.

4. Discuss the character and significance of the lithosphere and the asthenosphere.

5. Indicate the probable composition and structure of the core and the evidence that supports this theory.

6. Describe how and why isostatic adjustment takes place.

7. Describe the nature of Earth's magnetic field and the magnetic poles.

8. Explain the significance of the geothermal gradient and the flow of Earth's heat.

INTRODUCING THE LESSON

Have you ever wondered what Earth really looks like on the inside? From Lesson 2 you learned that Earth is layered, that it is mostly solid but that some layers – such as the asthenosphere within the upper mantle and the outer core – are non-solid. But what color is it? What kind of rocks are down there? This is no easy problem to solve since the deepest wells ever drilled barely scratch the surface of the thin crust. Remember, where you are at this moment is about 6,370 kilometers (3,900 miles) from the center of Earth. But the deepest well ever drilled is only about 12 kilometers (7 miles)!

Scientists for many years pondered not only the nature of Earth's interior but how they could find ways to explore beneath the surface. They recognized that Earth's grumblings and shakings resulted from movements within the restless planet. In the last few decades, geologists and geophysicists learned to translate these Earth movements into detailed information about the inaccessible deep interior. *Seismology* – the study of earthquake waves – became the primary tool for such exploration. When a major earthquake strikes, Earth rings like a gong and reverberates with *seismic waves* that pass through it, race around the surface, and reflect or refract from rock layers buried deep under the crust. From studies of thousands of *seismograms*, the depth, density, state of matter (solid or molten), and the type of rock within each layer can be inferred.

There is very little *direct* evidence about Earth's interior, but some does exist. Scattered *inclusions* of rock called *xenoliths* are brought to the surface by erupting lavas. Some mantle rock is thrust up with oceanic crust in plate collision zones. *Kimberlite pipes* – the source of most of the world's diamonds – provide some clue in their composition to the abundance of dissolved gases in the upper mantle. (Look ahead to Lesson 12 on "Minerals" for the case study on diamonds and kimberlite.) A few deep wells have been drilled into the continental crust. But these are all rare and scanty occurrences.

Indirect evidence is provided by modern geophysical methods that use the physical laws of nature to reveal the internal structure of the planet. In addition to studies of seismic waves, analysis of Earth's magnetism discloses the waxing and waning of the magnetic field and, most astounding, the periodic reversal of the magnetic polarity. Magnetic analysis of sea floor rocks, for example, provides some of the most convincing evidence of sea floor spreading.

Other geophysical methods include measuring the variations in Earth's gravity from place to place and using the small differences to determine the kinds of rock or geologic structures that lie below the surface. Variations in the heat flowing from Earth's interior identify rising hot magma in spreading oceanic ridges. Variations in heat also help identify cold slabs of sea floor diving into the deep sea trenches – additional evidence that strongly supports the theory of plate tectonics.

In this lesson, you will discover what geophysical data reveal about Earth's crust, mantle, and core. You will learn about two important layers: the *lithosphere*, (stone sphere), the outer 100 kilometers (60 miles) of Earth that includes the solid crust and part of the upper mantle, and the *asthenosphere* (weak sphere) which is partially molten and lies below the lithosphere between 100 to 350 kilometers (60 to 210 miles) beneath the surface. The rocks in the asthenosphere seem to have the consistency of toffee or tar and are easily deformed. These two zones are important because in plate tectonic theory the rigid moving plates of the lithosphere are believed to be slipping on the plastic partially molten asthenosphere below.

This newfound knowledge directly affects human affairs because it is related to the cause and prediction of earthquakes and volcanoes, the possible effects of a magnetic polar reversal, and also to all geologic features influenced by Earth's internal heat engine. (This includes just about everything we see on our surface terrain!)

Completing the following eight steps will help you master the lesson objectives and achieve the goal for this lesson:

Step 1: Read the TEXT ASSIGNMENT, Chapter 2, "The Earth's Interior."

Step 2: Study the KEY TERMS AND CONCEPTS as noted in the study guide.

Step 3: Watch the VIDEO portion of the lesson using the VIEWING GUIDE in the study guide.

Step 4: Read the PUTTING IT ALL TOGETHER section in the study guide, which will help you summarize and integrate all the information in this lesson.

Step 5: Complete any assigned LAB EXERCISES.

Step 6: Complete any assigned ACTIVITIES found in the study guide.

Step 7: Review the material for the lesson and complete the SELF-TEST found in the study guide.

Step 8: Go back to the LEARNING OBJECTIVES and make sure you have learned each one.

1. Read Chapter 2, pages 25-43. Be sure to include the chapter's "Introduction" and "Summary" in your reading.

2. The diagrams in this chapter will take a little study, but from them you will gain much valuable information about *how* we learn about Earth's interior.

The key terms listed under "Terms to Remember," page 43 in the text, will aid your understanding of the lesson's material. Make sure you look up the definitions in the glossary or in the chapter. One other important term should be noted:

Xenolith: A *stranger stone* from the Greek words *xenos* meaning stranger and *lithos* meaning stone. When magma rises in a volcanic vent, bits of the surrounding rock may be torn loose and carried to the surface

LESSON ASSIGNMENT

TEXT ASSIGNMENT

KEY TERMS AND CONCEPTS

with the erupting lava. Some xenoliths – such as those found in certain Hawaiian volcanoes – are composed of peridotite and dunite and may be bits of mantle, lower crust, or walls of the deep magma chamber.

Dunite can be a handsome green rock composed of the mineral *olivine*, known in the jewelry trade as *peridot*. Garnets are also found in some peridotites. So we know that some of the minerals of Earth's interior are colorful – grassy green olivine or red garnet. And, of course, there are the extremely rare diamonds found in the kimberlite pipes. Peridotites generally are dark green or greenish black. Xenoliths are relatively rare, one of the very few samples of lower crust or mantle that provide direct information about Earth's interior. (See Box 2.2 "Mantle Xenoliths – A Peek at the Deep," page 31, including Figure 1 for dunite xenoliths in basalt.)

VIEWING GUIDE

Once you have read the text material for the lesson and studied the terms, you are ready to view the lesson's video portion. Review the questions that appear in this section to help you watch for important points and to prepare you for what you will see in the video.

Before viewing, answer the following questions:

1. What kind of *direct* evidence do we have about Earth's interior?

2. Name four *indirect* methods used to determine the nature of Earth's interior. Which is the most important?

After viewing, answer the following questions:

1. In the video, did you see how seismic waves react when encountering boundaries between layers within Earth? This is how seismologists build up a picture of the general structure throughout Earth's interior.

2. How do seismologists generate seismic waves artificially to obtain information about Earth's interior?

3. From seismic information, what is the basic structure of Earth's interior? What are the characteristics of the continental crust?

4. How does the ocean crust differ from the continental crust? Why are *ophiolites* of special interest to geologists?

5. From the sequences shown in animation, what do the special properties of the P and S waves reveal about Earth's core?

6. Explain the workings of a gravimeter. What can it detect within Earth?

7. Explain the causes of the beautiful Aurora Borealis. What conditions must be met on other planets to produce this phenomenon? Do you think any other planets have similar Auroras?

8. How do rocks become magnetized? What has rock magnetism revealed about Earth's magnetic field?

9. Apparently, Earth's magnetic field is declining in intensity. What might be the significance? If Earth temporarily lost its magnetic field, can you predict how it might affect human affairs and life on the planet?

You have read the text assignment and seen the video portion of the lesson. This section, made up of a SUMMARY, a CASE STUDY, QUESTIONS, suggestions for ACTIVITIES, and a SELF-TEST, will help you pull the information together.

With our dawning understanding of the varied and dynamic aspects of the surface of our planet came the need to know more about Earth's interior. The turbulence within the planet that is reflected in violent upheavals such as earthquakes and volcanoes, as well as in the slow growth of mountains and continents and the spreading of the sea floor, required exceptional methods of study because the interior was so inaccessible. In this lesson, you will learn how diligent research has revealed the hidden interior, layered like an onion, slowly moving, hot enough to melt rock, and still an enigma in spite of decades of investigation.

Seismic Exploration: A Key to Earth's Interior

Earthquake vibrations occur when Earth is suddenly jolted, like a bell when struck by a hammer. Volcanic explosions, sudden slipping of rock masses along a fault, thumping or pounding of the surface during exploration for petroleum, and even nuclear bomb blasts all set up vibrations that travel into Earth. These vibrations can be recorded by delicate instruments called *seismographs*. The vibrations are shown on *seismograms* which show the time of arrival of the first waves and the intensity. With some calculations, seismograms also indicate the distance to the epicenter and the depth to the point of focus where the first release of energy or movement took place. (Look up *epicenter* and *focus* in your glossary.)

Seismic Waves

Vibrations – called *seismic waves* – radiate in all directions from the point of focus with a spherical wave front. Remember – the energy of the wave is carried by solid rock which moves slightly as the wave passes through and then returns to its original shape. Rock is more or less elastic, which means it has a resistance to being deformed. Don't think rubber bands, but think steel ball bearings to understand rock elasticity.

Seismic waves are of two kinds: *body waves* that pass through Earth and *surface waves* which are limited to Earth's surface. The body waves are also of two kinds – *primary* and *secondary* (or P and S waves, respectively) – and it is these waves that provide most of the information about the deep interior of Earth. The primary waves are similar to sound waves and are a series of compressions and expansions. They are sometimes known as *push-pull* waves because they cause a pulsing, back-and-forth motion of the rock particles. Like sound, the P waves are transmitted by solids, liquids, and gases. P waves have the greatest velocity of all seismic waves, 6 kilometers (3.6 miles) per second through the continents and about 7 kilometers (4.2 miles) per second through oceanic crust. They are the first to be recorded by seismographs.

The S waves are slower, about 2 to 5 kilometers (1 to 3 miles) per second and the second to arrive. They are shear or shake waves that cause the rock to vibrate from side to side at right angles to the direction of wave movement. They can be transmitted by solids only – an important property to remember. The surface waves are the last to be detected by the seismograph because they must pass around Earth. (Seismic waves will be discussed further in Lesson 9 on Earthquakes.)

Seismic Waves and Earth's Interior

If Earth's composition were uniform, the velocities of P and S waves would increase proportionately with depth, because the higher pressure would cause an increase in density and in the rigidity and elasticity of the solid rock. But observed travel time differs greatly from predictions. Earth is layered, and the layers of rock have different properties. The waves are reflected back from some of the layers. And by knowing the total travel time, the depth to a boundary between layers can be calculated. The seismic waves also bend or refract just as a stick lowered into a fish tank seems to bend at the air-water interface. (See Figures 2.1, 2.2, 2.3, and 2.4 on pages 26 and 27. Visualize Figure 2.4 B in three dimensions, spreading out in all directions.)

The Crust: Continental, Oceanic, and the Moho

What has the analysis of seismic waves shown about the deep interior of our planet? We have learned that Earth's crust is not only thin (proportionately thinner than the skin on a grape) but it varies in thickness – being thinner under the oceans and thicker beneath the continents. Also, based on the velocity of P waves and on drilled or dredged samples of sea floor rock, the ocean crust is known to be primarily formed of basalt – a dark, dense, igneous rock. The P waves move more slowly through continental crust, indicating the rock is less dense and less rigid than that of the ocean basins. The velocity of the P waves approximates those traveling through granite, also an igneous rock but differing in texture and chemical composition from basalt. The continents, however, are older and geologically far more complex than the sea floor. They also include many different types of rock. Perhaps "granite-like" or "granitic" are better terms to describe their composition (Figure 2.6 and Table 2.1, page 29).

The boundary separating the base of the crust from the mantle is called the *Mohorovicic discontinuity* (*Moho* for short), where the seismic waves speed up to 8 kilometers per second (Figure 2.7, page 30). From a world-wide analysis of the depth of the Moho, it was determined that the continental crust has an average thickness of about 40 kilometers (24 miles) while the oceanic crust is only about 7 kilometers (4 miles) thick.

The Mantle

The mantle is the thickest layer and contains most of the mass of Earth. Within the upper mantle – at a depth of about 100 kilometers (60 miles) – lies the low velocity zone or *asthenosphere*. It's defined by the decrease in velocity in both P and S waves. Based on the observation that seismic waves slow down in less rigid and less elastic materials, we can infer that the asthenosphere must be partially molten and more plastic than the mantle rock above and below.

Data from both seismic reflection and refraction indicate that the rest of the mantle is solid and layered. Based upon the increase in velocity at the Moho, we know the mantle is more dense than either the continental or oceanic crust (Figure 2.7). From studies of xenoliths, rocks, and minerals in the laboratory, combined with seismic evidence, it is believed that the mantle rocks are *peridotites* – iron magnesium silicates rich in the minerals olivine, pyroxene, and garnet. Peridotites are igneous rocks that differ in mineral composition from both oceanic and continental crustal rocks.

The Core: Outer and Inner

The existence of a core was surmised from a rather surprising turn of events. The P waves were known to be transmitted through solids, liquids, and gases. It was discovered that after a major earthquake, there was a belt between 103 and 142 degrees from the epicenter – on the opposite side of the world – where the P waves disappeared from the seismograms. This was named the *P-wave shadow zone* (Figure 2.8, page 31). Since the P waves should have been carried through any layer – solid or liquid – the interpretation of the shadow zone was that the P waves were bent or refracted upon reaching this deep boundary and could not pass unimpeded through the interior.

The second unusual finding was that the S waves are not recorded at all in the entire region more than 103 degrees away from the epicenter of the earthquake. This is now known as the *S-wave shadow zone* and it covers almost half of Earth. Since liquids do not transmit the S waves, it is assumed that the outer core is a liquid or at least acts like a liquid and absorbs the S waves. This is a pronounced boundary and indicates the depth to the core is 2,900 kilometers (1,740 miles) (see Figure 2.9, page 32).

At a depth of about 5,150 kilometers (3,090 miles), another strong reflecting and refracting boundary occurs, consistent with a change from a liquid to a solid. On this basis, the inner core is assumed to be in the solid state even though it is hotter than the surrounding molten outer core.

Based on the composition of meteorites and on seismic and density data, the core is believed to be metal – probably iron with minor amounts of nickel, silicon, and sulfur. The chemical composition of both the solid and liquid core is likely the same, but the pressure on the inner core is so great it cannot exist as a liquid. The core at the center of Earth is very dense, perhaps 13 or 14 grams per cubic centimeter (Figure 2.10, page 32). (The density of water is 1 gram per cubic centimeter while the overall density of Earth is 5.5 grams per cubic centimeter). Based on the density of the crust and mantle, and using the calculations for the whole Earth, the density of the core could be deduced.

Isostasy

Isostasy (from the Greek *isos* or equal, and *stasis*, meaning standing still) contends that there is a balance or equilibrium among different blocks at Earth's surface. Crustal rocks weigh less than mantle rocks and are, therefore, essentially "floating" on the plastic upper mantle. If the equilibrium is disturbed, through tectonic movements or erosion, the blocks will slowly move up or down. Over long periods of time, the denser material at depth will deform until the crustal rocks are balanced by the weight of the displaced mantle rocks. This slow rate of rising or sinking provides a measure of the viscosity of the mantle (Figure 2.11, page 33). All the blocks at some depth of equal pressure will have the same weight – whether it is a thick column of light continental crust or a column containing sea water, a thin oceanic crust or a section of heavy mantle rock.

Repeated seismic studies have shown that mountains and continents have *roots* much like floating icebergs. Under the Sierra Nevada mountains in California, the root extends downward about 60 kilometers (37 miles). As the mountains are eroded down, the Sierra range undergoes slow but continual isostatic uplift. The eroded sediments will be deposited in a basin that will slowly sink under the added weight (Figure 2.12, page 33). Isostasy explains why the continents stand high and the ocean basins are low. It also accounts for the depression of the land during the latest ice age when ice covered much of the continents and explains why the land slowly rebounded after the ice melted (Figures 2.13, 2.14, and 2.15, pages 34 and 35).

Gravity Measurements and Earth's Interior

Gravity, another indirect method of gaining information about Earth's interior, is one of the great forces in the universe and is just as mysterious as it is pervasive. It can be measured, described, and understood to a point, but exactly what causes this force is still not well known. Sir Isaac Newton, in 1666, explained that the force of gravity between any two objects is based on the *mass* of the objects and the *distance* between them. The more massive the objects, the greater the attraction; the greater the distance between them, the less the attraction (Figure 2.16, page 35). To get a feeling for Earth's gravity, try jumping straight up. It's easy to believe that gravity is a very strong force. But only if a body is very large – such as Earth – does the gravitational attraction become evident.

Earth acts as if all its gravity were concentrated at the center; in other words, all objects fall down in a straight line pointing to the center of the planet. Earth is not a perfect sphere, however, but has an *equatorial bulge* due to rotation. The gravitational pull on objects at the equator is lessened because they are farther from the center while the pull at the poles is greater because they are closer to the Earth's center. When you travel from the equator to the poles, you will gain about one-half pound in weight. But you will weigh less on the top of Mount Whitney than at the bottom of Death Valley since you will be farther from Earth's center on top of the mountain.

Variations in gravity are measured with a *gravity meter* consisting of a known mass of metal suspended on a sensitive spring. Some meters are so sensitive they can detect the difference when moved about 1 meter from the floor to a table top. Theoretically, if you carry a gravity meter from the equator to the poles, the readings should increase gradually after a correction is made for the height of the surface.

But careful measurements do not fit the predictions because, as we noted before, Earth is not homogeneous. Concentrations of massive, high density ore bodies or dense igneous rocks in the crust, for example, can be located by their *positive gravity anomaly* or readings higher than predicted. Low density rocks such as sedimentary rock will show lower readings (see Box 21.1, page 472, on gravity meter exploration for salt domes). The greatest *negative gravity anomalies* are found over oceanic deep sea trenches and are interpreted to mean that the trenches are actively being held down and are out of isostatic balance (Figures 2.18, 2.19, and 2.20, page 36). What kind of reading would you get – negative or positive – if there were a large underground cavern in the region you were studying (Figure 2.17, page 36)? Why do you think so?

Earth's Magnetic Field

Another indirect approach to the study of Earth's interior involves its invisible magnetic field (Figure 2.21, page 37). Known to navigators since the early days of exploration and the use of *lodestone* as a compass, the magnetic field surrounds Earth and deflects certain iron minerals and magnetized objects. The magnetic field is *dipolar* – having a north and a south magnetic pole. Your magnetized compass needle does not point to true north – it lines up with the *north magnetic pole* which is displaced at this time from the *geographic pole of rotation* (true north) by about 11½ degrees. The position of the magnetic poles wobbles slowly and irregularly around the poles of rotation. The changes in location have been documented for hundreds of years and are constantly revised on all modern navigational charts.

The source of the magnetic field is believed to be electric currents generated within the slowly circulating fluid of the outer core. This hypothesis requires that the core be a good electrical conductor which further implies that the core is metallic and probably iron. Certain minerals – the most important of which is magnetite, a form of iron oxide – can become permanently magnetized and record the direction of Earth's magnetic field at the time the mineral crystallized (Figure 2.22, page 38).

All magnetism is destroyed above a certain temperature – called the *Curie Point* – which is about 500° C for magnetite. When the magnetite occurs within cooling lava and the temperature drops below 500° C, the magnetite grains in the rock become like tiny permanent magnets recording the orientation of Earth's magnetic field at the time of cooling. Based on studies of lava rocks on the continents, a remarkable discovery was made – some layers contained a record of reversed magnetic polarity. The south magnetic pole was where the north magnetic pole is today and vice versa! Since the age of the lava can be determined using radiometric dating techniques, it is now possible to state when the magnetic polarity reversals occurred (Figure 2.23, page 38).

Analysis of variations of magnetic field strength can indicate the presence of certain buried iron ores or iron-bearing igneous rocks, which show a *positive magnetic anomaly* (Figures 2.15 and 2.26, pages 35 and 39). Thick sedimentary rocks may show a *negative magnetic anomaly* or a decrease in field strength. Important variations in magnetic field strength were discovered in sea floor basalts where a record of both the changing magnetic polarity and repeated volcanic eruptions proved to be confirming evidence of sea floor spreading.

Earth's Internal Heat

Variations in the heat flow or loss of heat through the crust also provide information about Earth's interior. The geothermal gradient – or the increase in temperature with depth – averages about 25° C per kilometer (about 75° F per mile). But on complex Earth, it varies slightly from place to place. The rate of temperature increase in the mantle was thought to lessen with depth (Figure 2.27, page 40), but recent studies question previous calculations. The temperature at the core-mantle boundary may be as high as 4,800° C and 6,900° C at Earth's center.

The source of the heat may be residual from the origin of the planet or more likely from radioactive minerals in Earth's interior which are considered responsible for supplying most of Earth's present-day internal heat. But early studies of heat flow gave an unexpected result – the average heat flow through the continents is about the same as through the sea floor, even though there is a greater concentration of heat-producing radioactive material in continental rock. This was interpreted to mean that the heat flowing through the thin ocean crust must come from the hot mantle below. There are regional differences in heat flow – especially in the ocean basins. These variations enable scientists to map the elusive convection currents in the mantle that are believed to move the crustal plates.

Through years of study, scientists have learned much about the hidden interior of Earth – mostly through the indirect methods you learned about in this lesson. But our interpretations of the data may change slightly as new techniques and results appear (Box 2.1, page 29). This is a fascinating study and a tribute to the persistence and creativity of scientists working together in many fields to reveal that deep dark frontier – our planet's interior.

For well over a century, scientists have observed a steady and significant weakening in the strength of Earth's magnetic field. If this trend were to continue at the present rate, the field would vanish altogether in about 1,500 years. Does this mean that Earth will soon experience a phenomenon that has recurred throughout geologic time . . . the reversal of the geomagnetic field? The answer lies concealed 3,000 kilometers (1,800 miles) below Earth's surface within the outer core – a slowly churning mass of molten metal sandwiched between the mantle of Earth and the hot solid inner core.

How have geophysicists determined that the polarity has reversed? First, let us examine the nature of Earth's magnetism. The magnetic field is defined as the lines of magnetic force surrounding Earth. If a bar magnet is suspended from a string so it can swing freely, the north-seeking end of the magnet will point towards Earth's north magnetic pole, while the south-seeking pole of the bar magnet will turn toward the south magnetic pole. But the bar magnet is usually tilted at an angle to horizontal. At the north magnetic pole, the lines of force enter the ground vertically and the north-seeking end of the bar magnet will point down. At the South Pole, the north-seeking end of the bar magnet will point skyward since the lines of force emerge from the south magnetic pole, but the south-seeking end will point down. Only at the magnetic equator will the bar magnet hang horizontally. The angle of dip – called *magnetic inclination* – has become a powerful tool for determining the latitude (degrees north or south of the magnetic equator) where ancient lavas erupted and cooled (Figure 2.21, page 37).

Magnetic declination is the angle measured in a clockwise direction between a compass needle and true north. By analyzing the inclination (dip) and the declination, the ancient magnetic fields may be reconstructed from the orientation of iron particles in both igneous and sedimentary rocks.

In the 1950s, evidence began to accumulate that Earth's magnetic field has periodically reversed its polarity. But it is only in the last few years that detailed measurements have revealed the events during the important transition period (Figure 2.23, page 38).

Paleomagnetic records show that the geomagnetic field does not reverse instantaneously from one polarity state to the other but involves a transition period that seems to span a few thousand years. Theoreti-

CASE STUDY:
Magnetic Polar Reversals – Clues and Concepts

cally, from the inclination and declination of magnetic particles in ancient rocks, we should be able to follow the geomagnetic pole as it swings from one polarity to the other. Evidence indicates that about 98 percent of the time, the magnetic field is stable and can be well documented. But for the remaining two percent of the time, the field strength declines and the poles move erratically and fluctuate widely around the globe.

It was found that the paths traced by the magnetic pole positions from the last *reversed* episode to the present *normal* episode ranged over the surface of Earth. When the positions were determined from different localities such as Japan, Maui, California, and the North Atlantic, the positions were widely scattered. This scattering indicated the lack of a strong dipolar field. (If the pole positions coincided, the reversing field would be dipolar.)

Sometimes an intermediate pole position between the reversed and normal states lasted for long periods. But for other times, the field changed very quickly as during the cooling of one lava flow. In some studies, it appeared that the magnetic field underwent several unsuccessful attempts to reverse. Normal polarity was approached and then lost as the field direction returned to an intermediate position.

Recordings at Mono Lake, California indicate that as recently as 28,000 years ago a strange looping excursion or rapid change occurred in Earth's magnetic field. It changed to a pole position halfway between the two polarity states. This might have been a failed attempt at reversing the polarity.

Polarity reversals have now been dated back to the mid-Jurassic period – about 162 million years ago. Reversals are found in even older rocks, but their duration has been difficult to determine. The period and duration of reversals has been very irregular for the past 4.5 million years as you can see in Figure 2.24, page 38. The average reversal seems to be about every 440,000 years. The present normal orientation has lasted for the past 700,000 years. But since the magnetic field seems to be declining, is a reversal on the way?

What then? Will the magnetic field start to twist and turn with the poles pausing somewhere near the present magnetic equator? Will ultraviolet rays from the sun become more intense at Earth's surface and affect genetic material in some organisms? Has life on Earth survived previous polar reversals? Of course life survived!

Organisms that have bits of magnetite in their brains for navigational purposes during migration – and that includes such differing forms as dolphins, birds and bacteria – will likely be affected by the next polar reversal. Also, the beautiful Aurora Borealis – the result of the interaction among the magnetic field in polar regions, solar particles, and Earth's atmosphere – would probably fade with the disappearance of the magnetic field. Or will it glow in the Tropics as the poles shift position? It will be a very interesting time.

1. **Why aren't more deep wells drilled – such as the well in the Kola Peninsula, U.S.S.R. (Box 2.1, page 29) – to give us direct evidence about the interior of Earth?**

   When such an obvious answer to a problem is not actively pursued, there is probably a very good reason. The conditions of tempera-ture, pressure, and extreme depth are so severe that a completely new drilling technology had to be developed. Work at depths of 10 to 15 kilometers (6 to 9 miles) proceeds very slowly and is extremely costly. Continental drill sites that may be the most instructive scientifically may be in regions that are not considered the most promising commercially. Private companies may be less than en-thusiastic about embarking upon a project that may run for years without the assurance of finding valuable resources. Nevertheless, the scientific interest in the Kola well has encouraged deep drilling in several countries, and the results may eventually clarify some of our concepts at least about Earth's continental crust. For evidence about the deep mantle and core, the indirect evidence of geophysical methods is still our best source.

2. **What is the Moho? Why were the first wells that attempted to reach the Moho drilled from specialized ships at sea rather than from existing wells on the continents? Did they succeed in reaching the Moho?**

   The Moho is the boundary between the crust and the mantle. The first wells were drilled at sea because seismic studies indicated that the oceanic crust was very thin – about 7 kilometers (4 miles) – compared with the 30 to 50 kilometer (18 to 30 mile) thickness of the continental crust. In other words, the Moho (at the top of the upper mantle) was much closer to the surface under the ocean. However, the Moho has never been penetrated by a drill ship or by the deepest continental well.

3. **What type of indirect methods of research provide the most information about the deep interior of Earth?**

   While all of the indirect studies – such as Earth magnetism, heat flow measurements, and variations in Earth gravity – reveal differ-ent facets of the interior, it is the analysis of seismic waves that has enabled us to "see" deepest inside the planet to discover the layers and the strange nature of the outer and inner cores.

4. **What are the major units of Earth?**

   The major units into which Earth is divided are, from the surface down: the *crust*, the thick *mantle* which contains most of the mass of Earth, and the outer and inner *core*.

5. **What are some of the divisions of each of the major units?**

   The *crust* consists of the thick, low density *continental* crust which is comprised of a complex of igneous, metamorphic, and sedimentary rocks – with granitic-type rocks probably dominant. The crust also includes the *oceanic* crust which is thin, dense, and basaltic in composition.

The *lithosphere* is a solid, rigid zone that includes continental and oceanic crust. It also consists of upper mantle down to the non-solid *asthenosphere*.

The *mantle* is generally solid rock consisting of dense minerals such as pyroxene, olivine, garnets, and other iron-magnesium silicates. The temperature and density increase downward.

Within the upper mantle is a layer termed the *asthenosphere* which is plastic or non-solid or partially molten. The temperature is sufficiently high that the rock is in a molten state despite the increased pressure. The presence of the asthenosphere is of crucial importance to the theory of plate tectonics as it provides a layer on which the plates are able to move. Also, the molten material rises in the central rift valleys of the mid-ocean ridges where the sea floor is being pulled apart.

There are other concentric layers in the mantle in which the velocity of the seismic waves increases – indicating changes in the structure of the minerals at depth. The *core* consists of a molten outer core and a solid inner core. The temperatures there are very high, estimated to be about 6,600° C at the boundary between the outer and the inner core and 6,900° C at the center.

6. What is the importance of seismic reflection?

Seismic reflection occurs when seismic waves penetrate Earth and bounce off rock layers having differing densities. The reflected waves are recorded on a seismogram. And from a study of the time spent to travel down to the boundary and return to the surface, the depth to the boundary can be calculated. The layers within the mantle were discovered from the reflected seismic waves.

7. What is the importance of seismic refraction?

Refraction of seismic waves is similar to refraction of light waves as they pass through glass or water. Waves are bent (refracted) as they pass through substances having differing densities – glass and water are denser than air. This is the basic principle of correcting vision defects by bending the light rays as they pass through eyeglasses.

Within Earth, seismic waves are bent as they pass through the boundaries of rock layers where the wave velocity slows down in molten or partially molten layers such as the asthenosphere and speeds up in denser layers (Figures 2.1, 2.2, 2.3, and 2.4, pages 26 and 27). Refraction thus identifies rock boundaries and indicates the comparative density of the various layers.

8. What is the P-wave shadow zone, and why is it significant?

The P-wave shadow zone is the region in which P waves disappear. It is located between 103 and 142 degrees on the opposite side of

Earth from the epicenter of an earthquake. This zone offers a clue to the size and nature of Earth's core. The P waves which can travel through solids, liquids, and gases are bent or refracted as they enter the molten outer core so that no waves reach the surface within the shadow zone (Figure 2.8, page 31).

9. What is the significance of the S-wave shadow zone?

The S-wave shadow zone is the region in which S waves are lacking, and it extends over the entire surface more than 103 degrees away from the epicenter (Figure 2.9, page 32). It is believed that the S waves do not pass through the core at all. And since the S waves are not transmitted by liquids, the outer core must be liquid or at least act as a liquid.

10. Why do scientists believe that Earth has an iron-nickel core?

The evidence for the chemical composition of the core comes partly from meteorites which may be remnants of the basic material of the original solar system and are often composed of iron mixed with small amounts of nickel. The density of iron with some nickel is the correct density calculated for the core. The presence of a magnetic field around Earth also suggests a metallic or iron core.

11. If you wanted to build a large office building but were concerned about possible caves in the bed rock below, how could you gather data without drilling exploratory wells?

A valuable method of exploring beneath Earth's surface without drilling is to use a gravity meter. Dense rock bodies – such as metallic ores or dense igneous rock – or the presence of shallow mantle would give a positive gravity anomaly or an excess of gravity compared to the regional readings. If the underlying rock is a low density material such as the salt in salt domes, piles of low density sedimentary rock, or the suspected cave or caverns, the gravity meter would show a negative gravity anomaly or a decrease in gravitational pull compared to the regional average. Remember, gravity is related to mass. A dense mass in the crust added to the expected gravitational pull of Earth will be noted as a positive anomaly.

12. Why does your weight increase slightly as you travel from Hawaii to northern Alaska? And what happens to your weight as you climb up Mount Whitney – the highest peak in continental United States?

Your weight equals your mass times local gravity; therefore, wherever the gravitational pull is stronger you will weigh a little more. Since Earth has a greater diameter at the equator than at the poles – the equatorial bulge – you are farther from the center of Earth at the equator; hence you will weigh slightly less. Hawaii is closer to the equator, and Alaska is closer to the North Pole, so you will gain slightly as you travel from Honolulu to Fairbanks.

If you are on a mountain top, you are also farther from the center of Earth and will weigh a little less. Theoretically, your lowest weight in the continental United States will be on top of Mount Whitney, California. (You may also lose a few pounds making the climb to over 14,000 feet, but that is another story.)

13. What is a magnetic anomaly? How is it measured?

A *magnetometer* is used to measure the strength of Earth's magnetic field. When it was discovered that the strength of the field varies from place to place, geophysicists realized that here was another important source of indirect evidence about Earth's interior. Any variation from average magnetic readings is called a *magnetic anomaly*. Positive anomalies indicate a reading of greater magnetic field strength, and negative anomalies indicate a field strength that is lower than the regional average. Positive anomalies will be recorded over bodies of iron ore, over certain rocks that contain large amounts of iron minerals, and where certain basement rocks have been uplifted. On the sea floor, rocks having the same magnetic orientation as Earth's present field will show a positive anomaly. Rocks with a reversed polarity – measured against the normal polarity of the present time – will show a lower or negative anomaly.

14. What is paleomagnetism, and why is it important?

Paleomagnetism is the study of ancient magnetic fields. Many rocks contain a record of the strength and direction of Earth's magnetic field at the time the rocks formed. When the mineral magnetite is crystallizing in a cooling lava flow, the atoms will respond to Earth's magnetic field and like tiny compasses will point towards the north magnetic pole. Analysis of the magnetic orientation of a sequence of lava flows has revealed that some of the layers have a magnetic orientation that is parallel to that of the current magnetic field and are termed normal. But some layers record an orientation in which the magnetic lines of force are *reversed*. From a study of the magnetic polarity of rocks both on land and on the sea floor, scientists have strong evidence regarding the growth and movements of continents and about the opening of new ocean basins in the far geologic past.

15. When was the last magnetic reversal, and when can we expect another one?

The present normal orientation has lasted about 700,000 years. Periods of primarily normal polarity – or primarily reversed polarity – are called magnetic *epochs*. Within each epoch, there may be one or more short-term polarity reversals that are called magnetic *events*. For example, during the present long-lasting normal epoch, there have been some short events of polar reversals. If the present magnetic field continues to decline in intensity at its present rate, there should be some indication of an impending reversal in about 1,200 years and a complete reversal in about 1,500 years. We can't tell, however, if a new magnetic epoch or just a small magnetic event is on the way.

1. P waves are similar to sound waves. Can you *prove* they are transmitted by solids, liquids, and gases? Every time someone speaks to you, his or her vocal cords set the air molecules vibrating back and forth. This sets your ear drum vibrating at the same frequency, proof that sound waves are transmitted by gases. How about solids? Put your ear flat on a table, and have someone tap the other end of the table. In doing this, you will be like the native Americans of the Old West who learned to put their ears on the railroad tracks to hear the approach of the railroad train. Sound actually travels faster through steel than through air.

   But what about liquids? The next time you are underwater in a swimming pool, have a friend click two rocks together. Anticipate the result. Actually, the excellent transmission of sound waves through the sea is the basis for the songs of the whales, the grunts, croaks, and clicks of some fish, and the mapping of the sea floor with sonar.

2. Take a magnet and a small bottle to the beach. Draw the magnet through the sand. What are you picking up on the magnet? Magnetite, of course, a form of iron oxide. Fill a good part of your bottle with the black grains. Then put a piece of paper or a piece of plate glass over your magnet, and scatter the magnetite grains evenly over the paper. The magnet will move the grains into specific patterns that define its magnetic field. Earth acts as if there were a bar magnet inside. And Earth's magnetic field resembles the field you made around your magnet. Turn the magnet around slowly. Make a magnetic reversal, and watch what happens to your magnetite grains.

3. How far out into space does Earth's gravitational field extend? Clue: Go out and look at the moon – almost 400,000 kilometers (240,000 miles) away. Does the moon mark the limits of Earth's gravitational field? What holds all the planets in their orbits? (The sun's gravitational field.) But why are the tides on Earth influenced primarily by the moon, which is a speck compared to the mass of the sun? Another clue: the distance from Earth to the sun is about 155,000,000 kilometers (93,000,000 miles).

1. The best evidence about the *layered structure* of Earth's interior is from

   a. variations in heat flow.
   b. magnetic polar reversals.
   c. reflected and refracted seismic waves.
   d. xenoliths.

2. The presence of the asthenosphere is indicated by

   a. high heat flow through the sea floor.
   b. slowing of both P and S waves.
   c. positive gravity anomalies.
   d. the P-wave shadow zone.

3. The Mohorovicic discontinuity marks the boundary between the

   a. crust and the mantle.
   b. crust and the asthenosphere.
   c. lithosphere and the mantle.
   d. mantle and the outer core.

4. Most of Earth has properties of the

   a. continental crust.
   b. outer core.
   c. mantle.
   d. lithosphere.

5. The nature of the outer core is convincingly demonstrated by the

   a. samples brought up in volcanoes.
   b. S-wave shadow zone.
   c. positive gravity anomalies.
   d. calculated temperature at depth.

6. One way to search for hidden chambers in the Pyramids might be to use a

   a. gravity meter.
   b. magnetometer.
   c. compass.
   d. heat probe.

7. One way to prospect for buried iron deposits is to

   a. set up a seismograph in a likely spot and wait for an earthquake.
   b. drill exploratory wells.
   c. use a magnetometer and look for a positive magnetic anomaly.
   d. use a gravity meter and look for a negative gravity anomaly.

8. The first attempt to reach the Moho was made from a drilling ship at sea because

   a. the oceanic crust is relatively thin.
   b. the oceanic crust is hot and easier to drill through.

c. the high geothermal gradient on the continents posed many technical problems.

d. drilling at sea is inexpensive because most of the depth is through water.

9. Based on evidence from meteorites and certain magnetic effects, the core of Earth is believed to consist of

   a. dunite, an olivine-rich rock.
   b. peridotite, a rock composed primarily of olivine and pyroxene.
   c. basalt.
   d. iron-nickel.

10. The base of the crust is defined by

   a. a marked increase in temperature.
   b. the slowing of the P and S waves.
   c. an increase in velocity of seismic waves.
   d. a negative gravity anomaly.

11. All of the following statements apply to the lithosphere except it

   a. is rigid and brittle.
   b. includes the moving plates of plate tectonic theory.
   c. consists of the crust and the upper mantle.
   d. is a zone of low velocity seismic waves.

12. The flow of heat through the crust of Earth is

   a. about the same through the continents as through the sea floor.
   b. greater near the deep sea trenches.
   c. greater through the continents due to radioactive elements.
   d. greater through the sea floor due to the presence of basalt.

13. Your weight should increase slightly as you move from

   a. the bottom of the Grand Canyon to the top.
   b. the equator to the North Pole.
   c. the South Pole to Panama.
   d. a location over a buried iron deposit to one over a salt dome.

14. From geophysical studies, we have evidence that

   a. continental crust is much denser than oceanic crust.
   b. radioactive elements are rare in continental rocks.
   c. continental crust is composed primarily of sedimentary rocks.
   d. continental crust is about four times thicker than oceanic crust.

15. The observed rebound of lands that had been buried under ice-age continental glaciers

   a. is a result of shifting crustal plates.
   b. indicates an impending magnetic polar reversal.
   c. is an effect of isostatic adjustments.
   d. is a result of motions within the liquid outer core.

4

THE SEA FLOOR

The goal of this lesson is to help you to understand the nature of major sea-floor features such as the mid-oceanic ridge, oceanic trenches, and fracture zones, and the surprisingly young age of sea floor rocks.

GOAL

After reading the textbook assignments, completing the exercises in this study guide, and viewing the lesson's video portion, you will be able to:

LEARNING OBJECTIVES

1. Describe how the sea floor is studied.

2. Describe the general topography of the ocean floor.

3. Sketch and label a cross-section showing: continental shelf, slope, rise, abyssal plain, mid-oceanic ridge, abyssal hills, and guyots.

4. Describe the typical features of submarine canyons and current theories on canyon formation.

5. Indicate the relationship between oceanic trenches, Benioff zones, andesitic volcanism, and island arcs or continental edges.

6. Describe the geologic nature and significance of the mid-oceanic ridge.

7. Differentiate between seamounts, guyots, and aseismic ridges.

8. Describe the formation of three major types of coral reefs.

9. Discuss the origin of the principal sea floor deposits and explain their geographic distribution.

10. List the fundamental differences between continental and oceanic crusts in terms of their structure, age, and rock type.

**INTRODUCING
THE LESSON**

It is quiet there and bitter cold. The pressure of 4 kilometers (2.5 miles) of water above is crushing. The blackness is broken only by flashes of living light from strange soft-bodied creatures that shine, blink, flash or glow, while attracting prey, avoiding a predator, looking for a mate, or just lighting up the scenery. This is the realm of the deep sea, the abyss.

Learning about the deep sea floor is no easy matter. It is only in the last few decades that humans with their *samplers, dredges, seismic profilers*, deep-sea television cameras, drilling ships, and various other methods have explored this vast forbidding frontier. People in *submersibles* dive deeply to visit this unknown world and scurry back with tales of incredible wonders.

Ocean scientists have learned that the water in the ocean basins is very ancient, perhaps almost 4 billion years old. But analysis of the age of rocks of the sea floor has revealed some very surprising and disturbing data: the oldest rocks are only 200 million years old – on a 4.6-billion-year-old planet! If you want to find Earth's primitive crust and the first evidence of life, for example, you must seek them on the continents, not on the deep sea floor.

Today, the oceans cover about 70 percent of Earth's surface. The configuration of the basins and the continents, however, has changed drastically since the waters first appeared on the planet. Oceans have opened and closed while the waters sloshed from one basin to another; continents have grown and rifted apart and gyrated over the surface of Earth in response to sea floor movements. What we see today is merely the way it looks now. By tomorrow, changes will have taken place – the Atlantic will be a little wider, the Mediterranean will be a little narrower, and East Africa will move away a little from the rest of the continent as a new ocean basin continues to open.

In this lesson, you will see that the remarkable landscape of the sea floor – under 4 kilometers (2.4 miles) of water – is a result of the mobile crust of Earth. The crust moves the great plates around with some colliding and some separating, creating the features of the ocean floor as well as those of the continents.

To understand the sea floor, you must leave the continents and enter the realm of the oceans. You will study the submerged *continental margins*, with the gently inclined *continental shelf*, the steeper *continental slope*, and the apron of sediments at the base of the slope called the *continental rise*.

Proceeding seaward, you will discover the vast, flat, almost featureless *abyssal plains*. Blanketed by unconsolidated sediments, these cold deep plains – about 3,000 to 4,550 meters (10,000 to 15,000 feet) below sea level – are quiet, below the limits of light penetration, perhaps 3 to 5° C, and under incredible pressures from the weight of the water above.

Sometimes the stillness of the abyss is interrupted by an underwater avalanche called a *turbidity current*. These turbidity currents are a bottom flow of mud, sand, gravel, sometimes shells and other shallow water debris, moving rapidly down the continental slope at great speeds and depositing sediment far beyond the continental margin on the abyssal plains (Figure 3.11, page 51).

Rising above the deep sea floor are the mid-oceanic ridges and rises, volcanic mountains that are the dominant features of every ocean basin and quite unlike anything seen on the land. The high ridges form where the sea floor is pulling apart, opening a deep rugged central valley in which hot molten basalt lavas periodically emerge, quench in the frigid water, and fill the widening rift. Also rising from the sea floor are various islands and submerged seamounts and guyots, all born of volcanic activities associated with the moving oceanic crust.

Descending into the sea floor are the deep sea trenches, at depths that may be over 10,000 meters (33,000 feet) below sea level. Here, great slabs of sea floor are diving down either beneath a continental margin or beneath an arc of volcanic islands.

The ocean is not just a body of water filling some low spots on the continents. From your study of the vast sea floor, you will know that continents and ocean basins are very different in origin and geologic history – in the thickness of crust, in rock type, and in geologic structure. This lesson will introduce you to an exciting study of a largely unseen and unfamiliar landscape but one that covers most of the face of our planet. You will understand how the discoveries made in the ocean basins led to the verification of the new theory of plate tectonics that have revolutionized earth science. And you will also learn why living on the "Pacific Rim of Fire" is such a moving experience.

Completing the following eight steps will help you master the lesson objectives and achieve the goal for this lesson:

Step 1: Read the TEXT ASSIGNMENT, Chapter 3, "The Sea Floor."

Step 2: Study the KEY TERMS AND CONCEPTS as noted in the study guide.

Step 3: Watch the VIDEO using the VIEWING GUIDE in the study guide.

Step 4: Read the study guide's PUTTING IT ALL TOGETHER section, which will help you summarize and integrate all of the information in this lesson.

Step 5: Complete any assigned lab exercises.

Step 6: Complete any assigned ACTIVITIES found in the study guide.

Step 7: Review the material in this lesson and complete SELF-TEST found in the study guide.

Step 8: Go back to the LEARNING OBJECTIVES and make sure you have learned each one.

LESSON ASSIGNMENT

TEXT ASSIGNMENT

1. Read the paragraph introducing Chapter 3, page 45.

2. Read Chapter 3, pages 45-63, paying particular attention to the "Summary."

3. Make sure you study the photographs and diagrams in this chapter.

4. Try the "Questions for Review" and "Questions for Thought" at the end of the chapter.

KEY TERMS AND CONCEPTS

Read "Terms to Remember," page 62 in the text, and try to become familiar with the topographic features of the ocean floor. Determine which features are characteristic of *passive continental margins* and which are related to *active continental margins*.

In addition, look up the following terms and concepts in the text and glossary. They have been re-arranged from the alphabetical listing in the text to aid your understanding of the material in this lesson.

Continental slope	Terrigenous sediment
Continental rise	Continental shelf
Submarine canyon	Fringing reef
Oceanic trench	Barrier reef
Active continental margins	Atoll
Passive continental margins	Island
Abyssal plain	Seamount
Mid-oceanic ridge and rise	Guyot
Turbidity current	Aseismic ridges
Pelagic sediment	Ophiolite

VIEWING GUIDE

Once you have read the text material for the lesson and studied the terms, you are ready to view the lesson's video portion. Review the questions that appear in this section to help you watch for important points and to prepare you for what you will see in the video.

Before viewing, answer the following questions:

1. List at least five methods by which the deep sea floor is being studied.

2. Describe the origin and features of the mid-oceanic ridges.

3. Differentiate between seamounts, guyots, volcanic islands, and as-eismic ridges.

After viewing, answer the following questions:

1. How does the video explain the origin of the waters of the ocean?

2. How is sonar used to map the sea floor?

3. What is the importance of the Exclusive Economic Zone that makes it worthwhile to study in detail?

4. What are some of the sea-floor features described by Dr. Tanya Atwater that you would see as you travelled from the beach to the deep sea?

5. You are peeking out of the heavy glass window as you descend in a submersible over one of the hydrothermal vents on the sea floor. Describe what the landscape looks like and what organisms live there.

PUTTING IT ALL TOGETHER

You have read the text assignment and seen the video portion of the lesson. This section, made up of a SUMMARY, a CASE STUDY, QUESTIONS, suggestions for ACTIVITIES, and a SELF-TEST, will help you pull the information together.

Summary

During the Age of Discovery form the 16th to almost the 20th century, detailed exploration of the deep sea floor was beyond the capabilities of the technology of the time. Soundings were laboriously taken by hand (later with a steam winch) using a weight on the end of a cable. One sounding in the deep sea might have taken half a day or more. It is no wonder that early reports described the ocean basin as a "a featureless plain". With the use of sonar and other methods, a fascinating picture evolved of the hidden 70 percent of our planet's surface. In this lesson, you will learn of the techniques used by modern research vessels, and you will become acquainted with the strange unfamiliar yet captivating landscape of the abyssal depths.

The Continental Margins: The Continental Shelf, Slope and Rise, Submarine Canyons, and Deep Sea Trenches

If you could sail out to sea in a glass-bottom boat, you would discover that the continent extends seaward many kilometers as a submerged platform called the *continental shelf*. The shelf – except for local irregularities – slopes seaward at an angle that averages about 0 degrees, 7 minutes (60 minutes=1 degree of angle). The smallest slope that the human eye can detect is about 17 minutes; therefore, the continental shelf would appear to be as flat as a billiard table.

The average width of the shelf world-wide is between 50 and 100 kilometers (30 and 60 miles) but varies from about 160 kilometers (100 miles) off the East Coast of the United States to a little over 1.5 kilometers (1 mile) in places off the West Coast. In parts of the Arctic Ocean, the shelf reaches a width of about 1,500 kilometers or 930 miles! (See Figure 3.7, page 49.) The outer edge of the shelf is typically under about 130 meters (426 feet) of water.

The topography of the shelves was profoundly affected by the periods of rising and falling sea level during the Pleistocene Ice Age. As the water from the oceans became locked up in the ice sheets on the continents, sea level dropped 100 to 150 meters (about 330 to 490 feet) and exposed the shelves of the world as dry land. Ice Age mammals probably roamed the shelves seeking food during the cold glacial periods as did the early humans. The last maximum extent of the ice and of sea level lowering was about 18,000 years ago. The present sea level is only 6,000 to 7,000 years old – a mere blink of an eye on a geologic time scale.

Beyond the edge of the continental shelf, the angle steepens and forms the *continental slope*, producing the confining rim of the ocean basins. Around much of the Pacific basin where deep-sea trenches border the continents and the islands, the slope may plunge from about 5 to 9 kilometers (3 to 6 miles) – one of the most dramatic topographic features of our planet. Not everywhere is the scenery so dramatic, but nowhere is the continental slope merely a smooth featureless scarp. It is incised by numerous valleys, canyons, and smaller gulleys.

At the base of the slope – especially in the Atlantic – is the *continental rise* which is a great apron of thick sediment that has washed down the slope to the sea floor. The sediments of the slope and rise mask the junction between the thick granite continental crust and the thin basalt oceanic crust, an important transition zone but one that has proved difficult to explore and thus is not well understood (Figures 3.7 and 3.8, page 49).

Other topographic features found on all continental margins are submarine canyons. These are V-shaped valleys on the continental shelf and slope, perpendicular to the coast, and incised into both unconsolidated sediments and solid bed rock. They have been found in various shapes and sizes with some approaching the Grand Canyon in depth and form (Figures 3.9 and 3.10, page 50).

The deep-sea trenches are unlike any feature found on the land. While continental valleys are usually the result of erosion by streams or glaciers, the origin of the trenches is apparently unrelated to any erosional geologic processes. The trenches are elongated V-shaped troughs, some hundreds of kilometers in length. The Peru-Chile trench and the Aleutian trench are two examples. Most of the trenches circle the Pacific and are generally parallel to the continental margins or on the seaward side of volcanic-island arcs. Trenches form as a result of great slabs of sea floor sinking into Earth (Figure 3.13, page 52). The rocks of the diving sea floor become partially molten in the rising temperatures within Earth and the resulting molten magma ascends buoyantly through the upper crust to form the volcanoes of the island arcs or portions of the Andes Mountains. The rocks formed near the trenches are not the basalt of the sea floor or the mid-oceanic ridges. Instead, these rocks are called *andesite* and are found around the Pacific Rim (hence the name Andesite.)

According to modern theories, it is within the trenches that *subduction* of the old sea floor – sliding of the sea floor beneath a continent or island arc – begins. This would account for the lack of ancient sea floor rocks in the ocean basins – they have all been subducted. Many trenches lie at the boundary between continental crust (granite) and oceanic crust (basalt). This area, called the *Benioff zone*, is also the site of both shallow to deep earthquakes. The zone, which traces the plane of the descending sea floor into Earth's interior, begins at the trench and dips landward under continents or island arcs (Figure 3.14, page 53). The trenches, which are the deepest parts of the sea, are close to land – contrary to the ideas of the ancient mariners that the deepest parts of the ocean must be in the center of the basins.

The Deep Sea Floor: Abyssal Plains, Mid-Oceanic Ridges and Rises, Seamounts, Guyots, and Volcanic Islands

Beyond the continental margin on the deep sea floor are the vast *abyssal plains* – the flattest features on Earth. The abyssal (meaning deep or unfathomable) plains are at depths ranging from 4,000 to over 5,000 meters (13,120 to 16,400 feet) – far below the limits of light penetration – where the water temperature hovers just above 0° C. On these plains, even the sluggish waters are undisturbed by the even fiercest storms of the surface waters.

The plains are blanketed with fine sediment which is composed of mostly mud and red clays and, locally, volcanic ash or shells of microscopic planktonic (floating) organisms. Strange black potato-like lumps called *manganese nodules* are abundant in places – especially on the deep floor of the Pacific. In the Pacific, the deep sea trenches that rim much of the basin capture the sediments washed off the land. The abyssal plains are not as widespread as in the Atlantic.

The veneer of sediments blankets an even older sea floor – completely volcanic in origin – consisting of basalt pillow lavas, lava flows, and other deep-sea eruptive materials. Towards the center of the ocean basins beyond the reach of the turbidity currents, various abyssal hills and other volcanic features emerge through the thinner sediments.

Rising from the sea floor in every ocean basin are the great mid-ocean mountains called *ridges* in the Atlantic or *rises* in the Pacific (Figure 3.15, pages 53-54). These mountains are interconnected like the seam on a baseball. They are the greatest ranges on Earth – over 80,000 kilometers (48,000 miles) long, 1,500 to 2,500 kilometers (900 to 1,500 miles) wide, and rising 2 to 3 kilometers (about 1 to 2 miles) above the ocean floor.

The oceanic ridges and rises have a very different origin and history from the folded and eroded sedimentary rocks of such major continental mountains as the Alps, the Himalayas, or the Appalachians. The mid-oceanic mountains are rugged piles of igneous black basalt that welled up from the crust and mantle into the faulted central rift valley where the sea floor is being pulled apart (Figure 3.16, page 55). The mountains stand high above the sea floor because of the continuous upwelling of hot, expanded rock materials. This is a zone of volcanic activity and of shallow-depth earthquakes.

The central rift that runs the length of the mid-Atlantic ridge is another unique feature, for no mountain range on any continent has such a valley on the summit. In a few places – Iceland, for example – the peaks of the ridge rise above sea level (Figures 3.17 and 3.18, page 56). There, you can walk on the mid-Atlantic ridge, see the central rift, and not even get your feet wet!

Deep sea sampling of the ridges revealed that the rocks at the rift valley are the youngest on the sea floor (and may be, in some places, forming at this instant) and that the age increases equally on both sides away from the central rift. The oldest sea floor rocks are closest to the continental margins. This was an amazing discovery. It means that the entire basalt floor of all of the ocean basins – over 70 percent of our planet – originated from volcanic activity in the central rift systems of the mid-oceanic ridges. The basins grew as the sea floor spread away on either side. In the Atlantic, considered a relatively young ocean, not only is the sea floor still spreading, but the continents which are attached to the same crustal plates are also moving apart.

Also rising from the sea floor – especially in the western Pacific Basin – is a jumble of literally thousands of submarine conical hills called *seamounts* (Figure 3.19, page 57). Most are believed to be extinct volcanoes. *Guyots* (gee-ohz) are a special class of seamounts having flat tops that are now between 1,000 and 1,500 meters (3,300 and 5,000 feet) below sea level. While their exact origin is still in question, it is believed that guyots were once volcanic islands that became extinct and were deeply eroded by waves and streams. These islands slowly sank beneath the sea as the underlying sea floor aged and subsided (Figure 3.20, page 57).

Most of the true oceanic islands such as Hawaii, Tahiti, the Galapagos, and the Azores are also volcanic in origin, building their cones of lava on the sea floor – flow by flow over a million years or more until their peaks broke through sea level. The islands are just the summits of high mountains; in fact, the peak of Mauna Kea on the island of Hawaii stands about 4,180 meters (13,710 feet) above sea level, on a sea floor that is deeper than 5,000 meters (16,400 feet) below sea level. Measured from base to summit, Mauna Kea is the tallest mountain on Earth – over 9,000 meters (30,000 feet)!

Surrounding many of these volcanic islands are *fringing reefs* constructed primarily by living coral (animals) and calcareous algae (plants), plus other reef organisms living adjacent to the island shore (Figure 3.22, page 58). As the volcanic islands eventually became extinct and slowly subsided, the living reef grew upward, forming a *barrier reef* with a protected lagoon between the reef and island. If the island subsided below sea level and the coral continued to grow upward, enveloping the remains of the old volcano, then this would form an *atoll* consisting of a circular pattern of coral islands enclosing a quiet lagoon.

Reefs and atolls are characteristic of the tropics. Reef-building corals are limited to shallow seas where the salinity is normal and the temperature stays above 18° C.

You have now been introduced to some of the major topographic features of the sea floor and have learned they are very different from the familiar topography of the land. You will study more about this remarkable seascape in the next lesson when you relate the origin of the sea floor to the fascinating theory of plate tectonics.

The ocean floor is still the new frontier of Earth exploration. The discoveries of deep-sea hot springs – called *hydrothermal vents* – with their associated oases of life were some of the most exciting events of the past few decades. The vents and the strange animals that live around them were discovered by deep-diving geologists, descending in submersibles to depths over 2,500 meters (8,250 feet) to explore the mid-oceanic ridges and to study the volcanic processes that create new oceanic crust. To their surprise, the floor around the vents was teeming with animals, some of which were previously unknown. The photographs they brought back captured the imagination of scientists and the public alike – *black smokers, tube worms* 1.5 meters (5 feet) long, giant clams, *dandelions* – colonies of animals that live together on a stalk and resemble dandelions – and strange food chains based on chemical synthesis rather than the photosynthesis of the land and sun-lit portions of the sea.

After many dives on the rift zones on the East Pacific Rise and near the Galapagos, it was determined that seawater continually seeps into the highly permeable crust at the spreading centers through faults and cracks. The water is heated by circulation through the young volcanic rocks below and exits through vents in the sea floor (Figure 21.14 D, page 486).

In one study site on the East Pacific Rise south of the Gulf of California, springs were found that were gushing forth water at about 350° C! The central vents were *chimneys* – tall, hollow spires formed by mineral deposits that had precipitated from emerging hot fluids into the cold surrounding sea water. These were the famous black smokers. They are not really smoking but appear dark from the finely divided particles of sulfides of iron with lesser amounts of copper, zinc, lead, and silver in the outflowing water.

The chimneys themselves were sizeable features – up to 20 meters high (almost 70 feet) and several meters in diameter. They had varied shapes and surface textures and seemed to be surrounded by mounds of debris produced by chimney collapse, possibly as a result of frequent local earthquake activity.

Between the chimneys were fields of black pillow basalts and, in some places, pale-colored patches of furry bacterial mats that covered soft sediments. With the forests of creamy white tube worms with scarlet gills, acres of giant clams, white scuttling crabs, and an assortment of mollusks, it was an incredible landscape.

The water around some vents was sampled and found to be somewhat acid. The water was rich in hydrogen sulfide – toxic to most living things and the source of a strong rotten egg odor emitted when the

CASE STUDY: Deep-Sea Hot Springs, Black Smokers, and the Ecology of the Hydrothermal Vents

samples were first opened in the ship's on-board lab. The presence of sulfides from the deep circulating sea water is a factor in the formation of potentially valuable ore minerals found around the smokers. However, not all the vents visited were acid. Some were alkaline, with few ore metals but high concentrations of ammonia and dissolved hydrocarbons. Those water samples smelled of diesel fuel!

The deep sea organisms, some previously unknown to scientists, were recovered from dredge samples of the abyssal plains. The organisms' growth was slow and the population densities and amount of living material (biomass) were very low in these areas. This was attributed to the low food supply, the high pressure, and the low temperature of the "typical" deep-sea floor.

At the vents in waters that are warm but high in toxic hydrogen sulfide, the ecosystem is completely different. Here, there is a high population density and biomass as well as rapid rates of growth of organisms. This discovery of rich biologic communities at the hydrothermal vents presented a fascinating enigma – what was the food source for so many different species of animals? How could they live in the toxic waters?

The answer that emerged within the last few years was equally surprising. Bacterial chemosynthesis is the major source of food for the vent communities. The primary producers – similar to the plants of the terrestrial environment – are bacteria that derive energy from the oxidation of sulfur compounds, especially the abundant hydrogen sulfide. Furthermore, there is evidence of a symbiotic association between the bacteria and several of the vent organisms such as clams, mussels, and giant tube worms.

A new area full of thermal vents and unique animal life was identified in the Mediterranean Sea in 1989. These vents were at a depth of about 700 meters (2,300 feet) between Sicily and Naples, Italy, near an underwater volcano called the Marsili Seamount. This vent appears to be the residence for a peculiar type of tube worm never identified before at other underwater vent sites.

The discovery of the hydrothermal vents raised probably more questions than answers. Why are the communities so similar at vents that are located so many kilometers apart? How do these attached animals, such as the giant tube worms, migrate so far? How do they locate a suitable hot spring for sustenance and growth? And how did these communities, which are the only ecosystems on Earth supported entirely by Earth heat and energy, ever get started? Are the fauna a very ancient community or is it a fairly recent development? And how many of the great spreading centers of the sea floor ridges and rises have hydrothermal vents yet to be discovered? Most of the seas are not yet explored in any detail. New discoveries we cannot even anticipate are ahead and will continue to shed light on the dark unknown abyss that covers so much of our planet.

1. What is the source of all of the water on Earth?

When you stand on the beach gazing out at the endless sea you are looking at a very ancient body of water – a relic from the time of the birth and growth of a young Earth. The still hot planet erupted enormous quantities of water vapor and other gases to form thick billowing clouds in the upper atmosphere in a process known as *degassing*. As Earth slowly cooled, the vapors condensed into liquid water, and for countless millennia the rain poured down on the barren crust. Thus, the seas were born – perhaps as long ago as 4 billion years. A very small amount of water is still being added through volcanic activity, but the total volume of water in the oceans and in the hydrologic cycle seems to remain essentially unchanged.

2. What is the source of the salts in the sea?

The salts in the sea came from the volcanic emissions containing chlorine (among other things), substances that were readily dissolved in the water. These salts also came from the chemical weathering of minerals from the primitive rocky surface of Earth, providing sodium, calcium, and magnesium. Weathering and erosion are still going on, but the amount of minerals entering the sea is about equal to the amount being removed by various geological and biological processes – creating a steady-state system.

3. Describe and give examples of some of the features of active continental margins.

Active continental margins occur where plate movements have brought a continent into contact or collision with, usually, an oceanic plate. The result of this collision is a *subduction zone* where the denser sea floor descends beneath a continent, producing deep sea trenches or volcanic island arcs. Active margins are the sites of narrow continental shelves, steep slopes, frequent earthquakes, volcanoes, and rugged mountainous coasts such as are found on the west coasts of North and South America. The west coast of South America, where the Nazca oceanic plate is colliding with and subducting under South America, is considered the classic example of an active margin. The deep Peru-Chile trench borders a narrow continental shelf below the soaring Andes mountains. Earthquakes are frequent and often cause devastating land or mud slides from the high glaciated peaks.

4. Describe and give examples of some passive continental margins.

Passive continental margins occur at some distance away from the edge of crustal plates that are diverging or moving away from one another. These margins are characteristic of the opening Atlantic basin, parts of the Gulf of Mexico, much of the Arctic Sea and the

Indian Ocean, as well as the coasts of Antarctica, Africa, and parts of Australia. Passive margins are noted for thick sediments deposited on very broad, flat continental shelves. The volcanoes and earthquakes of the active margins do not occur here. Trenches are lacking, and wide continental rises form the transition between shelf, slope, and the abyssal plain. Ancient passive margins such as the Gulf of Mexico frequently contain abundant deposits of oil and gas in the thick sedimentary layers.

5. Why is there a renewed interest in the geology of the continental shelves?

As we have discussed, continental shelves are the submerged areas of continents but cover about 7.5 percent of the total area under water. The renewed interest in continental shelves is a result of their economic potential during a period of increased awareness of the scarcity of natural resources.

The most important resources recovered from shelf areas are, of course, oil and gas, which far exceed the value of all other minerals put together. Second in value, surprisingly, is sand and gravel. Gold has been found off the mouths of rivers in the continental United States and Alaska in shallow-water placer deposits. Diamonds have been recovered on the shelf of South Africa. Tin is found in sands off Malaysia. Calcium carbonate has many uses including the making of cement. It is widespread, especially in coral-reef and limestone deposits. Phosphate rock is an important potential resource for use in fertilizers. Phosphates occur on shallow banks off the coast of Florida and Southern California (Box 3.1, "Geologic Riches in the Sea," page 61).

6. What geologic processes have contributed to the features found on continental shelves?

Most of the detailed features can be attributed to seven processes: glaciation, sea-level changes, waves and currents, deposition of sediments from the land, coral-reef building, faulting, and volcanism.

When the sea level stood 100 to 125 meters (328 to 410 feet) below its current level during the Ice Age, large parts of what is now the continental shelf were exposed as dry land. Fossil mammoth teeth discovered in dredgings off New England and the North Sea floor provide convincing evidence of such exposure. Glaciated U-shaped troughs and valleys and the hills of rocky debris that were terminal moraines have also been charted – especially on shelves in higher latitudes. This is evidence that the ice sheets pushed their way on to the exposed land during periods of maximum glaciation.

The submerged shelves today receive most of the rock debris from the weathering and erosion of the continents. Deep soundings indicate thousands of feet of accumulated sediments and sedimentary rock piled on the edge of the continents. Major rivers such as the Mississippi, the Nile, the Ganges, the Indus, and many more have deltas that extend many kilometers beyond the shoreline. Beneath the sediments on the shelves is the thick and granitic continental crust which confirms that the continental shelf is actually part of the continental mass even though it is now under sea water.

The bottom sediments of most shelves consist of sand in shallow water grading into finer mud in deeper water. Sometimes coarse sediments are dredged up near the edge of the shelf. These sediments are the relic deposits of waves that were breaking on what was the ice-age shoreline. While waves caused by wind influence only the shallow near-shore water, storm surges and the great tsunamis or seismic sea waves stir the unconsolidated sediments on the shelf. This results in complex depositional patterns and sometimes causes turbidity currents that wash down the slopes or through the submarine canyons. Tides too – especially the swiftly moving tidal currents in narrow inlets – may sweep the shelf clean of any sediments down to bed rock.

In the warm, shallow, sunlit tropical waters off Texas, Florida, and near the Great Barrier Reef of Australia, carbonate deposits associated with rich marine life and coral reefs may reach great thicknesses on the continental shelf. In Florida, for example, some deposits are more than 3,000 meters (9,900 feet) thick. Carbonate deposition must have been much more common in the past than it is today. The thick sequences of carbonate limestone that are found in mountains of North America, the Alps, and the Himalayas are evidence that these great mountain ranges were once part of some ancient continental shelf.

Since continental shelves are the surface expression of the transition zone between continents and ocean basins, they tend to be geologically unstable. There is evidence of faulting on every continental shelf explored to date, and many show evidence of subsidence. The Southern California Borderland, for example, is a complex faulted shelf with deep down-dropped basins and uplifted islands and shallow banks. All of these features are related to plate movement on a *translational (sliding) plate boundary.*

7. **What are some of the features of submarine canyons, and what are the current theories on canyon formation?**

Since World War II, various research institutions have been conducting intensive studies of the continental margins because of their military importance and their commercial potential. These studies have revealed great V-shaped submarine canyons that are incised into the continental shelf and slope on all continental margins (both passive and active). Some extend many kilometers from shore and can be traced to depths of over 1,800 meters (almost 6,000 feet). Among the best known are those off California, New York, Baja California, Colombia, southern France, the Mediterranean shore of North Africa, and the Congo River. For instance, the Monterey

Canyon in central California resembles the Grand Canyon in profile and depth.

Some of the canyons occur off the mouths of rivers, but many occur on shores lacking entering streams. Some canyons are branched with tributaries, but others – such as the great canyon off the Hudson River in eastern New York – have a straight valley that continues 50 kilometers (about 30 miles) offshore to a depth of 3,000 meters (9,900 feet) on the floor of the Atlantic Ocean. The Hudson River Canyon seems to be controlled by the fault system that extends from Canada into the continental shelf south of New York.

Some canyons are incised into sediments, but others are cut into solid bed rock. Most canyons have a fan-shaped deposit spreading out from the base of the canyon (Figure 3.9, page 50) composed of near-shore sediments and shells of shallow-water organisms.

Although many submarine canyons have been charted, photographed, and described, their exact origin is still a matter of controversy. On land, most canyons are created by stream erosion. However, other processes must be operating in the sea. It seems likely that some of the submarine canyons were first eroded into the continental shelf by rivers flowing over the land that was exposed during the Ice Age. Sediments washing down from the shoreline could have abraded the walls and floor of the developing canyons. Sand has actually been photographed flowing down the canyons off the southern tip of Baja California, Mexico. Longshore currents that flow parallel to the coast in shallow water also move great quantities of sand which can be deflected seaward by headlands or underwater topography. These currents may account for erosion of the canyons on coasts lacking major rivers. Many marine geologists believe that landslides or slumping walls and underwater turbidity currents carrying sand, mud, and rocky debris are responsible for cutting and enlarging the submarine canyons – processes still going on today.

8. **What is the origin of some of the principal sea floor deposits? What factors determine their geographic distribution? Why are sea floor sediments of particular interest to marine geologists?**

Classification of sea floor sediments. Although the rough lava surface of much of the ocean floor has not been gentled by erosion, it has been smoothed and modified by multicolored carpets of sediments – unconsolidated materials derived from various sources. These sediments have been classified in various ways: organic and inorganic, *lithogenous* (from rock material), *hydrogenous* (precipitated from the water), and *biogenous* (formed from living organisms). Your text divides sea floor sediments into *terrigenous* (derived from the land) and *pelagic* (sediment that settled slowly through the water).

Terrigenous sediments. Following the system of classification used in your text, terrigenous sediments are the products of weathering and erosion of the continents carried to the shore primarily by rivers. They generally consist of sands and muds and are found on the continental shelves, slope, and continental rise. In many places, the coarser deposits on the abyssal plains – such as sands and materials from shallow water sources – were carried by bottom-flowing turbidity currents to their deep-water resting places.

Pelagic sediments. Pelagic deposits have various sources – some inorganic and some organic. The most widespread deposits in the deeper ocean are the fine-grained muds and clays, usually brick red, chocolate brown, and occasionally dark olive-green in color. These particles were carried to the shore by streams. But because of their small size and low density, they floated out to sea and settled slowly to the sea bed. The rate of accumulation of these fine clays located far from land may be only 1 or 2 millimeters (less than $\frac{1}{16}$ of an inch) per thousand years!

On the peaks of the ridges and rises, a creamy white deposit rests like a blanket of snow. Microscopic examination of this light-colored "ooze" reveals countless tiny shells of floating plants and animals (plankton) that have settled to the sea floor. These include the tiny green plants (the *diatoms* with glassy shells of silica SiO_2) and the *forams* (related to the well-known amoeba of high school biology with their intricate and beautiful shells of calcium carbonate $CaCO_3$).

Because the shells tend to dissolve under pressure in the cold, slightly acid water of the deep sea, the oozes are generally restricted to shallow waters. They are also most abundant in areas of high biologic productivity such as the nutrient-laden polar seas and in the equatorial Pacific (Figures 3.23 and 3.24, page 59). Ancient oozes which are now dried and compacted are found in many places on the land, evidence of previous periods of seawater flooding. The distinctive white cliffs of Dover and the sticks of chalk found in every classroom originated as carbonate oozes on some distant sea floor. And diatomaceous Earth forming a silica ooze is the main ingredient in many other products including swimming pool filters.

Manganese nodules. Deep-sea photographs and dredging have revealed another widespread deposit, especially on the floor of the Pacific – the enigmatic manganese nodules (Figure 2, Box 3.1, page 61). Spread out like a field of sooty black potatoes, the nodules apparently were not washed in from the land like the terrigenous clays, nor did they settle through the water like the pelagic red clays or the biologic oozes. Some marine geologists feel they formed in place from slow chemical reactions on the sea floor or may be related to volcanic activity. In addition to manganese, these nodules contain appreciable percentages of iron, copper, cobalt, and nickel

– all metals that will someday be in short supply on land. Problem: the nodules rest in deep water, over 5,000 meters (16,500 feet). Who owns them, and how can they be mined economically?

Geologic importance of sea floor sediments. The sea floor sediments provide powerful clues to many aspects of the history of the ocean basins. By examining the layers in deep cores of the sea floor, we can see the changing forms of marine organisms through the ages and use the tiny shells to determine the age of the sea floor deposits. We can plot long-range climatic changes such as the repeated advances and retreats of the ice sheets during the Ice Ages. We can locate islands that may have literally blown themselves into oblivion from the remains of their volcanic debris. And we can obtain invaluable information about the birth and growth of the ocean basins themselves which is further support for the theories of sea floor spreading and plate tectonics.

9. **What are some of the differences between continental and ocean crusts?**

The oldest rocks ever found – about 3.9 billion years old – were found on the continents, which are complex patchworks of rocks of many ages, many origins, and many structures. Here we can see the effects of Earth's internal heat engine, raising land surface to great heights, moving continents around, and allowing surface processes to repeatedly reduce the continents to almost sea level. The continents also record most of Earth's geologic history in the widespread marine sedimentary rocks that cover about 75 percent of the land.

The basement rock of the continents is granite. Granite is an igneous rock that is termed *intrusive* since it rose from deep within the crust, intruded older rocks, and slowly solidified at some depth below the surface. It is exposed only after uplift and long erosion of the overlying materials. Granite is a low-density material and tends to float high in Earth's mantle.

The rocks of the deep sea floor – in sharp contrast to the continents – proved to be younger than 200 million years old. This is less than one-twentieth of Earth's history! The sea floor which covers about 70 percent of the planet is almost entirely volcanic in origin. It consists principally of basalt rock that is also igneous but termed *extrusive* because it flowed out on the surface. Basalt is considerably denser than granite and hence floats lower in the mantle.

The basic difference in elevation of the continents and ocean basins is a result of the different densities of granite and basalt. In addition, the continental crust is thick (averaging about 35 kilometers) while the oceanic crust is thin (averaging about 7 kilometers). As we mentioned before and from what you have learned in this lesson, the oceans certainly are not just low spots on the continent (Figure 3.25, page 60).

1. Mentally, drain off the ocean basins and take a walk on the sea floor. Start at the beach and walk out on the flat continental shelf. If you were in the northern high latitudes, what would the shelf look like? If you were in the tropics, what would you find on the shelves? If you slide down the slope off the west coast of South America, about how far would you go?

2. Mentally, take a dive aboard the "Alvin" to a hydrothermal vent. What would you see there not usually seen on the open sea floor? The clams you would see there are about 6 inches across but probably inedible. Why? Why does the water smell like rotten eggs? Why aren't plants found near the vents?

3. You have been offered a contract to mine manganese nodules on the deep sea floor. What are some of the problems you will face in retrieving this valuable resource? And what are your solutions? (Note: The nodules are non-magnetic, so forget the super-magnet idea. Also, some kind of super-scoop will kill the organisms living on the sea floor, and this will not be acceptable to environmental protection groups. And at depths of about 5,000 meters (16,500 feet), you will have a hard time finding divers who are ready to fill buckets of the little black potato-like nodules.) Be creative.

4. Go to an aquarium shop, a shell shop, or a nature store and look for specimens of coral. Many are exquisite in form and color. Consider how these colonies of animals grew and expanded on warm continental shelves, forming the largest biologic structures on Earth. Better still, go to Tahiti and snorkel for a hands-on appreciation of coral reefs.

1. Of the following, the best method to learn the history of the deep sea floor would be

   a. towing a rock dredge.
   b. lowering a deep sea camera.
   c. descending in a submersible.
   d. taking cores from a deep sea drilling ship.

2. The source of the original waters in the oceans is probably from

   a. rivers and streams.
   b. degassing of Earth's interior.
   c. water in the solar nebula.
   d. photosynthesis.

3. Which of the following does NOT apply to the ocean basins?

   a. The oceanic crust is relatively thin, about 7 kilometers.
   b. The basement rock has a relatively high density.
   c. The basement rock is granite.
   d. The basement rock consists of igneous rock.

4. It appears that at the present time the

    a. volume of water in the oceans is increasing.
    b. salt in the sea is increasing.
    c. volume and salts are in a steady state.
    d. salts are increasing but the water may be decreasing.

5. The continental shelf may be described as a

    a. flat surface of sediments overlying a granite basement.
    b. steep slope leading to the deep sea floor.
    c. flat surface made up of andesite lava flows.
    d. flat surface with a thin veneer of sediments overlying pillow lavas.

6. Most of the sediments eroded off the continents will be deposited

    a. in submarine canyons.
    b. on the continental shelf.
    c. on the abyssal plains.
    d. on the ridges and rises.

7. During periods of maximum glaciation

    a. the continental shelves were exposed as dry land.
    b. the continental rises were dry land.
    c. sea level rose and inundated continental shelves.
    d. shells of deep water organisms were deposited on the shelves.

8. Of the following, evidence of sea floor spreading is most convincing in the

    a. andesitic volcanism of the island arcs.
    b. thick sediments of the shelves, slope, and rises.
    c. rift valleys of mid-oceanic mountains.
    d. active continental margins of the Pacific Rim.

9. Submarine canyons

    a. are always located off the mouths of major rivers.
    b. result from the sea floor diving beneath a continental margin.
    c. are erosional and are found on both passive and active continental margins.
    d. are formed where the sea floor is pulling apart.

10. The Benioff zone is considered evidence of

    a. an ocean basin rifting apart.
    b. a sea floor diving beneath a continent.
    c. a line of active volcanoes forming an aseismic ridge.
    d. the subsidence of reefs to form atolls.

11. Evidence of sea floor subsidence is best seen in

    a. fringing reefs.
    b. flat abyssal plains.

    c. peaks of mid-oceanic mountains.

    d. guyots.

12. The ocean basins lie at lower elevations than the continents because

    a. basalt is denser than granite.

    b. the basins are depressed by the weight of the water.

    c. the sea floor sediments are thicker and heavier than the continental sediments.

    d. andesite is denser than basalt.

13. The oozes on the sea floor consist mostly of

    a. red clays and other fine muds from the continents.

    b. black potato-like lumps.

    c. microscopic shells of planktonic organisms.

    d. rotting and decaying fauna.

14. Of the following, the least likely origin of submarine canyons is erosion by

    a. streams during periods of sea level lowering.

    b. water flowing off glaciers during periods of warm climate.

    c. turbidity currents.

    d. sand in longshore currents.

15. New hydrothermal vents are most likely to be discovered

    a. within deep sea trenches.

    b. on the abyssal plains.

    c. within submarine canyons.

    d. within rift zones of mid-oceanic mountains.

5

THE BIRTH OF A THEORY

This lesson will help you understand the origins of the ideas of continental drift and sea-floor spreading and how they contributed to the theory of plate tectonics.

GOAL

After reading the textbook assignments, completing the exercises in this study guide and the lab, and viewing the lesson's video portion, you will be able to:

LEARNING
OBJECTIVES

1. Explain the theory of continental drift.

2. List the lines of evidence used by Alfred Wegener to support his theory of continental drift and the existence of a supercontinent.

3. Discuss the objections to the theory of continental drift.

4. Cite the evidence that contributed to a revival of the idea of continental drift.

5. Explain methods used to determine the drift history of the continents.

6. Summarize Harry Hess's theory of sea-floor spreading.

7. Describe the causes of magnetic patterns on the sea floor and how they allow us to measure the rate of sea-floor motion and predict the age of the sea floor.

8. Describe the rates of plate motion.

The theory of plate tectonics, as you learned in previous lessons, is a revolution in earth science. It is in fact the most recent of three great revolutions in our understanding of Earth. The first was the realization that the planet was round, a fact known to the ancient Greeks but not widely accepted until after the voyages of the early navigators. (A few

INTRODUCING
THE LESSON

"Flatlanders" still resist the evidence.) Another revolution – also anticipated by the Greeks 22 centuries ago – was that Earth circles the sun instead of being central to the universe.

Today, as evidence accumulates that Earth is a dynamic body and that the continents and ocean basins are in a state of constant change, we must look again at our planet home in a new and unaccustomed way. Just as the older revolutions were unifying concepts that brought together many apparently unrelated events, so plate tectonics enables us to explain such diverse phenomena as earthquakes and volcanoes, mountain building, the shapes of the continents, the youthful age of the sea floor, and even the unusual distribution of both fossil and living animals and plants.

The emergence of the theory of plate tectonics brought together ideas that had been simmering for a long time. With the far-flung voyages of discovery in the 16th century, early mapmakers noted with curiosity that the outline of the west coast of Africa seemed to match that of the east coast of South America. In 1620, Francis Bacon commented on this remarkable fit. But it was not until 1858 that Antonio Snider-Pellegrini of France published a book suggesting for the first time that a great continent had once broken apart and that the Atlantic Ocean had been formed by powerful forces acting within Earth.

Many traditional geologists, unable to abandon their notions of "terra firma," clung to their ideas that large blocks of continental crust could not possibly move such great distances over Earth's surface. They preferred the idea that long narrow *land bridges* had at one time connected separated landmasses. This could account for the distribution of certain living and extinct animals and plants, they said, adding that the land bridges had later sunk beneath the sea. As you now know, continental crust has a lower density than oceanic crust, and there is no way that a light "land bridge" could sink without a trace into the dense ocean floor. On the other hand, the similarities in shape – while intriguing – did not *prove* that the continents were ever together.

During the late 19th century, explorers and naturalists roamed Earth and new information became available to the scientific community. For example, a strange similarity was found between the fauna of the island of Madagascar, off the east coast of Africa, and that of India, separated by nearly 4,000 kilometers (2,400 miles) of ocean. (This really stretched the land bridge hypothesis!) Missing from Madagascar, however, were the lions, giraffes, elephants, zebras, and other animals that characterized nearby Africa.

Another puzzling discovery was the similarity of certain rock formations such as *tillites* that provided evidence of ancient shared periods of glaciation on the continents of Africa, South America, Australia, Antarctica, and India (which is far north of most of the other landmasses). Tillites of the same age (Paleozoic) are absent from North America, Europe, and Asia.

The great expeditions of the H.M.S. *Challenger* between 1872 and 1876 discovered the mid-Atlantic Ridge. Frank Taylor, an American geologist, suggested in 1908 that the ridge was the site at which an ancient landmass had split apart. The stage was now set for Alfred Wegener's first detailed argument for continental drift.

In this lesson, you will learn about Alfred Wegener, a German meteorologist, who studied the fit of the continental margins and the evidence of Paleozoic glaciation in southern hemisphere continents now widely separated from each other. He also studied such diverse topics as the occurrence of an unusual group of fossil plants called the *Glossopteris flora*, found in coal seams of the same southern hemisphere continents, the trace of ancient mountain ranges, and the apparent wandering of the poles. As a result of his research, in 1915 Wegener proposed a giant supercontinent of all the continents joined together. He named this supercontinent *Pangaea* (*pan* meaning "all" or the "whole of a diversified group" and *gaea* (jee-ah) from the Earth goddess of Greek mythology). In Mesozoic time, about 170 million years ago, Pangaea split apart, and the Atlantic ocean opened as the continents began to migrate away from one another.

Wegener's theory of *continental drift*, as he called it, was generally rejected because his evidence lacked a plausible mechanism to explain the movement of the continents. There was a resurgence of interest following new discoveries in the 1960s that led to the theory of *sea-floor spreading* and eventually to the all-encompassing *plate tectonics*. You will learn of the many new lines of evidence that include, among other things, distribution of earthworms and dinosaurs, changing climatic zones, the youthful age of the sea floor, variations in heat flow, and what is considered by some geologists as the "clincher" – the magnetic striping of the sea floor.

In this lesson, the birth and development of the great theory will be presented with relevant evidence. Further investigations of plate dynamics will continue in the next lesson.

Completing the following eight steps will help you master the lesson objectives and achieve the goal for this lesson:

LESSON ASSIGNMENT

Step 1: Read the TEXT ASSIGNMENT, Chapter 4, "Plate Tectonics."

Step 2: Study the KEY TERMS AND CONCEPTS as noted in the study guide.

Step 3: Watch the VIDEO portion of the lesson using the VIEWING GUIDE in the study guide.

Step 4: Read the PUTTING IT ALL TOGETHER section in the study guide, which will help you summarize and integrate all of the information in this lesson.

Step 5: Complete any assigned lab exercises.

Step 6: Complete any assigned ACTIVITIES found in the study guide.

Step 7: Review the material for the lesson and complete the SELF-TEST found in the study guide.

Step 8: Go back to the LEARNING OBJECTIVES and make sure you have learned each one.

TEXT ASSIGNMENT

1. Read the first part of Chapter 4, pages 65-79, up to the section entitled "Diverging Plate Boundaries." The last part of the chapter will be discussed in the next lesson.

2. Review the "Introduction" to the chapter. Reading the "Summary" at the end of the chapter will also help, even though you haven't read the whole chapter yet.

3. Review the text and study guide Lessons 3 and 4 to help you understand the basis for the theories of continental drift, sea-floor spreading, and finally plate tectonics.

KEY TERMS AND CONCEPTS

Here are some key terms that will aid your understanding of the material in this course. Carefully study the text or the glossary for the "Terms to Remember" listed at the end of the chapter on page 98. Here are some other key terms:

Pangaea: The single supercontinent, named by Alfred Wegener, in which all the large continental areas of the modern world were united (Figure 4.2, page 67).

Gondwanaland: The southern land mass consisting of Africa, South America, India, Australia, and Antarctica. Gondwana is a place in India where seams of coal yield fossils of the *Glossopteris flora* (Figure 4.3, page 67).

Laurasia: The northern land mass consisting of all of North America, Greenland, Europe, and Asia. Laurasia and Gondwanaland formed from the rifting of Pangaea and subsequently were further rifted; this movement led to the distribution of the present continents.

Tectonics: The study of the origin and development of the broad structural features of Earth. The term originates from the Greek word *tecton*, meaning "to build."

Paleomagnetism: Magnetism remaining in ancient rock. This magnetism records the direction of the magnetic poles at some time in the past and is an important method of tracing the movements of the continents and sea floor.

Magnetic stripes: Alternating bands of strong and weak magnetic fields on the sea floor. Equidistant from the central rift, the bands are mirror images of each other, with similar magnetic polarity. The magnetic stripes record the episodes of changing magnetic orientation between reversed and normal as well as the spreading of the sea floor.

Once you have read the text material for the lesson and studied the terms, you are ready to view the lesson's video portion. Review the questions that appear in this section to help you watch for important points and to prepare you for what you will see in the video.

Before viewing, answer the following questions:

1. List four lines of evidence used by Alfred Wegener to explain his theory of continental drift.

2. List three sea floor discoveries that led Harry Hess to formulate his theory of sea-floor spreading.

3. If paleomagnetic striping of the sea floor is a "clincher" as evidence for continental movement, why didn't Wegener use it in his arguments?

After viewing, answer the following questions:

1. How did the concepts of James Dana change the way people regarded Earth?

2. According to Dr. Gary Ernst, how did Wegner explain his theory of continental drift?

3. Why did most geologists of the 1960s and mid-1970s object to Wegner's theories?

4. What discoveries were made by Harry Hess that supported the theory of sea-floor spreading?

5. What did the discovery of magnetic polar reversals have to do with the theory of sea-floor spreading?

6. How did the new concepts of plate tectonics discussed by Dr. Jason Saleeby alter the way we look at Earth?

You have read the text assignment and seen the video portion of the lesson. This section, made up of a SUMMARY, a CASE STUDY, the QUESTIONS, suggestions for ACTIVITIES, and a SELF-TEST, will help you pull the information together.

These are the most exciting and rewarding times to be a geologist. Since Wegener, Hess, and many others stirred our senses, we have a new way of looking at Earth and understanding its restless, dynamic nature. In this lesson, we followed the evidence and discoveries that led to the acceptance of plate tectonics. In so doing, we developed an appreciation of how observation, gathering and testing of data, and formulation of hypotheses led to the development of a scientifically acceptable theory. (Study Figure 4.1, page 66. Which is your "home plate?")

What Led Wegener to his Belief in Pangaea?

Wegener was intrigued by the jigsaw-puzzle fit of continents now separated by ocean basins (Figure 4.2, page 67). He found there were geologic similarities between eastern South America and western Africa, such as ancient rock sequences containing evidence of Paleozoic glaciation. The same glacial event was discovered in rocks of Australia, India, and Antarctica. It is important to remember that the younger rocks and fossils in these five localities are very dissimilar, having been formed after the breakup of Pangaea (Figures 4.4, 4.5, and 4.14, pages 67, 68, and 71).

He also looked for shifting climatic zones such as ancient coral reefs that would indicate a position near the tropics or glacial scratches on the rocks that would indicate a position in polar latitudes. From his research, he found that the climatic belts had indeed shifted from their present latitudes. He interpreted this to mean that the continents had moved rather than the geographic poles (Figures 4.6 and 4.7, page 69). Based on paleoclimatic data, Wegener plotted the curves of apparent polar wandering, which, as we know today, trace the wandering of the continents (Figure 4.8, page 69). He also recognized that the rift valleys of east Africa might be newly forming rifts or the beginnings of future ocean basins.

Later Developments

Wegener's arguments were more fully developed by South African geologist Alexander du Toit. Among the groups of living animals that supported the concept of Gondwanaland were *earthworms*! One genus of earthworm was found to be restricted to the southern tips of South America and Africa. Another genus was encountered only in southern India and southern Australia. How could soil-dwelling animals have such a strange distribution in the modern world unless the land areas had once been connected? A small early Permian reptile, *Mesosaurus*, was found in *freshwater* deposits in both South Africa and southern Brazil and could not have made its way across such a broad expanse of ocean as the Atlantic. During the Mesozoic, some of the continents were still connected, and various kinds of dinosaurs and other animals spread over large areas of Earth. After Gondwanaland and Laurasia separated, each was colonized by its particular set of dinosaurs.

Early Skepticism

Many of the southern hemisphere geologists were accepting of Wegener's ideas and of the hypothetical landmass Gondwanaland. Geologists of the northern hemisphere, however, continued to be skeptical. The problem was the absence of a mechanism that could move continents horizontally over long distances. Geophysicists of that time knew that the continental and oceanic crust were continuous above the Moho (Lesson 3) and could not see how the continents could plow through oceanic crust. In addition, Wegener's timing of the breakup of Pangaea placed it into the more recent Cenozoic Era, about 50 million years ago. This error misled paleontologists. The rifting of Pangaea actually occurred near the start of the Mesozoic Era, about 170 or more million years ago. Also, the geologists of the time may have been reluctant to accept Wegener's theories because he was a meteorologist.

Revival of Interest

During the late 1950s and 1960s, evidence from studies of the sea floor and paleomagnetism renewed interest in continental movements. (Unfortunately, Wegener died in 1930 and never saw the full flowering and acceptance of his theory.)

From studies of magnetic polar reversals and of the inclination and declination of magnetized iron oxide particles in ancient lava rocks, geologists had much stronger evidence for the opening of the Atlantic and the movement of the continents. There were still objections from many earth scientists, however, based on the lack of precision of paleomagnetic data and on the seemingly insoluble problem of how continents could push their way through the ocean crust. (Review Chapter 2 and Lesson 3. Note Figures 4.10 and 4.11, page 70).

Harry Hess and Sea-Floor Spreading

In 1962, an American geologist, Harry Hess, published a novel solution to this problem in his paper entitled "History of Ocean Basins." Hess had been a naval officer during World War II and used his ship's sounding equipment to chart many areas of the sea floor. He discovered *guyots*, the flat-topped submarine peaks (review Lesson 4) that were evidence of sea floor subsidence. He sampled deep sea sediments, and, using the known rate of deposition as evidence, theorized that the deepest ocean floor could be no more than 260 million years old. We now know that the sea floor is less than 200 million years old, but Hess's projection was amazingly close. (The youth of the sea floor was a new concept; many geologists believed the ocean basins were among the most ancient features of Earth's crust.)

Hess was well aware of the Mid-Atlantic Ridge and was curious to know why the crest exhibited such a high heat flow. Careful mapping of the ridge revealed the central rift valley – a deep furrow on the summit unlike any feature on a continental mountain range.

Hess suggested that mid-oceanic ridges were narrow zones where hot rock – part of a mantle convection system – was rising into the ocean crust. The crust is expanded from the rising molten magma and forms the ridges and rises. (Review Figures 3.15, 3.16, and 3.18., pages 53-55 and 56) Below the central rift the rising circulation pattern within the mantle splits, diverges, and horizontally moves apart slabs of the ocean crust and upper mantle. (Study Figure 4.16 A and B, page 73 to understand the process of sea-floor spreading.) Tension cracks form along the crest of the rise, allowing basalt lavas to seep up from below. Sometimes, the molten rock overflows onto the sea floor and builds mounds of pillow lavas, volcanic cones, and sheets of basalt. The lava solidifies to form many square kilometers of new ocean floor each year.

From these observations, Hess developed a hypothesis that contrasted with Wegener's idea of continents pushing through the ocean crust and solved that particularly difficult stumbling block of how the continents move. The sea floor was also moving, growing away on either side from the central rift and carrying with it the continents – like passive passengers – attached to the same lithospheric plates.

Hess also made the observation that if the sea floor were spreading, theoretically Earth must be increasing in size. As there was no evidence that this was occurring, he believed the sea floor, moving away from the ridge crest, cooled and increased in density sufficiently to sink back into the mantle. The deep sea trenches provided the solution to this puzzle. The trenches are known for their low heat flow and a large negative gravity anomaly. The cold oceanic crust sinking into the trenches accounts for the low heat flow. The relatively less dense oceanic crustal rock diving into the denser mantle provides the negative gravity anomaly. Earthquakes of the Benioff zone and the andesitic volcanism associated with the trenches confirmed the validity of the theory. (Review Lesson 4 and take another look at Figure 4.16, page 73; it is all there!)

Next Step: Paleomagnetism and Polar Wandering Curves

As you learned in Lesson 3, the strength and direction of Earth's magnetic field is known from the orientation of tiny magnetite crystals preserved not only in solid igneous rock but in certain sedimentary rocks as well. Also, the crystals, by their inclination (dip) and declination, indicate where the rock formed in relation to the magnetic poles and the direction of the magnetic field at that time. As data accumulated, it began to appear that Earth's magnetic poles had wandered. (They actually do wander a little but stay close to the geographic pole.) Surprisingly, a plot of the pole's apparent positions over time in North America and in Europe didn't match (Figures 4.12 and 4.13, page 71); this raised the possibility of two magnetic poles. Alternatively, the poles didn't wander at all; instead, as Wegener predicted, the continents of Europe and North America moved in relation to the northern pole and to each other.

From the magnetic evidence and the polar wandering curves, it became possible to trace the movement of the continents since the breakup of Pangaea. Within the last 100 million years, Laurasia and Gondwanaland further fragmented into the land masses we know today (Figures 4.14 and 4.15, pages 71-72).

Paleomagnetism and Magnetic Striping of the Sea Floor

Despite strong circumstantial evidence that favored Hess's hypothesis of sea-floor spreading, his publication in 1962 still lacked a convincing test of his basic idea. Such a test was soon found. It was known that Earth's magnetic field had periodically reversed its polarity. In 1963, the British geophysicists Fred Vine and Drummond Matthews reported that newly formed lava rocks on the ridge crests of mid-ocean mountains were magnetized while Earth's magnetic field was in its present polarity. This was no surprise. It turned out, however, that the rocks on either side of the rift were magnetized in the reverse direction! The normal and reversed magnetic stripes of sea floor rocks have been likened to a tape recorder that shows the reversals of Earth's magnetic polarity as well as the spreading of the sea floor in both directions from the central rift.

Here was the "clincher" for sea-floor spreading. Vine and Matthews observed that the *magnetic anomalies* were symmetrical on either side of the ridge crest and continued in matched pairs parallel to and away from the ridge axis. (Study Figures 4.18, 4.20, and 4.21, pages 75-77.) This magnetic "striping" of the sea floor rocks could be correlated in both the Atlantic and Pacific sea floors. It actually consisted of measured strong and weak magnetic fields as preserved in the lava rocks that had been extruded in the central rift of the ridges. The rocks containing "normal" polarity (in the present orientation) would strengthen the magnetic field readings, while the rocks containing reversed polarity would subtract from the present magnetic field and would weaken the magnetic field readings.

During the 1960s, a time scale was developed, originally based on measurements of the magnetic polarity of continental lava rocks of known age. The pattern and age of magnetic events found in sea floor rocks proved to be the same as those recorded for continental lava flows, since polar reversals are world-wide events. Using the time scale and the magnetic anomalies, it became possible to date any portion of the sea floor. As anticipated from the Vine-Matthews hypothesis, the age of sea floor rock increased equally on both sides of the rift towards the continental margins.

New Directions

The *hypothesis* of plate tectonics – developed from the discoveries of continental drift, sea-floor spreading, and many other lines of evidence – has advanced in certainty and acceptance until it is now considered a *theory*. Meanwhile, new discoveries are confirming our ideas about how lithospheric plates break up and reassemble. Alaska and Western Canada, for example, were apparently assembled from over 50 *exotic terranes*, bits of land, island arcs, and sea floor that travelled long distances, sometimes halfway across the globe, before attaching to these growing continents. In a future lesson, you will learn about mountain building, the growth of continents, and more about these interesting exotic terranes.

What would Wegener have thought about all of this? Not all geologists are sure what *they* think. These theories are more than just working models, but many observations remain unexplained. As we have learned, the data and observations do not change, but interpretations change as theories evolve. In the next lesson, we will continue the exciting story of plate tectonics.

CASE STUDY:
Sea-Floor
Spreading and
Green Sea Turtles

Sea turtles are common members of tropical marine communities. The green sea turtle, *Chelonia*, has feeding grounds on the east coast of Brazil but migrates to remote, isolated islands for nesting. These islands apparently lack many of the predators that would harass the turtles and raid their nests on mainland beaches. The best documented feats of island-finding by green sea turtles are migrations between the east coast of Brazil and Ascension Island. Ascension Island is a mere dot of land, only 8 kilometers (4.8 miles) wide, in the Atlantic Ocean midway between Brazil and Africa, 2,000 kilometers (1,200 miles) from the feeding grounds! How and why do the sea turtles return here? Plate tectonics may provide part of the answer.

The turtles lay their eggs in warm, sandy beaches along the north and west coast of Ascension. Immediately after hatching, the young turtles head directly out to sea. Once they are beyond the hazards of surf and shore, they apparently are picked up by the South Atlantic Equatorial Current flowing westward and are carried toward Brazil at speeds of 1 to 2 kilometers (.6 to 1.2 miles) per hour. Less than two months are needed to drift passively to Brazil, yet nothing is known of the young turtles' whereabouts or activities for their first year. Nor it is known how they get back to Ascension Island, but the adult turtles do show up during the nesting season. The females leave Ascension after laying their eggs and return to the coast of Brazil. Two or three years later, they return again to tiny Ascension to mate and lay eggs, thus completing another incredible odyssey. (Some green sea turtles in the Caribbean do not return to their nesting site for an estimated 30 or more years!)

Marine turtles similar to *Chelonia* inhabited the seas between North America and northwestern Gondwanaland by the beginning of the late Cretaceous, about 100 million years ago. The northern coast of South America was also a suitable habitat for these ancient turtles because it provided a tropical environment. Fossil evidence indicates that even then the turtles were herbivorous and had pasture and breeding grounds that were separate from one another.

About 80 million years ago, the opening of the equatorial Atlantic marked the final separation of South America and Africa. A rift and ridge system linked the spreading ridges of the North and South Atlantic and made it a single ocean. Volcanic eruptions are a frequent feature of mid-oceanic ridges, and some grow sufficiently to emerge as islands. As the sea floor opened and new islands formed, the sea turtles were apparently able to migrate the short distances to the new breeding grounds. It seems reasonable to assume that their pattern of travel – constant over a long period of time – would become an established inherited part of the turtles' behavior.

As the sea widened, some of the older islands may have subsided, but new ridge islands became part of the breeding grounds. It was necessary that the turtles extend their travel path as the sea slowly widened. It has been suggested that by about 70 million years ago, the colony of ancestral turtles was making seaward breeding migrations to other young islands about 300 kilometers (180 miles) away, consistent with the spreading rate of about 2 centimeters (1 inch) per year. Ascension is probably the last and youngest of the islands – less than 7 million years old.

How the turtles target the distant island is still not well known. They probably use the rising sun, equatorial currents, and chemicals in the water specific to Ascension. It is not known if the ancestors of *Chelonia* were even the same species, but it is believed that today's green sea turtles use migration patterns developed over millions of years.

The spreading oceans must have influenced the migrations of other animals as well. The incredibly long routes of migrating birds and the migrations of salmon, tuna, the eels of Europe and eastern North America, and the whales of both northern and southern hemispheres may some day be correlated with the changing patterns of land and sea since the break-up of Pangaea.

1. **What are the main points of Wegener's theory of continental drift?**

   Wegener believed that all the continents at one time fit together to form a giant supercontinent that he called *Pangaea*. He said that Pangaea broke apart and that the continental fragments moved to their present position.

2. **What evidence did Wegener use when he proposed continental drift?**

   Wegener tried to use as many different lines of evidence as were available to him. He used the remarkable correspondence of the continents on either side of the Atlantic basin but measured at the edge of the continental shelf rather than at the changeable coastline. He also noticed a distinctive sequence of ancient rocks in the southern continents that consisted of a tillite (a sedimentary rock deposited by glaciers) overlaid by shales and coals that yielded members of the Glossopteris flora. In addition, he noted that fossils of the freshwater Permian reptile *Mesosaurus* from 270 million-year-old strata (late Paleozoic) were found in South Africa and Brazil and nowhere else in the world. Dune sands had covered most of the layers and were followed by outpourings of Jurassic lava flows on all the Gondwana continents. The lava flows seemed to end the sequence and marked the beginning of the disruption of the southern landmass. Wegener's basic concept was that the present distribution of both rock sequences and plants and animals could only be explained if all the continents of Gondwanaland were at one time joined together.

Wegener also tried to reconstruct old climate zones (paleoclimatology) using sedimentary rocks that formed under different conditions, such as the cold polar regions or the warm tropical zones. Earth has such climatic divisions, but the belts Wegener discovered were unlike those found today. As with his analysis of possible polar wandering, there were two possible interpretations: either the continents remained stationary and the poles wandered, or the poles remained fixed and the continents drifted. (This is an excellent example of how valid data can have more than one interpretation – an important part of the scientific method.)

3. **What is the significance of the famed Glossopteris flora? Did it support the theories of continental drift, or was it used as evidence against continental drift?**

During the latter part of the 19th century, late Paleozoic coal deposits of India, South Africa, Australia, and South America were found to contain a group of fossil plants that were collectively designated the *Glossopteris flora* after their most conspicuous genus, a variety of seed fern. After the turn of the century, the flora were discovered in Antarctica as well. The occurrence of these fossil plants led one geologist, Eduard Suess, to suggest that land bridges once connected all of these continents. He even introduced the term *Gondwanaland* for the southern continents and their connecting land bridges. The distribution of the plants had been widely discussed, and means of dispersal such as wind, birds, and ocean currents were suggested, but seed ferns are not likely to be carried such great distances over oceans to so many diverse continents. Frank Taylor, an American geologist, proposed a new explanation: the continents had once been side by side as components of a very large landmass that eventually broke apart and moved across Earth's surface to their present position. This general idea is central to modern plate tectonic theory and supported Wegener and his contention of drifting continents.

4. **Why were geologists in southern hemisphere countries more impressed with the theory of continental drift than their colleagues in northern hemisphere countries?**

In southern hemisphere countries, the distribution of the Glossopteris flora, the similarity of the ancient glacial deposits, as well as the widespread occurrence of the great Jurassic lava flows were very visible and well known to the geologic community. The remarkable fit of South America and Africa was too close to be merely coincidental. Despite some of the apparent problems, many respected geologists such as Alexander Du Toit supported Wegener in his quest for answers. On the other hand, many United States geologists felt the idea of continents moving through solid ocean crust violated what was known about the strength of sea floor rock, and few took Wegener very seriously.

5. How did Hess's theory of sea-floor spreading eliminate the principal objections to Wegener's theory of continental drift?

The theory of sea-floor spreading demonstrated that new ocean floor was being created in the rift valleys of the mid-oceanic ridges and was spreading symmetrically away from the central crest. The moving sea floor also moved the continental masses that were part of the same lithospheric plate. This spreading eliminated the necessity for the continents to plow their way through the ocean crust.

6. How did Hess explain that Earth was not getting larger despite a spreading center in every ocean basin?

Studies of the deep sea trenches solved this enigma. Where plates containing sea floor converged with continents, the denser sea floor and the lithosphere *subducted* or subsided beneath the less dense continental mass. The low heat flow near the cold sinking sea floor, the negative gravity anomalies, the Benioff zone which started near the trenches and extended landward as a belt of increasingly deep earthquakes, and the andesitic volcanism associated with the diving sea floor were solid evidence of subduction of old sea floor. Earth was not getting larger, since the formation of new sea floor at the central rift system was balanced by the destruction of old sea floor within the subduction zone.

7. How were the magnetic anomalies of the rocks of the oceanic crust discovered?

During World War II, extremely sensitive airborne magnetometers were developed to detect submarines by their magnetic fields. These same instruments were modified by oceanographers for towing behind their research ships. Used in this way, the magnetometers measure two things: the main geomagnetic field of Earth and the local magnetic disturbances or *magnetic anomalies* caused by magnetized rocks on the sea floor.

Steaming back and forth across the mid-Atlantic ridge, the scientists discovered amazing magnetic anomaly patterns that formed linear bands and spanned hundreds of kilometers; they were almost perfectly symmetrical with respect to the central rift valley. The magnetic bands showing periods of normal and reversed magnetism on one side of the crest are almost mirror images of those on the opposite side. As the sea floor spreads away from the cracks and faults in the rift valley, about half the newly magnetized basalt in the rift moves to one side and the other half to the other side, forming two symmetrical magnetized bands. The process is repeated over and over again as the sea floor spreads and the magnetic polarity reverses.

8. Sea-going geologists can now determine the age of the sea floor rocks without even examining rock samples. How can this be done?

Predicting the age of the sea floor is an important part of the Vine-Matthews hypothesis. Magnetic reversals have probably occurred through much of geologic time and have been dated with some precision for the past 5 million years, with less exact evidence reaching back into the Mesozoic Era, over 160 million years ago (Figure 4.22, page 77). As expected, the youngest rocks on the crest of the mid-oceanic ridges are normally magnetized because they were extruded during the present normal magnetic epoch. If reversely magnetized rocks dated about one million years ago have been displaced from the ridge crest about 15 kilometers (9 miles) on each side, the spreading rate is about 1.5 centimeters (about ¾ of an inch) per year.

Since precise dating of the reversals and therefore of the spreading rates goes back only about 5 million years, it has been assumed that the spreading rate of the last 5 million years can be considered representative of rates of a much longer period. By comparing the magnetic anomaly patterns, it has been determined that the opening rates of the North Atlantic Ridge are the slowest, about 2 centimeters per year (about 1 inch), and the East Pacific Rise is one of the fastest, from 10 to 12 centimeters (4 to 5 inches) per year. This means that if you live in North America, you are moving westward about 2 centimeters (about 1 inch) per year — about the same rate as your fingernails grow!

Detailed magnetic patterns recorded in all the oceans are quite similar, with some variations due to differences in spreading rates. Marine geophysicists and their trailing magnetometers can now determine the age of any sea floor beneath them by comparing the magnetic patterns they are recording with those already obtained and interpreted for large sections of the world's oceans.

This hypothesis has been tested hundreds of times through the use of cores of sedimentary and igneous rocks recovered from drilling ships. The age of the sea floor has been verified usually by identifying the microfossils in the cored sediments lying above the basement rocks and generally corresponds very well to the age predicted by the magnetic data. Recently, the oldest sea floor rocks ever obtained were cored from a site in the far western Pacific basin, the farthest point from a spreading center.

Activities

1. Despite the convincing evidence of sea-floor spreading and plate tectonics, the exact mechanism that propels the crustal plates is still not well understood. Carefully read Box 4.3, pages 94-95, for a new method of studying convective currents in the mantle. Or go to a library and see *Scientific American*, October 1984, for Earth models derived from *seismic tomography*.

2. While you are at the library, look up some of the references listed at the end of Chapter 4. Your text and this telecourse are just the introduction to this fascinating theory. For your own interest and breadth of knowledge, do a little reading on your own.

1. The latitude where iron-bearing igneous rocks first solidified can be determined from the

   a. distance from the mid-ocean rift.
   b. plant fossils in the rock.
   c. inclination of the iron particles in the rock.
   d. declination of magnetite in the rock.

2. The main objection to Wegener's continental drift theory was

   a. the lack of correspondence of the continental coast lines.
   b. the absence of an obvious mechanism that could move continents.
   c. poor fossil evidence of connected continents.
   d. lack of rock continuity from one separated area to another.

3. The apparent fit of the continents on either side of the Atlantic was first noted by

   a. Alfred Wegener.
   b. Harry Hess.
   c. Francis Bacon.
   d. Antonio Snider-Pellegrini.

4. Wegener's studies on the location of the poles through geologic time

   a. indicated that the poles were originally located in Asia.
   b. showed the poles of rotation moved, causing the climatic belts to change.
   c. showed there were actually two poles, one for North America and one for Europe.
   d. concluded that the poles were stationary and the continents moved.

5. The name Wegener gave to his supercontinent was

   a. Laurasia.
   b. Gondwanaland.
   c. Pangaea.
   d. Panthalassa.

6. The time of breakup of the supercontinent can be estimated from

   a. comparison of Paleozoic life in different continents.
   b. the appearance of glacial tillites in rock sequences.
   c. the distribution of the Glossopteris flora.
   d. the magnetic anomalies and age of the sea floor.

7. The pattern of magnetic reversals at sea

   a. was found to differ in each ocean basin.
   b. occurred at different time intervals on the land.
   c. was the same for the land and the various ocean basins.
   d. was not found in land formations.

8. Land bridges were suggested mainly to explain

   a. the distribution of Paleozoic and early Mesozoic plants and animals.
   b. glacial deposits in now widely separated areas.
   c. the occurrence of coal in regions now in cooler climates.
   d. the fit of continental margins across the Atlantic.

9. The mid-Atlantic ridge is characterized by all of the following except

   a. an axial rift valley.
   b. young outpouring of basalt lavas.
   c. a great system of cross-faults.
   d. deep earthquakes and subsiding crust.

10. The major breakthrough towards the acceptance of continental drift came in the field of

    a. paleontology.
    b. paleoclimatology.
    c. paleomagnetism.
    d. paleobotany.

11. The late Paleozoic glaciation cited by Wegener as evidence for continental drift occurred in all of the following places except

    a. Europe.
    b. South America.
    c. Africa.
    d. Australia.

12. The theory of sea-floor spreading differed from continental drift by

    a. proving land bridges did exist.
    b. providing a mechanism that enabled the continents to move.
    c. showing tidal forces could shift the continents.
    d. demonstrating the sea floor was weak and continents could easily plow through it.

13. The Vine-Matthews hypothesis accounted for all of the following except

    a. polar wandering.
    b. sea-floor spreading.
    c. magnetic anomalies.
    d. prediction of the age of the sea floor.

14. When geologists say the sea floor is like a "tape recorder," they mean it contains a record of all of the following except

    a. reversals of Earth's magnetic polarity.
    b. the rate of spreading of the sea floor.
    c. the rate of expansion of Earth.
    d. the time of magnetic reversals.

15. According to oceanographic studies,

    a. the sea floor is approximately the same age everywhere in the Pacific.
    b. only in the Atlantic is new sea floor formed at the deep sea trenches.
    c. old sea floor in all oceans is subducted in the rift valleys.
    d. new sea floor is formed at the rift valleys in the ridges.

6

PLATE DYNAMICS

This lesson will help you understand the underlying theory and concepts in the movements of Earth's plates.

GOAL

After reading the textbook assignments, completing the exercises in this study guide, and viewing the lesson's video portion, you will be able to:

LEARNING OBJECTIVES

1. Differentiate between the lithosphere and the asthenosphere and state their significance to plate tectonics theory.

2. Describe and give examples of three types of plate boundaries, including their characteristic geologic features.

3. Describe the origin of island arcs and deep sea trenches.

4. Explain the process of backarc spreading.

5. Give examples of where continental collision has occurred.

6. Relate plate tectonic theory to the origin and distribution of most of the world's volcanoes, earthquakes, young mountain ranges, and major sea-floor features.

7. Summarize the various theories for the driving force behind plate tectonics.

8. Discuss the evidence relating the origin of the Hawaiian Islands to the hot spot theory.

9. Discuss the unresolved questions regarding plate tectonics.

It has taken 4.6 billion years to evolve that bit of land under your house. Think of the countless plate collisions, the rifting apart of continents and their reassembling into different shapes, the fiery volcanic birth and later subduction of the sea floor, the upward thrusting of great mountain

INTRODUCING THE LESSON

95

ranges and the weathering and erosion that wore them down, all part of the dynamics that created Earth's crust of today (and your backyard). When did all this action start? The exact date when tectonic movements began is not certain, but most likely it was well under way by 2 or 3 billion years ago.

What do we know about the plates of a billion or two years ago? Not much. The land masses apparently were numerous small micro-continents that grew over the eons by accretion and collision with other lands. The plates today are large, and they include both continents and oceans. The North American plate, for example, includes half of the North Atlantic Ocean as well as almost the whole continent of North America. (On what plate is the other half of the North Atlantic Ocean? See Figure 4.1, page 66). Most of the very ancient ocean floors have been subducted down trenches, but some are scraped off and preserved in belts of deformed rock that are seen in the Appalachians, the Urals of eastern Europe, the Himalayas of Asia, and the Alps. Careful examination of these and other ancient mountain belts enables geologists to locate some pre-Pangaea plate boundaries.

In this lesson, you will learn in greater detail how plate tectonics offers a new approach and solutions to the many anomalies that have mystified geologists for centuries. You will find that most major geologic activities take place at the boundaries between the slowly moving rigid plates. Each type of boundary produces distinctive and recognizable geologic patterns. The deep sea trenches and the island arcs, for example, are characteristic of continent-ocean and ocean-ocean convergence. The greatest continental mountains also are the result of convergence, specifically continent-continent collision. Divergent boundaries on the sea floor lead to the formation of the mid-oceanic ridges and the creation of all the oceanic crust on the planet. Divergent boundaries within a continent can produce rift valleys along which the continent breaks apart and opens a new arm of the sea. A third type of plate boundary is the transform fault – the best known example being the San Andreas fault, where part of California bumps and grinds northerly along the western edge of the North American plate.

The globe is continuing to evolve and change by mechanisms that remain difficult to decipher. The East African rift in central Kenya, for example, runs north-south through more than 6,000 kilometers (3,600 miles) and is a spectacular illustration of what may develop into continental separation and creation of new ocean floor. The volcanoes of the Hawaiian Islands occur within the great Pacific plate, 3,200 kilometers (2,000 miles) from any plate boundary, although most of the volcanism in the oceans takes place at boundaries such as the mid-oceanic ridges or above zones of subduction. Plate motion ranges from 2 to 10 centimeters (1 to 4 inches) per year, but the origin of the driving force that is shifting the plates remains uncertain. Most explanations appeal to convection within Earth's mantle, but the details of depth of convection, its rate and causes, the role of hot spots that result from mantle plumes and effects of the molten asthenosphere are still not known, and these factors may be important in propelling the plates.

Completing the following eight steps will help you master the lesson objectives and achieve the goal for this lesson:

Step 1: Read the TEXT ASSIGNMENT, Chapter 4, starting at "Diverging Plate Boundaries."

Step 2: Study the KEY TERMS AND CONCEPTS as noted in the study guide.

Step 3: Watch the VIDEO using the VIEWING GUIDE in the study guide.

Step 4: Read the PUTTING IT ALL TOGETHER section in the study guide, which will help you summarize and integrate all of the information in this lesson.

Step 5: Complete any assigned lab exercises.

Step 6: Complete any assigned ACTIVITIES listed in the study guide.

Step 7: Review the material in this lesson and complete the study guide's SELF-TEST.

Step 8: Go back to the LEARNING OBJECTIVES and make sure you have learned each one.

LESSON ASSIGNMENT

1. Review the first part of Chapter 4 and study guide Lesson 5, as needed, to refresh your memory of plate tectonics.

2. Read the second part of Chapter 4, starting at "Diverging Plate Boundaries," pages 79-98. This is a long chapter, with many concepts to learn. It will take more than one reading to absorb all the material, but the subject is fascinating and will be of great value in understanding *Earth Revealed*.

3. Read "Backarc Spreading," Box 4.1, pages 86-87, to understand the process and problems of interpretation.

4. The "Questions for Review" will help you prepare for the SELF-TEST and your other examinations. The "Question for Thought" is a good one that you should be able to answer.

5. Carefully study the text diagrams and photographs. They provide visual understanding of plate dynamics.

TEXT ASSIGNMENT

KEY TERMS AND CONCEPTS

Most of the "Terms to Remember, " page 98 in the text, should be very familiar to you by now. Look up those that still seem new and unclear. Here are some other key terms:

Aulacogen: See description in the glossary, and study Figures 4.42, 4.43, and 4.44, pages 93-94.

Hot spots: Volcanic areas that are the surface expressions of mantle plumes, which are vertical columns of rising molten mantle rock. There are at least 100 hot spots around the globe that have been active within the last 10 million years. Some hot spots appear to be very long-lived and fixed in location. The *relative* motion between the plates can be easily demonstrated. The importance of hot spots is that because they are fixed in position, they provide a direct means for measuring *absolute* motion of lithospheric plates.

Rift Valley: An elongate valley formed by the down-dropping of a block of Earth's crust between two parallel faults or fault zones. Rift valleys are usually steep sided and occur on the summits of the mid-oceanic ridges in East Africa, Asia, and wherever there is tension or divergence within or between plates. Also see *graben* in your glossary.

VIEWING GUIDE

Once you have read the text material for the lesson and studied the terms, you are ready to view the lesson's video portion. Review the questions that appear in this section to help you watch for important points and to prepare you for what you will see in the video.

Before viewing, answer the following questions:

1. What type of plate boundary is characteristic of the mid-Atlantic ridge? Between India and Asia? Between the South American and the Nazca plates in the eastern Pacific? Between the North American and the Pacific plates as seen in California?

2. Where do most shallow earthquakes occur? Where do deep focus earthquakes occur?

3. Where does most of the basalt volcanism take place?

4. Where are most of the andesite volcanoes found?

After viewing, answer the following questions:

1. How are the mid-ocean mountains and the rift valley of eastern Africa similar? How do they differ? The mid-Atlantic ridge is in the middle of the Atlantic Ocean. Does it have any effect on the North American continent?

2. What are the three types of plate boundaries? According to Dr. Jason Saleeby, there are also three basic types of converging plate boundaries. List them and tell where they occur.

3. How are the Cascade and the Andes Mountains related? How does the origin of the Himalayas differ from that of the Andes?

4. According to the host, why was the discovery of the asthenosphere of crucial importance to the theory of plate tectonics?

5. Based on the video demonstration, describe the way convection works and how it might work within Earth. How is this related to plate motion?

6. How did Dr. W. Jason Morgan describe his concept of mantle plumes?

7. What evidence was presented by Dr. Tanya Atwater that supported the concept of mantle plumes, especially in the Hawaiian Islands?

8. Now that plate tectonics is an accepted theory, what are some of the geologic features that can be explained, and what predictions can scientists make regarding changes on Earth's surface?

You have read the text assignment and seen the video portion of the lesson. This section, made up of a SUMMARY, a CASE STUDY, QUESTIONS, suggestions for ACTIVITIES, and a SELF-TEST, will help you pull the information together.

PUTTING IT ALL TOGETHER

Summary

The great value of plate tectonics is its ability of explain such diverse events as earthquakes and volcanoes, the changing face of the planet, and the evolving ocean basins. The theory grew slowly at first because there was much skepticism about Wegener's ideas. In the last decade however, intensive research, spurred on by the success of plate tectonic theory, has clarified many of the problems that puzzled scientists for years. In this lesson, you will share the insights geologists have about how the plates move and affect Earth's surface – that varied and wonderful landscape upon which all life subsists.

The Lithosphere and the Asthenosphere

When geologists speak of "plates," how much of Earth is really involved? The plates consist of the solid rocky *lithosphere*, the upper 100 to 150 kilometers (60 to 90 miles) that include continental and oceanic crust and the solid mantle lying below the Moho, down to the partially molten zone called the *asthenosphere* (Figure 4.17, page 75) upon which the plates are shifting.

Recent discoveries using the technique of *seismic tomography* are drastically changing our ideas about the boundary between the lithosphere and asthenosphere. (Refer to Box 4.3, "A CAT Scan of the Mantle," pages 94-95, and study Figures 1 and 2.) Geophysicists know that relatively cold, dense rock speeds up seismic waves, and hot, partially molten rock slows them down. (Review Lesson 5.) Using powerful computers to analyze data from thousands of earthquakes, seismologists have found an amazing complexity in the relationships between the cold, solid lithosphere and the hot asthenosphere. The colder rock extends deeply below most continents and under old sea floor such as the western Pacific. In fact, under North America, Asia, and Antarctica, the cold roots extend 400 to 600 kilometers (240-360 miles) downward. Hot rocks were found, as expected, along the crests of the mid-oceanic ridges and rises, and beneath the known hot spots, such as Hawaii and Yellowstone. But some areas that seemed to be hot at 100 kilometers (60 miles) were cold at 300 kilometers (180 miles); others that were cold at 100 kilometers (60 miles) were hot at 300 (180 miles) kilometers! The significance of these and other findings relative to the internal structure of Earth, to plate dynamics, and to mantle convection is still uncertain.

Plate Boundaries

Because a lithospheric plate is rigid and moves as a unit, most of the geologic activity – such as earthquakes, volcanoes, mountain building, creation of new sea floor and subduction of old – takes place within narrow belts that define and locate the plate boundaries. Three major types of boundaries have been identified: *diverging* (pull-apart), *converging* (colliding), and *transform* (horizontal sliding.)

Diverging plate boundaries. Grand scale divergence occurred with the breakup of Pangaea, forming Gondwanaland and Laurasia. Divergence – whether on the sea floor or on the continents – is accompanied by regional tension, shallow depth earthquakes, high heat flow, and eruptions of basaltic lavas along normal (tensional) faults. A divergent boundary is the new growing edge of a plate caused by the volcanic construction of new ocean crust. Both continental and oceanic divergence are characterized by *rift valleys* or down-faulted blocks (*grabens*). In east Africa, a series of grabens marks a divergent boundary; continued separation will move east Africa apart from the rest of the continent, and new sea floor will form in the rifted zone. Areas of ancient rift valleys include Lake Baikal in the Soviet Union, (Earth's deepest lake), the Rhine Valley of Europe, and the Rio Grande Valley of New Mexico.

The Red Sea and the Gulf of Aden are active divergent zones that meet the African Rift Valley at a triple junction (Figures 4.26 and 4.27, page 81). The African Valley is a failed rift (aulacogen) that some geologists believe will not develop into open ocean. Iceland, in the middle of the North Atlantic, is on a particularly interesting rift zone, where the great grabens and active volcanoes rise above sea level. Here you can walk on the mid-Atlantic ridge, peer into an active rift valley, and later take a hot shower with water heated by molten magma under the island. (See Box 11.2: "An Icelandic Community Battles a Volcano – And Wins", Figure 1 A and B, page 255.)

After the initial divergence, what happens to the trailing edges of the separated continents? The volcanic peaks created at the central rift will subside as the sea floor spreads and cools with increasing age. The rugged topography will be smoothed and finally buried by sea-floor sediments to become the continental shelves and slopes. Thick sedimentary deposits will build up on the shelves from debris carried to the ocean by streams eroding the land. These margins, now far from the plate boundary, are passive in that they lack active faulting, volcanoes or earthquakes (Figure 4.29, page 82). Seismic reflection studies, however, have found buried under the cover of marine sediments, evidence of extensive tensional faulting, basalt lava flows, layers of evaporite rock salt, and, if the shelf is within warm tropical waters, carbonate (coral) reefs. Each stage in the separation is documented in the rocks and sediments of the continental shelves.

Converging plate boundaries. To balance sea-floor spreading, collisions must occur either between plates of two continents, a sea floor and a continent, or between two sections of a sea floor. At convergent margins, two plates move toward each other, but depending upon the type of plates involved, the geologic effects differ.

- *Ocean-continent convergence.* Plate collisions between sea floor and a continent are well exemplified by the convergence between the Nazca plate in the south Pacific, and the west coast of South America (Figure 4.34, page 85). Named after the resulting high mountain range, this type of plate boundary is called an *Andean margin.* For hundreds of millions of years, the dense sea floor of the Nazca plate subducted (subsided) under the less dense South American continent, forming a deep sea trench parallel to but offshore from the continental margin. The descending slab of basaltic sea floor (which can be traced downward by the deepening focus of the earthquakes of the Benioff zone, Figure 4.16, page 73) and part of the overlying continental rock continue to undergo partial melting to produce the andesitic volcanoes of the high Andes. The molten magma is erupted through a chain of volcanoes along the edge of the continental plate and contributes to the growth of the mountains. Granite and other deep-seated igneous rocks are believed to form at the base of the volcanic belt. Between the mountains and the trench, a *subduction complex* forms where the sea floor materials scraped off by the descending plate will be deformed, metamorphosed, sheared and stacked, thus extending the young mountains seaward. Although the present cycle of mountain building in the Andes started during the Middle Cretaceous (about 100 million years ago), the earlier geologic history of the great "cordillera" (system of mountain ranges) of which the Andes are a part dates back about 700 million years.

  Other examples of ocean-continent convergence include subduction along the western edge of Mexico and the northwest United States, the latter producing the andesitic volcanoes of the Cascades, including Mount St. Helens. Ancient Andean margins created the

Peninsular Range, Sierra Nevada, and Idaho batholith of western North America.

- *Ocean-ocean convergence.* When two oceanic plates converge, the plate that is older, colder, and therefore denser subducts beneath the other. As with ocean-continent convergence, partial melting will occur, generating andesitic magma. *Island arcs* are formed that have a typical arc-shaped curvature resulting whenever a straight line is drawn upon a globe. These chains of active volcanoes form on the landward side and parallel to a deep sea trench (Figure 4.30, page 83). Examples of island arcs in the Pacific basin include the Aleutians, the Philippines, Japan, the Marianas, Tonga, and Fiji. The angle of subduction may vary and will determine the spacing between the island arc and the trench (Figure 4.31, page 83). The earthquakes in the descending Benioff zone vary in depth from 0 to 700 kilometers (0 to 420 miles) and indicate the angle of subduction (Figure 4.32 and 4.33, page 84). The great earthquakes and volcanoes of the Philippines, Japan, and the Aleutians, as well as the disastrous volcano Krakatoa – between Java and Sumatra in the Indian Ocean – are all consequences of active ocean-ocean convergent boundaries.

- *Continent-continent convergence.* When two continental plates move toward each other, neither plate will subduct because both are composed of low density continental granites or sedimentary rocks. Volcanism is not common, but folding of the crust, metamorphism, mountain building, and earthquakes occur over a broad region.

A classic example of continent-continent convergence is seen in the Himalaya Mountains. A slow but steady collision between India and Asia over the last 50 million years created the highlands of Tibet, the highest (almost 5,000 meters or 16,500 feet above sea level) and largest plateau on Earth (Figure 4.35, page 88). According to some geologists, this great convergence affected and rearranged the entire face of southeast Asia.

After the breakup of Gondwanaland, India was an island that moved northward about 3,000 kilometers (1,800 miles) from its southern hemisphere origin. The intervening ancient sea, *Tethys*, became narrow and subducted. The continents eventually collided and India pushed deep into Asia and is continuing to do so today. The convergence yielded the highest mountains on Earth (the Himalayas) and a continental crust that measures some 65 to 80 kilometers (39 to 48 miles) thick, about twice the thickness of crust elsewhere on the globe. Crumpled remnants of the old sea floor can be found (complete with fossils) on the top of Mount Everest, 8,796 meters (29,028 feet) above sea level!

Other results of the collision are now being hotly debated in scientific journals. Was the uplift and the thickening of the crust the only result of the immense collision? Or, as the new theory of *continental escape* suggests, are parts of the Tibetan plateau "escaping" by sliding horizontally eastward as India puts pressure on Asia? Satellite images show an enormous strike-slip fault with horizontal movement that runs for a length of more than 2,000 kilometers (1,200 miles). Note in Figure 4.1, page 66, the south and eastward bulge of Indonesia. Also note that India and Australia are on the same plate along with a large portion of the Indian Ocean. What are the boundaries of this unusually-shaped tectonic plate? And why are these two lands, widely separated by a sea floor, placed upon the same plate?

Transform or sliding plate boundaries. Transform boundaries are the third type of plate interaction and represent areas where plates are moving horizontally along a single vertical fault or a group of parallel faults. The plate movements are neither compressional nor tensional. These are sliding margins that can abrade each other to such an extent that the plate boundary is a zone of intensely shattered rocks as much as 100 kilometers (60 miles) wide. The San Andreas is a major transform fault system that connects the northern end of the East Pacific Rise in the Gulf of California to spreading ridges off the coast of northern California and Oregon. The fault can be traced from the Mexican border, through most of the state of California, under the Golden Gate Bridge and out to sea north of San Francisco (Figure 4.36, page 89).

The shifting ground and the earthquakes along the San Andreas fault system arise from movement between the North American and Pacific plates. Baja California and all of California west of the fault are on the Pacific plate, which is moving northwest at a rate of a few centimeters per year. In about 10 million years, Los Angeles will approach the present location of San Francisco. At the present rate and direction of movement, in about 60 million years, the continental crust west of the San Andreas will become separated from the rest of North America, and with the already separated Peninsula of Baja California, will become a long, northward-sliding island heading straight for the Aleutian trench! Will it subduct, or will it accrete to Alaska, with a slow but mighty collision, as so many other bits of continental and oceanic crust have done before?

Another transform system is the Dead Sea fault zone in the Middle East. It connects the divergent plate boundary in the Red Sea to the convergent boundary in the Taurus Mountains of southern Turkey. The crust east of the fault is moving north with respect to the crust to the west. Within the fault zone are some "pull-apart basins" in which lie the Dead Sea, the Sea of Galilee and the Gulf of Aqaba. (Look carefully at the north end of the Red Sea in Figure 4.1, page 66, to find this transform fault.)

The sliding motion between plates causes many shallow-focus earthquakes, some of which are of high magnitude. Most transform faults do not have volcanic activity associated with them, although if there is a plate separation, volcanism may take place. On the Jordan side of the Dead Sea, for example, there have been extensive flows of basalt lavas, some of which appear geologically recent as they cascade down the faulted cliffs of the older basin. (See Figure 4.41, page 93, which shows several hot spots near the Red Sea and adjacent to the transform fault. Could this be an alternative hypothesis that explains the basalt flows near the Dead Sea in Jordan?)

Backarc Spreading

Plate boundaries may be rather narrow belts, but the effects of collision may reach hundreds of kilometers beyond a zone of subduction. Behind an island arc, the upper plate may be extended or compressed, or it may be relatively passive. If the *backarc* plate is under extension, a zone of stretching may thin the lithosphere and the crust, forming depressions such as the Aegean Sea of the eastern Mediterranean. If extension continues, new oceanic crust will be created in the backarc region, such as the Sea of Japan between the continent of Asia and the islands of Japan. The new sea floor does not have a ridge with a central rift valley but does exhibit a high heat flow (Box 4.1, Figure 1 A and B, page 86).

As you have learned, the subduction of oceanic lithosphere causes partial melting at depth. The molten igneous material rises into the overlying rock but does not always erupt at the surface. Backarc spreading is attributed to the rising and spreading of these intruded bodies, called *mantle diapirs*. The island arc and trench move away from the continent as the young ocean spreads.

But backarc spreading within a continent can also thin the continental crust, which may account for the features in the Basin and Range province of Nevada (Box 4.1, Figure 2 A and B, page 87). There is no subduction zone near Nevada today, however, and the high heat flow and tensional faulting are puzzling.

Cause of Plate Motions: Mantle Convection

Mantle convection has long been regarded as a cause of plate motion. As with any convection current, whether in a boiling pot of soup, in the atmosphere where heated air rises, or in the slow moving hot rocks of the mantle, heated material is less dense and more buoyant and will rise toward regions of less confining pressure. Cooler substances, being denser, will sink under the pull of gravity to be heated again, and the endless cycle of convection will be repeated.

One theory of mantle convection holds that it is the rising magma at the diverging plates along the mid-oceanic ridges that pushes the plates apart horizontally. Other geologists believe that subduction is the driving mechanism. As a plate descends into the mantle, partial melting will allow less dense minerals to be erupted as andesitic lavas, while the residual dense rocks will continue to descend into the mantle and pull the plate behind them. Perhaps both processes are working together but at different plate boundaries.

The depth of mantle convection is still not known. Some recent theories suggest that convection is shallow and remains within the asthenosphere rather than within the entire mantle down to the molten outer core (Figures 4.38, and 4.39, page 91).

Mantle Plumes

A modification of the theories of mantle convection has been suggested by W. Jason Morgan of Princeton University. This interesting concept holds that the upward convection of hot mantle rock is confined to narrow *plumes* or vertical columns rather than long belts as previously visualized under the mid- ocean divergent boundaries. The plumes are expressed at the surface as *hot spots* in which the hot mantle material spreads out radially at the surface and tends to break up the lithosphere, moving the plates horizontally (Figures 4.40 to 4.44, pages 92-94).

A plume rising beneath a continent would heat the land and cause it to bulge upward into a dome. The stretched surface of the dome would fracture into a three-pronged pattern (Figure 4.42, page 93). Two of the fractures would eventually join to become the edges of the rifted continent and part of a spreading sea floor (Figure 4.44, page 94). The third rift, called an *aulacogen*, would become inactive, eventually filling with sediment. The triple plate junction at the Red Sea, the Gulf of Aden, and the African Rift Valley might well be above a mantle plume, with the African Rift acting as the failed arm or the aulacogen (Figure 4.43, page 93).

The future Atlantic Ocean, before Pangaea split, must have had several mantle plumes with three-pronged rifts on the surface. The failed arms are rifts on today's continents, mostly on the east coasts of North and South America and the west coast of Africa. Some are filled with sediments, but others serve as channel-ways for large rivers such as the Amazon, the Mississippi, and the Niger. But not all aulacogens are at present continental margins. An aulacogen has been proposed at the south end of Death Valley, in eastern California, far from a present plate boundary. This area might have been a very ancient divergent plate boundary about 140 million years ago, related to rifting of western North America and formation of the present Pacific Ocean.

Unresolved Questions

While plate tectonics, as a unifying theory, offers explanations for much recent geologic activity, many geologic phenomena remain unexplained. We understand the reasons that belts of earthquakes usually coincide with the plate boundaries. However, some major earthquakes, such as the Missouri New Madrid (1812), Charleston, South Carolina (1886), and Boston, Massachusetts (1755) earthquakes, cannot be explained in terms of present plate relationships but may be a reactivation of ancient faults created at former plate boundaries.

The deep focus earthquakes at the base of the Benioff zone landward of a subducting trench are also something of a puzzle. Many occur at depths of almost 700 kilometers (420 miles) where the temperature should be so high that the brittle lithosphere should flow quietly and not snap, crackle, or rumble.

The Rocky Mountains are a great assemblage of mountain ranges, all formed within a few million years of each other, that extend northward from near the Mexican border, through the United States, and up to western Canada. These are youthful mountains, but the time of uplift is problematical. Some have suggested that the tectonic forces leading to their initial uplift were collisions and island arcs – about 60 million years ago – along the western edge of the Cordilleran mountain belt. Despite the intense folding and thrust faulting seen in the Rockies, the nearly horizontal sedimentary layers in the nearby uplifted Colorado plateau show very little deformation – another situation that has not been satisfactorily explained.

The Midcontinent rift system is a 1.1 billion-year-old graben that extends from Lake Superior southwestward into Kansas. The rift is filled with huge thicknesses of volcanic and well-preserved sedimentary rocks. The area is puzzling but is clearly related to the rifting of the North American continent. (As a side note, in the last few years, the rift has been drilled to 3,424 meters (11,300 feet) in the exploration for oil; so far no hydrocarbons have been encountered.) Although there are still many geologic features that are not well understood, the overall picture of plate tectonics remains a powerful theory for understanding the origin of the major features of Earth's surface.

CASE STUDY:
Hawaii, the Pearl
of the Pacific

The Hawaiian Islands form a chain that extends northwest-southeast for 3,360 kilometers (about 2,000 miles) in the central Pacific Ocean. The chain makes a sharp bend to the north and continues as the submerged Emperor Seamounts for about another 2,400 kilometers (1,440 miles).

Until a few decades ago, the origin of the island chain and the seamounts was explained as a series of eruptions along a 3,200 kilometer (1,920 mile) fault at the bottom of the Pacific Ocean that started in the northwest and proceeded to the southeast. Early explorers recognized that the individual islands increased in age from the youngest, the Big Island of Hawaii at the southeast end to the oldest, Kauai at the northwest end. Toward the northwest, the erosion was deeper, the soils more developed, the variety of plants was greater and the summits of the islands were lower, all evidence of increasing age. In addition, the only active volcanism was limited to the island of Hawaii; the last eruption on another island was a small flow in 1790 on Maui (Figure 4.46, page 96). Volcanoes on all the other islands in the chain are considered extinct.

The eruptions from the island of Hawaii's five volcanoes also vary, with the greatest current activity limited to the southeast vents, which includes Kilauea and its subsidiary craters on the flanks of the volcano. Mauna Kea, 4,180 meters (13,796 feet), and Hualalai, 2,506 meters (8,271 feet), are classified as dormant volcanoes. Mauna Loa (Figure 4.46, page 96, the western red dot on Hawaii) last erupted in the 1950s. While some observers believe that Kilauea is declining in activity, the almost continuous and voluminous eruption from the southern flank of Kilauea during 1989 and 1990 was one of the most active episodes in the history of the volcano.

With the acceptance of plate tectonics theory, especially with the concept of rising plumes, the geologic history of the Hawaiian and Emperor chain now has an elegant explanation. Instead of the eruptions proceeding from northwest to southeast along a massive fault, it is now believed there is a plume rising under Hawaii in the middle of the Pacific plate. As the plate slowly moves northwestward over the stationary plume or hot spot, large volumes of molten lava flow out on the deep sea floor, building layer upon layer of basalt pillow lavas. After perhaps half a million years, the new volcano reaches sea level, and continues to build a shield volcano whose summit may rise to 9,091 meters (30,000 feet) above the ocean floor. As the Pacific plate continues to move a few inches a year, the new volcano will be carried slowly northwest, away from the hot spot. Erosion will become dominant; the summit will be lowered; soils will develop, and coral reefs may form around the island. Eventually, as the volcano is uncoupled from the hot spot, it will become dormant and then extinct. After that it will subside, and perhaps become a guyot (Figure 4.45, page 96). As the plate continues to move, a line of volcanoes will form over the hot spot, the oldest being farthest from the eruptive center, exactly what we see in the Hawaiian and Emperor chain.

About 30 kilometers (18 miles) southeast of the Big Island is Loihi, a 2,424 meter (8,000 foot) high submarine volcano discovered in the 1950s. In 1971-72 and again in 1975, a swarm of earthquakes was recorded in the vicinity of Loihi. Recent dredgings conducted by a research ship in the vicinity of Loihi brought up young pillow-lava fragments, as would be expected from a growing volcano. Seismic studies of earthquakes indicate the mantle plume underlies the summits of Mauna Loa, Kilauea, and Loihi, which defines the plume's size and location. The plume is apparently fixed in position in the mantle. Its energy waxes and wanes as the lavas erupted build the individual volcanoes.

When will Loihi rise above sea level? Perhaps in 20 to 100 thousand years from now. In another few hundred thousand years, it might reach the size of Mauna Loa. At the present time, as Loihi continues growing, new rumblings on the sea floor may be signaling the birth of yet another submarine volcano to the southeast, as predicted by plate tectonics and hot spot theories!

1. Why was the discovery of the asthenosphere important to the theory of plate tectonics?

The partially molten, non-solid asthenosphere in the upper mantle provided the layer upon which the lithospheric plates were able to slide. Otherwise, neither the continents nor ocean basins would have been able to push their way through the rigid crust and mantle.

2. Why is most geologic activity limited to narrow belts on Earth's surface?

Earth's plates are rigid slabs of crust and upper mantle that move as a unit. It is uncommon to find earthquakes, volcanoes and young

Questions

mountain ranges in the center of a plate, although they do occur. Where the plate boundaries interact with each other, however, there will be tension and divergence – such as on the mid-oceanic ridges. Or there might be compression and convergence, leading to subduction or mountain building. Finally, there can be plates sliding or grinding past one another on the long transform faults. Since the boundaries between the plates are rather narrow belts, so will the geologic activity occur within relatively narrow zones.

3. **When two oceanic plates come together, how can you tell which plate is most likely to subduct?**

In this case of convergence, the denser slab will subduct, and the less dense section of crust will override. In ocean-ocean convergence, the older sea floor will be colder and denser, therefore it is more likely to subduct.

4. **What happens when two continental masses collide?**

This is still a controversial topic. One of the characteristics of continental collision is "thin-skinned tectonics" in which crustal sheets 10 to 20 kilometers (6 to 12 miles) thick slide over one another for tens or hundreds of kilometers. Such displacements stack and thicken different parts of the crust into a series of irregularly deformed and folded sheets, so that the rocks and structures at depth cannot be predicted from the rocks and structures exposed at the surface. This process can be seen in the Appalachians and the Alps.

The collision of India and Asia has caused Tibet, north of the Himalayas, to contract horizontally in the north-south direction while thickening in the vertical east-west dimension, accounting for the great thickness of crust under Tibet (almost like a piece of soft cheese squeezed in a vise).

Still another theory is that huge horizontally-moving faults allowed parts of Tibet to "escape" by sliding eastward. One crustal crack extends more than 1,000 kilometers (600 miles) from Tibet to the South China Sea and accounts for the eastward bulge of Indochina.

Working in Tibet has been a problem because entrance to the country was forbidden to western geoscientists until 1979. The plateau itself presents problems because it is vast and lacks good roads. The region is not completely unknown, however, but there just aren't enough facts yet to decide which of the theories is correct. In any case, it is clear that all continents do not follow the same simple behavior patterns during collision.

5. **What are some of the geological characteristics of diverging plate boundaries?**

Diverging boundaries usually have rift valleys, as seen along the summits of the mid-oceanic ridges or in east Africa. They also have shallow depth earthquakes and high heat flow as in the crest of the mid-Atlantic ridge, the East Pacific Rise, and the axis of the Red Sea. Finally, diverging plate boundaries have basaltic volcanism, which produced the sea floor of the ocean basins.

6. How do the Pacific island arcs differ in origin from such islands as Iceland in the Atlantic Ocean?

The Pacific island arcs owe their origin to ocean-ocean convergence, in which a slab of old, cold, dense sea floor subducts under a plate containing younger, less dense sea floor. As the older plate descends, a deep sea trench will form at the point of subduction. The area is marked by low heat flow where the cold slab sinks downward. The Benioff zone is a belt of descending earthquakes which dips downward under the overriding plate. As the descending plate reaches the hotter depths of the lithosphere, partial melting will occur, enhanced by the water in the sediments and in the old sea floor. The less dense minerals tend to melt first and, combined with partial melting of the overriding crust, will move upward toward the surface, to erupt as andesitic lavas. Thus, the volcanoes of the Pacific island arcs are usually formed of andesite, a rock lighter in color and density, differing chemically from the basalt of the ocean floor.

Iceland, in the Atlantic Ocean, straddles a divergent plate boundary and is formed of basalt that erupted in volcanoes and in the great rift valley that runs through the island. A high heat flow, frequent volcanic eruptions, and shallow focus earthquakes – all characteristic of a divergent sea floor – occur on Iceland.

7. Are the Hawaiian Islands part of an island arc system or part of a sea floor divergent boundary?

The Hawaiian Islands are neither island arcs nor situated on a divergent boundary, although they are in the middle of a vast oceanic plate – the Pacific – and they are predominantly constructed of basalt lava flows. The Hawaiian and Emperor chain of islands owe their origin to a "hot spot" or mantle plume – a vertical column of molten basalt rising from the mantle (Figure 4.40, page 92). The location of the plume has apparently remained stationary for millions of years, while the Pacific plate has passed over it, first moving northerly and creating the older Emperor chain, then changing to a northwestward motion and forming the Hawaiian group of islands. The age of the islands increases to the northwest, and a long line of older submerged volcanoes forms an aseismic ridge (the Emperor chain) northwest of Kauai. There are tens of thousands of other islands and aseismic ridges on the floor of the Pacific (Figure 3.21, page 58), some having similar changes in direction that have an origin similar to that of the Hawaiian Islands.

8. If volcanoes are characteristic of plate boundaries, why aren't there volcanoes in San Francisco, Los Angeles, or any other place along the San Andreas fault?

The San Andreas is a transform, or sliding, fault in which the west side is moving northwesterly. This fault is the boundary between the Pacific plate and the North American plate and is simply a margin along which two plates slide past each other. The sliding movement causes many shallow focus earthquakes that generally do not have any volcanic activity associated with them. Occasionally, however, plate separation does occur, and a "leaky" transform results in a small amount of volcanism. Transform faults on continents do influence the topography and produce ridges, valleys, ponds, belts of trees and other vegetation, offset streams, and offset rock units that enable geologists to trace the path of such long faults as the San Andreas.

9. Can tectonic plates change size?

The large size of the major tectonic plates of today is the result of many cycles of plate collision and plate growth. From evidence found in very ancient deposits on the continents, the early seas covered much of Earth, and the land areas were small, scattered and widely separated. As these islands or micro-continents approached each other, the sea floor between them must have subducted, and volcanism occurred and created new lands. In the billion or more years of tectonic cycles, the land masses grew, came together, and split apart in different ways, while whole oceans disappeared down deep sea trenches.

Today, as new sea floor is added to the North American plate at the spreading ridge in the Atlantic, the plate is growing. As the leading edge is made of continental rock, it is not being subducted. The Pacific plate, on the other hand, is growing smaller as North America is moving westward and the Eurasian plate is moving eastward (Figure 4.1, page 66). The anomaly is that the East Pacific Rise is a fast spreading center, diverging from 6 to 10 centimeters (3 to 4 inches) per year; thus, the Pacific basin should be getting wider! What is happening to all this extra sea floor? (It's being subducted, of course!)

10. What are some of the mechanisms suggested as causes of plate motions?

Causes of plate motions are probably related to slow convection movements within the mantle, in which hot, low density rocks slowly rise toward the surface, cool, become denser, and slowly sink back toward the hot center of Earth. The mid-ocean divergent plate boundaries are regions where heated basalt erupts as molten lava onto the sea floor. It was thought that the force of the rising basalt was *pushing apart* the plate boundaries, causing whole plates to diverge. The convection currents were visualized as broad rolling currents that rose to the surface under the length of the mid-oceanic ridges and spread the plates horizontally.

But more recent studies of central rift valleys showed many tensional cracks, as if the sea floor were being *pulled apart*. If the trenches are being pulled downward by some mechanism – and the strong negative gravity anomalies suggest this – the subduction must exert enough tension to open the rifts at the mid-oceanic ridges. Subducting plates become partially molten at depth; the minerals that are less dense and have lower melting points will melt and rise toward the surface. The rest of the subducting plate, made of heavier minerals, some of which collapse, would have greater weight and would increase the subduction. The question is whether these events are the prime movers of the tectonic plates.

The depth of mantle convection is uncertain. Seismic studies have shown that there are distinct layers having different physical properties within the mantle. Deep convection, if it exists, should have stirred up and homogenized these layers long ago. Shallow convection seems more convincing. Two-tiered convection with cells separated at the 670 kilometer (402 mile) boundary in the mantle has also been suggested (Figures 4.38 and 4.39, page 91).

Activities

1. For a lesson in new geography, on Figure 4.1, page 66, locate the mid-ocean divergent boundaries (double lines) including the mid-Atlantic Ridge, the Atlantic-India Ridge that extends into the Red Sea, the Southeast Indian Rise that forms one boundary of the Indian-Australian plate, and the Pacific-Antarctic rise that connects with the East Pacific Rise. Notice that the East Pacific Rise heads into the Gulf of California and joins the San Andreas fault but emerges offshore again adjacent to the northwest United States and British Columbia as a spreading center. (Note: Not all of these and succeeding boundaries are labeled on the chart in your text, but you should be able to figure them out from their location.)

2. Now locate convergent plate boundaries on the same diagram, shown as heavy single lines with triangles pointing down the subduction zones. Find the Peru-Chile trench and the adjacent Andes Mountains; notice the small Cocos plate north of the Nazca plate, and the mid-America trench (both unmarked), where subduction has caused devastating earthquakes as far inland as Mexico City. Follow the trenches around the north and west boundaries of the Pacific plate all the way to the Tonga trench in the South Pacific. The Philippine Sea plate is bounded on the east by the Mariana trench, the deepest spot on Earth. Is the Mediterranean an opening or closing sea?

3. Are there any transform faults shown in Figure 4.1 in addition to the San Andreas? Notice the transform fault (single line) that runs north from southeast Alaska into central Alaska: this is the major

Denali fault along the Alaska Range, topped by Mount McKinley, which at over 23,000 feet (6,900 meters) is the highest peak in the United States. Also notice the jogs in the mid-oceanic ridges; these are transform faults that offset the central rift valleys. Some recent maps show the zone through the Mediterranean as a transform fault rather than the convergent margin seen in your text, a clue that the controversies over the details of plate tectonics are still alive.

4. Notice from the arrows on Figure 4.1 the direction each plate is moving. From the direction the arrows on the Pacific plate are pointing, determine if this agrees with the direction of alignment of the Hawaiian Islands.

5. Turn to Figure 4.23, page 78, showing how the sea floor ages away from the central rift valleys. Notice how far the Eocene sea floor (deep orange color) is from the central rift in the Pacific as compared with the distance this belt has moved in the Atlantic. What does this tell you about the comparative rate of spreading in these two ocean basins? Why is the oldest sea floor only from the Jurassic Period, about 150 million years ago? This was in the middle of the Mesozoic Era, the Age of the Dinosaurs. Where is the sea floor from the time of the trilobites and crinoids and the wondrous armored fishes from the Paleozoic, which preceded the Mesozoic?

How was this sea-floor chart made? What technique was used? The amount of deep sea sampling, magnetic determinations, age dating, and field and laboratory work involved in producing such a world-wide chart is awesome.

Self-Test

1. Californians know that when the dishes jiggle, the plates are moving. In this case, which plates are involved?

   a. The Nazca and the Pacific plates
   b. The eastern and western North Atlantic plates
   c. The North American and the Pacific plates
   d. The North American and the Cocos plates

2. The Benioff zone is important in plate tectonic theory because it

   a. traces the subduction of slabs of sea floor under continents.
   b. marks the trace of transform faults on continents.
   c. follows the line of divergence in sea-floor rift valleys.
   d. locates aulacogens and other rifted valleys on continents.

3. The advantages of living in Iceland, such as plenty of inexpensive hot water and central heating, are due to its location

   a. on a transform fault.
   b. on a volcanic island arc system.
   c. near a deep subduction zone.
   d. on a divergent plate boundary.

4. Recent analysis of continent-continent convergence has revealed the boundaries can be

   a. wide subduction zones.
   b. areas of high heat flow.
   c. regions of extensive horizontal overthrusting.
   d. narrow belts of andesitic volcanism.

5. The Mariana trench – the deepest place on Earth – has its origins as a

   a. subduction zone near an island arc.
   b. rift valley in the crest of the East Pacific Rise.
   c. graben on an opening sea floor.
   d. backarc spreading center.

6. The difference between islands of the island arc system and those formed over hot spots is that

   a. hot spots are volcanic and island arcs are not.
   b. island arcs are limited to the Pacific, but hot spots occur over wide regions.
   c. island arcs are primarily andesitic, while hot spot islands are basaltic.
   d. island arc islands form on divergent plate boundaries, and hot spot islands form on convergent plate boundaries.

7. Subduction zones are important to the theory of plate tectonics as regions

   a. where new sea floor is being formed.
   b. of plate divergence or pull-apart.
   c. of plate convergence, where old sea floor is being "consumed."
   d. where continental masses are coming together.

8. According to plate tectonic theory, the plates are sliding on the

   a. Moho, at the base of the oceanic crust.
   b. lithosphere.
   c. boundary of the mantle and outer core.
   d. asthenosphere.

9. Of the following, which is NOT considered part of a young rifting sea?

   a. Red Sea
   b. Mediterranean Sea
   c. Gulf of California
   d. Atlantic Ocean

10. All of Earth's ocean floor originated

   a. in the subduction zones.
   b. in the rifts of the mid-oceanic ridges.
   c. as backarc spreading.
   d. from continental collision.

11. All of the following mark a plate boundary except the

    a. San Andreas fault.
    b. Peru-Chile trench.
    c. Aleutian Islands.
    d. Hawaiian Islands.

12. Mount Everest – the tallest mountain on Earth – apparently grew as a result of

    a. continent-continent collision.
    b. island arc formation.
    c. basaltic volcanic eruptions.
    d. location on a rifting continental margin.

13. The origin of the Hawaiian-Emperor chain is best explained as

    a. eruptions proceeding southward along a major fault.
    b. islands formed on a sea-floor divergent boundary.
    c. islands formed on an island arc trench system.
    d. islands formed on a plate moving over a mantle plume.

14. The Pacific Ocean basin is something of a puzzle because it

    a. is getting larger while the Atlantic is decreasing in size.
    b. has a spreading center that is becoming inactive.
    c. is getting smaller but is spreading the fastest of the ocean basins.
    d. is getting larger, but the rate of subduction around the edges is increasing.

15. One major area of uncertainty in the theory of plate tectonics is the

    a. cause of plate motion.
    b. origin of the sea floor.
    c. origin of island arcs.
    d. cause of earthquakes along the San Andreas fault.

7

MOUNTAIN BUILDING

GOAL

This lesson will help you understand how major mountain belts and continents have evolved.

LEARNING OBJECTIVES

After reading the textbook assignments, completing the exercises in this study guide and the lab, and viewing the lesson's video portion, you will be able to:

1. Discuss the characteristics of major mountain belts in terms of the following:

   a. Size and alignment
   b. Age
   c. Thickness and density of rocks and rock layers
   d. Patterns of folding and faulting
   e. Metamorphism and plutonism

2. Describe the evolution of a mountain belt from its inception on the sea floor.

3. Explain the concept of orogeny as it applies to ocean-ocean convergence, ocean-continent convergence, and continent-continent convergence.

4. Describe the ways in which continents grow.

INTRODUCING THE LESSON

The greatest peaks on Earth – Matterhorn, Everest, McKinley – tower above the horizon, thrill us with their serene majesty, and dwarf us with their incredible size. They lure painters and poets, explorers and sight-seers, but the most fascinating story of their long, turbulent, and frequently chaotic origin is revealed only to the inquiring geologist.

Mountains, as part of the evolving landscape, are not forever. Generally, they are born from the sea on a convergent plate margin; their ancient marine sedimentary layers, containing fossils of sea creatures, can be seen today on many a lofty peak. Remains from once explosive volcanoes and from deep igneous intrusions tell of fiery episodes in their past.

Mountains are also subject to erosion, to being slowly reduced to low hills – even flatlands – eventually exposing the ancient heart of the range. The movements of the lithospheric plates can change, and as new convergences develop, new mountains will form. Sometimes the older worn-down mountains will rise again, perhaps to greater heights than before. Even as mountains rise, erosion wears them away, but the fact that mountains reach such great heights tells us that uplift can proceed faster than erosion.

In general, the highest ranges are geologically the youngest; however, since the mountains we see today are in only one phase of their long history, the height of a peak may not be a reliable clue to its age. Some were raised less than 65 million years ago (the Alps and the Himalayas), and some are remnants of a barren landscape of over 2 billion years ago (the Precambrian Shield). Most of the present mountains were formed since the beginning of the Paleozoic Era, about 600 million years ago. The welding of younger and younger mountains to the borders of continents has increased the size of the land masses in a process roughly analogous to the formation of tree rings; the oldest rings are in the center, and the younger rings form on the outside.

Mountain belts are a mosaic of igneous, sedimentary, and metamorphic rocks in such complex relationships to one another that unraveling their history is a challenge for even the most enterprising geologist. In this lesson, you will apply your knowledge of plate motions and plate boundaries to the origin of mountain belts and the growth of the continental crust. You will learn that mountain belts form in stages: *accumulation* of sedimentary rocks; *orogeny*, an episode of intense folding and deformation; and *erosion*, the wearing away of the high peaks. You will recognize the significance of the great thickness of the folded sedimentary rock, of tensional faulting that can develop during uplift, and why earthquakes are frequent in geologically young mountains. And finally, you will understand how a growing mountain belt joins a *craton* (stable continental interior) and eventually becomes part of a growing continent.

Mountains are not to be taken lightly. They are complex, difficult to decipher, and frequently difficult to explore, but they reveal Earth processes on a grand scale. In contrast, the mid-oceanic ridges are the longest mountain systems on Earth, yet when compared to continental mountains, they appear to be the very essence of simplicity. Fortunately, plate tectonics provides a framework on which to build our understanding of the long history of the continents and of the soaring mountains that so lift our spirits.

Completing the following eight steps will help you master the lesson objectives and achieve the goal for this lesson:

Step 1: Read the TEXT ASSIGNMENT, Chapter 5, "Mountain Belts and Continental Crust."

Step 2: Study the KEY TERMS AND CONCEPTS noted in the study guide.

Step 3: Watch the VIDEO portion of the lesson using the VIEWING GUIDE in the study guide.

Step 4: Read the PUTTING IT ALL TOGETHER section in the study guide, which will help you summarize and integrate all of the information in this lesson.

Step 5: Complete any assigned lab exercises.

Step 6: Complete any assigned ACTIVITIES listed in the study guide.

Step 7: Review the material for the lesson and complete the SELF-TEST found in the study guide.

Step 8: Go back to the LEARNING OBJECTIVES and make sure you have learned each one.

LESSON ASSIGNMENT

1. Read Chapter 5, pages 101-121, making sure to read the "Introduction" and the "Summary."
2. Read the "Questions for Review" and try to answer them to see if you understood the material in the chapter.

TEXT ASSIGNMENT

Here are some key terms that will aid your understanding of the material in this course. Carefully study the "Terms to Remember," page 121 at the end of Chapter 5 and look them up either in the chapter or in the glossary. The following terms are explained in greater detail to help you understand the objectives of the lesson:

KEY TERMS AND CONCEPTS

Continental crust: This is the thick, granitic crust under which is under continents. It has two distinctly different kinds of structural units – *cratons* and *orogenic belts*.

Craton: A portion of the continent that has attained tectonic stability and has not undergone significant uplift for geologically long periods. The rocks within a craton are usually deformed, the result of tectonic forces acting in the very distant past. The cratons are the cores around which continents have grown (Figure 5.5, page 105).

Continental shield: The portion of a craton where very ancient metamorphic and igneous rocks are exposed at the surface. Within the craton of North America – called the *Precambrian Shield* or the *Canadian Shield* and seen over much of eastern Canada and around Lake Superior – most rocks are older than 2 billion years. There is no sedimentary cover over the remnants of the old mountain ranges because the surface has been scoured clean by Pleistocene glaciers (Figure 5.6, page 105).

Stable platform: That portion of a craton that is covered by a thin layer of relatively undeformed sedimentary layers. Surrounding the Precambrian Shield, for example, is a zone of former mountain belts ranging in age from about 1.0 to 1.9 billion years that are now covered by younger sedimentary rocks (Figures 5.5 and 5.7, pages 105 and 106). The Great Plains between the Appalachians and the Rockies are part of a vast stable platform of North America.

Orogens or orogenic belts: Elongate regions of the crust that have been intensely folded, faulted, and metamorphosed during cycles of mountain building. Orogens differ in age, history, size, and origin but all were once mountainous terrains.

Mountain range: An elongate series of mountains that may be closely spaced and parallel, and belong to a single geologic unit. Examples are the Sierra Nevada in eastern California and the Front Range in Colorado.

Mountain system: A group of ranges similar in general form, structure, and alignment that originated from the same general processes. The Rocky mountain system is an assemblage of ranges, all formed within a few million years of each other.

Mountain belt or chain: An elongate unit, thousands of kilometers long, consisting of numerous ranges or systems, regardless of similarity in form or age. A good example is *The American Cordillera* that runs along the western edge of the Americas from the tip of South America to northwestern Alaska. This belt at its widest portion includes the coast ranges of California and Oregon eastward to the Rocky Mountains of Montana, Wyoming, and Colorado (Figure 5.2, page 103). The Appalachian mountain belt on the east coast extends from eastern Canada southward through the eastern United States into Alabama (Figure 5.4, page 104).

Terranes: Geologically unique regions that are fault bounded and have characteristics unlike those of adjacent regions. Some terranes may cover thousands of kilometers, while others are smaller. Terranes may consist of fragments of oceanic crust (islands, plateaus, ridges, island arcs) and continental materials carried by subducting sea floor and sutured to the continental margins, an important process in continental growth.

Once you have read the text material for the lesson and studied the terms, you are ready to view the lesson's video portion. Review the questions that appear in this section to help you watch for important points and to prepare you for what you will see in the video.

Before viewing, answer the following questions:

1. What are the basic stages in the evolution of a mountain range?

2. What is a terrane, and how would you recognize one?

3. How do small cratons evolve into large continents?

After viewing, answer the following questions:

1. What are the processes that result in formation of mountains? Where does energy come from to raise mountains? Where does mountain building usually take place?

2. What does it mean when the host says that the North American continent has grown in a "concentric pattern"?

3. What are accretionary terranes and how are they recognized as being different from the rest of the continental crust? What marks the boundaries of the terranes?

4. Why do mountains stay high for so long after they form?

5. What are the basic concepts that explain isostasy?

6. How does Scott Bogue use a magnetometer to determine the place of origin of accretionary terranes?

7. The "Dance of the Continents" is a catchy phrase. To what does it refer? For example, how does it apply to the origin of Alaska?

8. What is happening to the Mediterranean Sea, and how is it related to the active volcanoes and earthquakes in the region?

You have read the text assignment and seen the video portion of the lesson. This section, made up of a SUMMARY, a CASE STUDY, QUESTIONS, suggestions for ACTIVITIES, and a SELF-TEST, will help you pull the information together.

The evolution of great continental mountain belts is written in the accumulation of thick layers of marine sedimentary rocks; in periods of volcanism and igneous intrusions; in intense deformation, metamorphism, and uplift; and finally in weathering and erosion that can reduce the most majestic peak to an insignificant nub. Mountains offer an unparalleled glimpse of the growth of continents, the shifting of plates, and the interactions at plate boundaries. For the geologist, mountains

VIEWING GUIDE

PUTTING IT ALL TOGETHER

Summary

offer the pleasure of working amidst the most beautiful scenery while solving some of the most intriguing geologic puzzles on Earth (Figures 5.1 and 5.3, pages 102 and 103).

General Characteristics of Major Mountain Belts

Despite the differences in age, origin, size, and height, the great mountain belts have certain features in common. They are all linear (longer than wide) and occur on the margin of the old core or craton at the heart of every continent (Figure 5.4, page 104). Roots of the older Precambrian mountains lie beneath the cratons and the sedimentary cover of the stable platforms (Figures 5.5 and 5.7, pages 105 and 106). In the Grand Canyon, for example, the almost horizontal (undeformed) Paleozoic sedimentary rocks of the Colorado plateau, consisting of limestones, sandstones and shales, overlie the complex igneous and metamorphic core of a vast Precambrian mountain range that was eroded flat almost a billion years ago (Figures 8.1 to 8.15, pages 177-182). It is well worth a raft trip down the turbulent Colorado River to experience first-hand this magnificent gorge that is cut into the colorful "layer-cake" of the Grand Canyon and into the ancient metamorphic and igneous rocks of the Precambian Eon.

The thickness of the sedimentary layers in the Grand Canyon is less than 0.6 kilometer (one mile), a small figure compared to the 10 kilometer (6 mile) thickness reached in some mountain chains. From these thicknesses, we can surmise that mountain building takes place where sediments accumulate over long periods of geologic time, such as on a passive, divergent continental margin.

The Evolution of a Mountain Belt

Most geologic events, such as the evolution of the Andes, the Alps, or the American Cordillera, happen over such long periods of time that on the scale of human history they seem to be everlasting and unchanging. We cannot easily imagine the gentle wooded hills of the Appalachians, for example, rivaling the Himalayas in height, as they have done more than once during their long evolution (Figure 5.8, page 106). In view of the complex history of a mountain belt stretching through vast lengths of geologic time, where does the inquiring geologist begin?

The Accumulation Stage

Mountain belts occur along existing or ancient plate boundaries. Thick sections of marine sedimentary rocks indicate an origin at a passive continental shelf and slope of a divergent boundary near an eroding land mass shedding rock debris into the sea. This environment of deposition existed about half a billion years ago on the east coast of ancient North America, and it led to the rise of the Appalachian Mountains during the last half of the Paleozoic era. A similar passive margin environment once existed along the western margin of North America, and today the sedimentary rocks that once accumulated there can be found among the lofty peaks of the Rocky, Sierra Nevada, Cascade, and other mountain ranges of the American West. These examples are close to home, but inspection of all of the major mountain belts of the world – the Alps, the Pyrenees, the Urals, the Andes, to name a few – leads to the same conclusion. They all have vast amounts of ancient sedimentary rock that once formed on a sea floor at the edge of a rifted continent.

The Orogenic Stage and Convergent Plate Boundaries

Convergence of sea floor and continent. After the long period of sediment accumulation, the next stage is orogeny and actual mountain building. Compression, folding, and faulting of the rock layers, igneous intrusion, volcanism, metamorphism, and uplift all take place then. Because of changes in the direction of plate movements, what was once a passive margin has evolved to an active convergent boundary where sea floor subducts beneath a continent or where there is continent-continent collision. (See Figures 5.12 and 5.15, pages 110 and 113, and review Lesson 6.) During subduction, a deep sea trench forms parallel to the continental margin, and a cold sea floor slab descends into the hot lithosphere. The descending sea floor will become heated, metamorphosed, and eventually partially molten; this will lead to the rise of magma, a portion of which will erupt to form a chain of volcanoes on the edge of the continental plate. The most common volcanic rock is called *andesite*, and the tectonic setting described here is called an *Andean margin*.

The Andes Mountains of South America (hence the name *andesite*) and the Cascades of northern California, Oregon, and Washington, (including the active Mount St. Helens) are prime examples of volcanic mountains formed on a continent's edge near a convergent boundary. Earthquake activity, volcanism, and deformed and metamorphosed rock are caused by the compression of the sea floor moving against and under the continent. The old continent itself is also involved in the mountain building process; further heating leads to more widespread (regional) metamorphism, melting of continental crust at depth, and construction of a magmatic arc. The rocks of the old continent and the accumulated sedimentary rocks yield to great compressive forces to become both folded and broken by reverse and thrust faults (Figure 5.13, page 110).

During this long orogenic episode, erosion of volcanoes and igneous intrusions of the magmatic arc will be the source of volcanic sedimentary debris that will be shed toward new basins on either side of the arc (Figure 5.12, page 110). As oceanic crust continues to be subducted beneath the craton, the sediments will be squeezed into folds and reverse faults both on the continental and oceanic side of the magmatic arc. Thus, as the mountains evolve, the continents grow through addition of magma, accumulation of sediment, and accretion of portions of the subducting slab onto the outer margin.

If a subducting plate is carrying a continent, then at some time in the future, the continent will arrive at the site of subduction and collide with an island arc. The continent is too buoyant to subduct, and the direction of subduction must "flip" to the opposite direction as the arc and the continent converge Figure 5.14, page 111). The wedge of sediments on the continent's former passive margin will be deformed. Sections of metamorphosed seafloor and sometimes mantle will be pushed upward, and the arc will be accreted to the continent.

Geologists believe that a subduction zone and a series of island arcs once existed along the western border of the Sierra Nevada. These arc materials are now part of California and are exposed in the Coast Ranges and the Sierra Nevada. Other collisions between continents and island arcs have taken place in Oregon and Alaska, as well as in the Andes of South America. (Read Box 5.1, page 108, about serpentine and talc in mountain belts. Serpentine is the California State Rock.)

The Appalachians: More of the story. Now let us return to the puzzle of the Appalachians, noted above. How can we account for the folding and faulting of the Appalachians, today on a passive divergent plate, far from any active plate boundary? Plate movements can change direction, and ocean basins that once were opening can later close. The Atlantic is an opening ocean now, yet it closed during the late Paleozoic Era. Northeast North America and Europe drifted together about 350 million years ago, closing part of the ancestral Atlantic Ocean, the Iapetus Ocean. During collision of the two continental masses, the marginal marine sediments were folded and uplifted, and the first Appalachian mountain belt was formed. Toward the end of the Paleozoic, southeastern North America collided with Africa, forming the second and more southerly Appalachian Mountains. This mountain building cycle was associated with the assembling of the supercontinent Pangaea.

About 190 million years ago, Pangaea rifted apart; the land masses separated along a margin roughly parallel to the old suture zone, and the Atlantic re-opened. Parts of the old mountain system were left in Europe (Norway, the British Isles, Spain), and parts were left in North America. The Appalachians were largely leveled by erosion during the Mesozoic Era, about 100 to 65 million years ago. Some of the moderate elevation we see today is a secondary uplift caused by isostatic adjustment during the Cenozoic Era. Much of the former great mountain system remains, eroded and buried, but remnants are exposed in Arkansas, central and west Texas, and northern Mexico.

An interesting discovery was made during a recent seismic study across the southern Appalachians. (See Box 5.3, page 112, "Thin Skinned Tectonics.") A major break was found several kilometers beneath the surface. The break, nearly horizontal in most places, is a trust fault along which an enormous slab of folded and fractured surface rock was shoved about 100 kilometers (60 miles) northwestward during collision. An ancient Precambrian basement of metamorphic and igneous rocks lies below the fault (Box 5.3, Figure 1). Large scale slippage of cover rocks over thrust faults is also seen in the Alps and in the Canadian Rocky Mountains.

Continent-continent convergence. You have just learned that the Appalachians resulted from continental collision that created an immense mountain system in the interior of a sutured continent stretching from present day Norway to northern Mexico. Likewise, the Ural Mountains resulted from collision of Europe with Siberia. The result was that former oceans closed and a supercontinent, Pangaea, was created, stitched together by the formation of large Paleozoic-age mountain belts, now long eroded. A series of collisions in the last 50 million years created the Alps and the Himalayas due to the northward movement of Africa against southern Europe, and India against Asia, respectively. A great east-west trending ocean, called the *Tethys Sea*, was closed by the collision, leaving the present day Mediterranean as a mere remnant. As Africa and India wedged under these northern land masses, the crust became doubly thick, and the highest mountains of today's world were created in the wake of folding, thrust faulting, and intense metamorphism. Major earthquakes continue to shatter regions along this youthful boundary – including areas in Armenia, Iraq, northern India, and Tibet.

After collision. Seismic studies have indicated that the highest mountains have roots of low density rock, usually granite, that extend deeply into the denser mantle. Long after the compressive forces of collision become inactive, and as erosion removes the upper part of the mountains, the mountain belt – being less dense than the mantle below – will slowly rise by isostatic adjustment and the continental crust under the mountains will become thinner (Figure 5.16, page 114). Uplift will continue until the crust beneath the mountains becomes as thin as the crust beneath the craton in the continental interior. This is evidenced by seismic data revealing that ancient mountains do not have the deep roots of the younger ranges.

Block Faulting

After the compressive stress ceases and the mountain range is still rising from isostatic movements, normal faults may develop from tensional forces (pull-apart movements) to form fault-block mountains (Figures 5.10, and 5.17 through 5.20, pages 108 and 114-116). There are literally hundreds of parallel, small block-faulted ranges separated by down-dropped valleys in the Basin and Range province of the western United States. This is one of the most extensive fault-block mountain systems in the world; the topography has been compared to an army of caterpillars crawling across the landscape. The famous Death Valley of eastern California is an example of an actively down-dropping valley bordered by fault-block mountains that are continuing to rise even today (Figures 19.9, 19.10, and 19.11, pages 439-440).

The origin of the Basin and Range structures (Box 4.1, Figure 2, page 87), may be the result of hot mantle rock beneath the crust causing regional tensional stress. Volcanism has accompanied the faulting, and it has been suggested that the source of the magma was a spreading ridge (originally part of the East Pacific Rise) overridden by the westward movement of North America. This is still a controversial area and a matter of open debate.

The Growth of Continents: The Story of Terranes

If all continents grew larger by accumulation and magmatic intrusion during subduction alone, then the age of rock units should become younger and younger away from the central craton. But mountain ranges are a complex amalgam or *melange*. They contain not only rocks that date back to the age of subduction and orogeny but also much older rocks that formed elsewhere and were rafted into position and accreted against the margin. These areas of displaced rock, unrelated to adjoining formations and having their own distinctive character in terms of age, rock type, or place of origin, are called *terranes* and have been classified by their mode of origin. *Accretionary terranes* represent continental or oceanic material welded together in the subduction zone. The oceanic rocks may have originated in an island arc, an ocean island or plateau, or a ridge system. *Suspect terranes* have an unknown place of origin, while an *exotic terrane* shows evidence of far travel. Paleomagnetic data are often used to verify the place of origin and to demonstrate that some terranes have moved as far as 5,000 kilometers (3,000 miles). There are at least 50 welldocumented terranes in western North America alone that have been accreted since the early Mesozoic, about 200 million years ago (Figure 5.21, page 117).

The Appalachians contain many terranes, large and small, that were welded onto the eastern North American craton during the subduction and orogenies that produced the Appalachians. Most of the blocks are themselves composites consisting of several smaller pieces of crust that were united before or during attachment to North America.

Geologists are beginning to discover that all convergent continental margins had periods of accretion of exotic rock formations, some from half an Earth away. This is a rather new and controversial idea, but with plates shifting all over Earth, and whole oceans being subducted down marginal trenches, it does seem likely that some of the features of the mobile crust should get scraped off, like peanut butter on the leading edge of a piece of toast (Figure 5.22, page 119).

CASE STUDY:
Terranes in My Back Yard?

In the last lesson, it was suggested that it took 4.6 billion years to evolve the bit of land in your back yard. The next concept to consider is that your land may have come from thousands of kilometers away, slowly heading your way as an ancient tropical island, a section of deep sea floor, or perhaps a microcontinent from a long-gone sea. These are the *terranes*, bounded on all sides by major faults and genetically unrelated to neighboring rocks. For example, if you live in Baja California or on that sliver of California west of the San Andreas fault, you are on a *future exotic terrane*. In about 50 million years, your strip of land, jolting north at a rate of about five centimeters (2 inches) per year, will accrete to Alaska – a land that has been constructed almost entirely of exotic terranes.

If you live near a continental margin or near a major mountain belt, it is likely that you also live on a suspect or exotic terrane. But how can terranes be recognized? And how can we establish where the terranes originated? These questions can be answered only by the close collaboration of scientists in many fields, especially geophysicists, geologists, and paleontologists. The evidence comes from different sources: the age and rock sequences in the suspect terranes, the fossil assemblages, and the detailed studies of the paleomagnetism in the displaced formations.

The fossils in the rocks could be from vastly different ages and from different climatic zones. As Baja California accretes to Alaska, imagine how the tropical fossils in the displaced rocks will differ from the cool-temperate forms in the Alaskan rocks. One of the most interesting examples of fossil evidence of accretion of exotic terranes comes from a group of tiny marine microscopic fossils known as *fusilinids*. Some are disc-shaped, and some look like grains of wheat. These organisms dating back to the Permian Period – about 250 to 290 million years ago – are widely distributed in rocks of westernmost North America but are totally unlike species found farther east in the Rockies and in the middle of the continent. The western forms belong to a species encountered in China, Japan, the East Indies, and the Malay Peninsula – while the fusilinids found in the rocks of Nevada, Texas, and Kansas belong to a separate eastern North American faunal assemblage.

This peculiar distribution was first explained by narrow *seaways* similar to the *land bridges* suggested before plate tectonics. Later, scientists believed that the Pacific Ocean had once closed, so that Asia and North America were in contact. On the reopening of the Pacific, scraps of Asia with their characteristic fusilinids were left behind, plastered to the rifted margin of North America. It is now assumed that rocks bearing the fusilinids were transported east and north on a moving oceanic plate and became accreted to the growing edge of western North America. Today, the exotic fusilinid rocks are found about 500 kilometers (300 miles) inland from the coast of North America. Therefore, it would seem that the rocks seaward must be exotic too, voyagers from distant climes, and this suspicion has been well-confirmed.

But most significant are the paleomagnetic characteristics of the terranes. As you learned from previous discussions, the magnetism – frozen into basaltic and other igneous rocks as they solidified – helps identify the places where they formed. The magnetic inclination of the iron oxide particles (the angle of dip from the horizontal) is an indication of the original latitude. At the magnetic equator, the inclination is horizontal, and at the poles it is steep. The declination or the angle between the magnetic particles and true north tells how the land changed orientation in relation to the poles.

From extensive studies of paleomagnetism, it appears that many of the terranes in western North America have traveled northward thousands of kilometers, some from the southern hemisphere perhaps as far south as Tahiti or Peru. One fascinating proposal is that a large portion of southeast Alaska may have originated in eastern Australia nearly 500 million years ago. This land mass, known as the *Alexander Terrane*, separated from Australia about 375 million years ago, traveled across the Pacific Ocean, stopped near the coast of Peru, then headed north. It collided with the North American continent about 100 million years ago. Three lines of evidence were used to deduce this history: (1) age dates found in igneous rocks; (2) measurements of the paleomagnetic characteristics; and (3) the fossil record of the terrane rock.

Other terranes have been discovered in central Asia, north of Tibet. This finding is a major shift from the original plate tectonic view that the Himalayas resulted from a single collision between northward drifting India and the southern rim of the Asian continent. India is a single great terrane. Some of its individual rock formations have ages exceeding a billion years; nevertheless, over the past 100 million years India has acted as a single mass. In contrast, Asia is now considered to be a *continental mosaic* made up of many blocks of diverse origins.

It has also been discovered that many of the terranes near the western margin of the North American continent rotated clockwise as much as 70 degrees in a geologically short period of time. The cause of rotation is not well understood but may be related to differential westward movement of the North American plate and the northerly motion of the Pacific plate.

Some geologists believe that during the early stages of the formation of continents, land masses were small – probably originating from subduction, igneous intrusion, and andesite volcanism – and moved in a vast sea that covered most of the planet. The microcontinents grew by accretion of fragments of volcanic islands, slicedoff margins of continents, and segments of subducted sea floor. Thus, the ancient cratons would have been the first lands to be assembled from exotic terranes. Many Earth scientists now believe that accretion is a major factor in continental growth. They propose that western North America has grown by more than 25 percent through addition of mostly oceanic terranes during the last 200 million years. We may find that most mountain belts, in addition to the American Cordillera, the Appalachians, and the Himalayas, have grown by piecemeal addition rather than by great plate collisions alone.

So depending on where you live, your back yard may have originated in some far distant land, may be full of unusual fossil plants and animals, and may be part of a mosaic of rocks of all ages. And your yard may have made a sharp right turn before docking at a continental margin. So much for *terra firma* and the everlasting hills!

1. What are some of the characteristics of major mountain belts?

Major mountain belts are always linear, longer than they are wide. They consist of many smaller mountain ranges, which may differ from one another in structure, rock type, and origin. The rocks within a mountain belt are folded, faulted, intruded by molten rock, and metamorphosed. Many of the rocks in mountain belts are old sea floor sediments, oceanic crust, island arcs, seamounts, or volcanic eruptions. Mountains have been raised slowly over great spans of time by tectonic movements such as plate convergence, with uplift that was generally spasmodic and accompanied by earthquakes, both shallow and deep. Sometimes the rate of growth was greater than the rate of erosion, but ultimately, as the cycle slows, erosion wears away even the greatest peaks.

2. What is the general anatomy of a mountain chain?

The anatomy of a mountain chain will depend somewhat on the origins of the chain. Where an oceanic plate is being subducted under a continent at a convergent plate margin, there will be partial melting of the sea floor and lower portion of the continent, producing granitic intrusions and andesite volcanoes inland from the continental margin. The intrusions form from molten magma deep within the crust and cool into huge volumes of lowdensity igneous rock, such as granite, within a *magmatic arc*. This low density rock also thickens the lithosphere, which in turn causes the mountain range to lift because of isostatic adjustments.

On both sides of the magmatic arc will be a high-temperature, regional, metamorphic belt of crustal rocks that have been altered by pressure and heat from the intruding granites. Metamorphism dies out away from the igneous core of the range and gives way to what is called a *fold and thrust* belt. The folds are usually overturned away from the core, indicating that the thrusting is coming from the magmatic and metamorphic region. The sedimentary rocks are folded less severely, but enormous thrust faults can develop. Along these faults, large masses of crust move away from the core. Folding and thrust faulting also shorten and thicken the crust. Closer to the trench, a separate high-pressure, low temperature metamorphic belt will develop and will be characterized by metamorphosed ocean sediments and basalt (changed to *blueschist*) and metamorphosed mantle rock (changed to *serpentinite*).

3. How old are mountain belts? Are the highest mountains always the youngest?

It is frequently true that the highest mountains are the youngest, but since a mountain range may undergo several episodes of erosion and renewed uplift, height alone cannot be used to determine the age of the range. The best evidence is from an analysis of the radio-active age and the fossils included in the rocks.

Geologists know, for example, that the Appalachians formed from collisions leading to the construction of Pangaea and are far older than the mountain belts of the western United States. The Alps and other mountains of the Mediterranean region are considered relatively youthful, since they were uplifted when the African plate moved northward against the Eurasian plate during the Eocene and Oligocene Epochs, about 60 to 25 million years ago. The lofty Himalayas are considered youthful and boast the highest peaks on Earth. They made contact with Eurasia about 15 million years ago, but most of the major uplift has occurred during the last 5 million years. The high Andes are considered a youthful range, although their geologic history extends back into the early Paleozoic Era, almost half a billion years ago. The present Andes are in a phase of active growth as evidenced by volcanic activity and fierce destructive earthquakes. The highest peak in the continental United States is Mount Whitney, in the Sierra Nevada. The granites of this range were emplaced about 100 million years ago into much older Paleozoic rocks at what was then an active continental margin; the present cycle of uplift started about 25 million years ago. Thus, the question "How old is this mountain range?" does not have a simple answer but is usually related to the age of the rocks and the last episode of deformation and uplift.

4. **What does the great thickness of sedimentary rocks indicate about the origin of major mountain belts?**

The layered sedimentary rocks found in mountain belts are usually of marine origin and several thousand meters thick, indicating that the original deposition was on a shallow sea floor such as a continental shelf and slope. In 1785, James Hutton was the first to recognize that mountain building began with a depositional cycle on the sea floor. About 100 years ago, however, the term *geosyncline* was introduced to describe a subsiding trough where sediments could accumulate for long periods of geologic time. This explained the great range of age and the thickness of sedimentary rocks found in the Appalachians (Box 5.2, page 111).

With the new view provided by plate tectonics, geologists realized the concept of a geosyncline needed modification. The site of deposition was not just in a trough but in a broad continental shelf and slope environment on a divergent plate boundary where near-shore sediments merged with sediments that accumulated in deeper water offshore. The new term *geocline* was given to this environment, where shales, sandstones, and limestones would be deposited along the continent's margin, derived from materials eroded off the adjacent land mass. As plate motions changed and convergence occurred, the ancient sedimentary deposits would undergo compression, folding and thrust faulting, intrusions of magma, and volcanic activity, depending on the type of convergence (Figures 5.12 and 5.13, page 110).

5. Which mountains formed from ocean-continent convergence?

The formation of the Andes of South America, the longest continuous mountain chain in the world and the highest range in the western hemisphere (including Mount Aconcagua at 7,021 meters [23,170 feet]), is a good example of mountains resulting from ocean-continent convergence and a tectonically active plate margin. The present pattern of mountain building in the Andes began at the end of the Paleozoic, and since then enormous volumes of igneous rocks have been produced from the subducting oceanic plate.

As a result of compression and stacking of slices and slabs of rocks, there has been uplift and thickening of the crust locally to more than 70 kilometers (42 miles). Volcanic eruptions and earthquakes are still frequent along this active plate boundary. The southern Andes have a more complicated history including a period of Alpine-type continent-continent collision that produced *cordilleras* (chains of parallel mountain ranges) in the late Precambrian and early Paleozoic; since the late Paleozoic, Andean ocean-continent convergence has dominated the mountain building regime.

Charles Darwin, incidentally, during the voyage of the *Beagle* in 1835, noted the presence of relatively youthful Cenozoic marine fossils at high altitudes in the Andes, proof that the range had been elevated greatly during recent geologic time. He observed firsthand an earthquake in which land along the seacoast was suddenly raised several feet. Darwin produced the first scientific description of what we now call a tectonically active plate margin.

6. What is arc-continent convergence, and where has it occurred?

Sometimes an island arc, originally formed by ocean-ocean convergence, will be approached by a continent riding passively on ocean crust being subducted at the trench. The arc, being more buoyant than ocean crust, will not subduct but will collide with and become sutured to the continent (Figure 5.14, page 111). Note in the diagram the reversal of direction of subduction and that the descending plate is still supplying volcanic material to the arc. During the late Paleozoic and early Mesozoic, island arc systems once existed on the west side of the Sierra Nevada (central California) and now occur accreted to the magmatic arc. Mantle-derived serpentinite often becomes tectonically emplaced by faulting during collisions (which is why the rock type is common in California).

7. Why are some mountain belts created within a continent rather than at the margins?

You have learned about the origin of the great Himalaya Mountains, as an example of *continent-continent collision*, caused by the collision of northward-drifting India with the continent of Asia (Lesson 6). Thus, the Himalayas owe their origin indirectly to the breakup of Gondwanaland. That rifting event slowly moved India northward between 10 to 80 million years ago. The Himalayan crust is extremely thick, and the mountain system has deep roots. The thickening developed from a compressive shortening of the crust, thick igneous intrusions, and the wedging of a large segment of the Indian craton beneath Eurasia (Figure 4.35, page 88). Great thrust faults form the borders of the high range, and numerous strikeslip faults (with horizontal movement) throughout the region seem to have permitted the Asian crust to squeeze eastward, as the Indian peninsula pushed northward. Subsequent extension of the crust in this region during a pull-apart/fault-block stage accounts for the grabens in the Baikal rift system, which holds the deepest lake on Earth (Lesson 6).

As mentioned above, the Appalachian Mountains formed from Paleozoic continental collision and, prior to the rifting of Pangaea, existed in the middle of a supercontinent. The greatest continent-continent suturing event in the assembly of Eurasia also occurred during the Paleozoic. The Siberian and Russian platforms, which had been huge separate cratons, were united by collision, forming the long north-south Ural Mountains in the middle of eastern Europe.

8. When does an orogeny end?

The evolution of a mountain belt is complete when the belt becomes part of the craton. Tectonic activity has ceased, and the forces of erosion have reduced the once high peaks to a flat low-lying plain. The crust is now isostatically balanced in relation to the underlying mantle. The thick folded rocks of the ranges are gone, and the igneous and metamorphosed rocks that were deeply buried during the accumulation stages are now exposed at the surface. The crust is no longer undergoing uplift; the continent has grown, and the cycle has ended.

9. To summarize, how do continents grow? Are they still growing today?

The very earliest continents were small, probably andesitic island arcs that continued to grow by accretion and magma intrusion resulting from subduction of ocean crust at their margins. Ocean crust originates from the eruption, at the mid-oceanic ridges and rises, of basalt derived from the mantle. It is in subduction zones that continents grow by accretion processes and by partial melting of subducted basalt ocean crust. The net result is the generation of both ocean and continental crust, either directly or indirectly, from the mantle.

Eventually all ocean crust is destroyed, either being accreted to continents, becoming partially melted with the magma intruding into arcs or continents, or returned to the mantle by subduction.

Hence ocean crust is never very old (usually less than 190 million years). Yet, because continental crust is too buoyant to be subducted, the continents represent giant reservoirs of all material ever generated from the mantle, with ages that range back to 4 billion years.

Are continents growing today? Of course. As young mountains rise and as ocean basins containing thousands of seamounts, islands, and plateaus move toward the continental margins, continents will continue to grow. For example, part of California and its adjacent ocean crust are sliding northward along the San Andreas fault to enlarge parts of Alaska.

Thus we can see that the story of mountain building is part of continental origin and growth. We can also understand that the present arrangements of land and sea are geologically temporary and that the face of Earth has been changing, probably through all of its history. It's certainly something to think about.

Activities

1. Study Figure 5.22, page 119 in your text. Start at the bottom, at example A. If fragments of a Siberian terrane finally accreted to Alaska, what happened to the ocean between? Remember, the small terranes are not *drifting* through the ocean crust, in Wegener style, but the ocean floor is spreading and carrying them along. What was the evidence that the terranes came from south of the Equator? And from the other side of the Pacific?

2. Study Figure 5.21, page 117 in your text and locate the terrane boundaries and the plate boundaries in North America. As North America moves westward, why isn't the leading (western) edge subducting? Why are so many terranes accreting? Do you live on an exotic terrane?

Self-Test

1. If you lived in the middle of a craton, the rocks in your back yard would probably be

   a. recent continental shelf and slope deposits.
   b. island arc andesites.
   c. part of a youthful mountain range that was still growing.
   d. metamorphosed remnants of a very old, former mountain range.

2. If there were pillow lavas in your back yard, you would know that you were living on

   a. the core of a youthful mountain range.
   b. part of an accreted ocean floor.
   c. old continental shelf deposits.
   d. part of an ancient magmatic arc.

3. The development of a mountain belt occurs in which of the following orders?

   a. Accumulation, isostatic adjustment, and orogeny
   b. Orogeny, accumulation, and erosion
   c. Accumulation, orogeny, uplift, and erosion
   d. Block faulting, orogeny, and uplift

4. If you wanted to take a vacation and see an orogeny in action, which of the following would be the best place to visit?

   a. East coast of North America
   b. West central Asia
   c. The Grand Canyon and the Colorado plateau
   d. West coast of South America

5. In trying to unravel the evolution of a mountain system, we see that the great thickness of sedimentary rocks is an indication that

   a. accumulation of sediments took place near a passive divergent margin.
   b. deposition was within a magmatic arc system.
   c. orogeny resulted from ocean-ocean convergence.
   d. accumulation was a result of plate convergence.

6. If the rocks you encountered in your study of a mountain range were generally lacking in volcanic materials, you might guess the origin of the range was

   a. ocean-continent convergence.
   b. continent-continent convergence.
   c. a sea-floor divergent boundary.
   d. an arc-continent convergence.

7. All of the following are useful in determining the previous location of a suspected exotic terrace except

   a. radiometric dating techniques.
   b. gravity anomalies.
   c. paleomagnetic orientation.
   d. analysis of fossil remains.

8. The folds and thrust faults that characterize many mountain ranges are the result of

   a. long cycles of erosion.
   b. isostasy and normal faulting.
   c. compression during convergence.
   d. uplift and divergence.

9. The last great mountain building episode of the Appalachian Mountains took place

    a. when Africa became sutured to North America.
    b. during the break-up of Pangaea.
    c. before the accumulation of thick sediments.
    d. when the Atlantic Ocean rifted apart.

10. When Earth scientists are speaking about "thin-skinned tectonics," you know they are referring to

    a. uplifting of thin oceanic crust.
    b. accretion of long thin terranes.
    c. thrusting of thin rock layers.
    d. subduction of thin continental margins.

11. Which one of the following statements regarding young mountain ranges is generally true? Young mountain ranges

    a. are usually higher than older mountain ranges.
    b. are closest to the central craton.
    c. are sites of frequent earthquakes.
    d. usually have deep roots.

12. The Atlantic Ocean is spreading and the western portion is moving toward the east coast of North America. As a result,

    a. a subduction zone is starting under the east coast.
    b. the Appalachians are being uplifted from this ocean-continent convergence.
    c. the North American continent is on the same plate as the western Atlantic Ocean and is moving with it.
    d. the subduction zone is now in the failed rift of the Mississippi Valley.

13. One of the following assemblages of mountains is considered a result of regional tension in the crust. It is the

    a. Basin and Range Province.
    b. Ural Mountains.
    c. Alps of Switzerland.
    d. mountains of Alaska.

14. Materials accumulating near a converging boundary are most likely to be any of the following except

    a. andesite flows.
    b. limestone layers.
    c. sandstone from weathered igneous rocks.
    d. sedimentary shales.

15. Of the following, the thickest sediments accumulating at the present time are

    a. in the down dropped valleys of the Basin and Range.
    b. off the east coast of North America adjoining the continent.
    c. off the west coast of North America west of the San Andreas fault.
    d. off the west coast of South America in the Peru-Chile trench.

8

EARTH'S STRUCTURES

This lesson will help you recognize how bed rock responds to tectonic forces originating deep within Earth.

After reading the textbook assignments, completing the exercises in this study guide and the lab, and viewing the lesson's video portion, you will be able to:

1. Compare and contrast stress and strain.

2. Distinguish among compressive, tensional, and shear stresses; and between plastic strain, elastic strain, and fracturing.

3. Describe how geologic structures provide clues that allow us to decipher the geologic past.

4. Illustrate how strike and dip are used to denote the attitude of inclined strata.

5. Explain how folds are classified.

6. Discuss why knowing the geometry and rock composition of folds is economically important.

7. Distinguish between joints and faults.

8. Explain how faults are classified.

9. Give reasons why characterizing a fault as active or inactive may be difficult.

10. Compare and contrast the three kinds of unconformities.

11. Explain what each type of unconformity implies about the sequence of geologic events.

INTRODUCING THE LESSON

A drive along any mountain road is like reading a tale of the turbulent history of Earth. It tells of the forces and upheavals that created the landscape. *Road cuts* or excavations into the hillsides made during road-building expose the hidden bed rock that lies beneath the vegetated surface of Earth. The originally horizontal sandstones, shales, or lime-stones may be standing vertically or tilted at some other unlikely angle, or they may be a series of upfolds called *anticlines* and downfolds called *synclines*. A dramatic *angular unconformity* may reveal a long series of events of the distant past. Photographed by satellites, the intensely wrinkled areas of continents give mute evidence of millions of years of plate collisions, tension, and shearing or transform faulting. The persistent movement of the plates creates *stress*, which is a measure of the magnitude and direction of a force that causes *deformation* – or change in the shape – of rock. The resulting change – described either in length, volume, or shape – is termed *strain*.

The actual process of rock deformation, however, is rarely observed in action because it is incredibly slow or may take place below the surface, hidden from view. Faulting, of course, is an exception, as the displacements can be abrupt and are felt as earthquakes.

In this lesson, you will learn about *structural geology* and how the detailed, painstaking work by geologists in the field enables them to construct a geologic map. Such maps use standard symbols to indicate the aerial extent and characteristics of the rock formation and the folding, faulting, or other deformation of the exposed rock layers. The structure of the crust, the sequence of events, and the direction and strength of the forces that shaped the area can be inferred. From this study, you will be able to identify the different folds and faults and become familiar with *unconformities*, ancient erosional surfaces important in deciphering the sequence of geologic events. You will learn *how* the geologist is able to interpret the deformed rocks in terms of tectonic movements, and thus you will add to your understanding of the evolution and *architecture* of Earth's crust.

LESSON ASSIGNMENT

Completing the following eight steps will help you master the lesson objectives and achieve the goal for this lesson:

Step 1: Read the TEXT ASSIGNMENT, Chapter 6, "Geologic Structures."

Step 2: Study the KEY TERMS AND CONCEPTS noted in the study guide.

Step 3: Watch the VIDEO portion of the lesson using the VIEWING GUIDE in the study guide.

Step 4: Read the PUTTING IT ALL TOGETHER section in the study guide, which will help you summarize and integrate all of the information in this lesson.

Step 5: Complete any assigned lab exercises.

Step 6: Complete any assigned ACTIVITIES found in the study guide.

Step 7: Review the material for the lesson and complete the SELF-TEST found in the study guide.

Step 8: Go back to the LEARNING OBJECTIVES and make sure you have learned each one.

1. Read Chapter 6, pages 123-145, making sure to include the "Introduction" and the "Summary" in your reading.

2. Read the "Questions for Review" and try to answer them.

3. Study the diagrams and photographs for visual understanding of rock structures.

TEXT ASSIGNMENT

There is an extensive vocabulary included in this chapter. For easier understanding, rearrange the "Terms to Remember," page 145 in your text. Place the words referring to *folding* in one list (see example), the words describing *faulting* in another, and *unconformities* in a third list. Some words, such as *geologic map*, will not fit into any category but could be listed separately. Your text, the words in italics in the "Summary," and the glossary will help with your definitions.

KEY TERMS AND CONCEPTS

EXAMPLE	TO BE COMPLETED	
Folding	Faulting	Unconformities
Angle of dip		
Anticline		
Direction of dip		
Fold		
Open fold		
Overturned fold		
Plunging fold		
Recumbent fold		
Hinge line		
Isoclinal fold		
Limb		
Structural basin		
Structural domes		
Syncline		

VIEWING GUIDE

Once you have read the text material for the lesson and studied the terms, you are ready to view the lesson's video portion. Review the questions that appear in this section to help you watch for important points and to prepare you for what you will see in the video.

Before viewing, answer the following questions:

1. Under what conditions will rocks fold or bend rather than break?

2. How does a dip-slip fault differ from a strike-slip fault?

3. List some of the types of information recorded on a geologic map.

After viewing, answer the following questions:

1. What do geologic structures tell us about Earth, and why do they have economic importance?

2. Why is original horizontality important in working out geologic structure?

3. Describe "dip" and "strike" to someone, demonstrating the concepts with two books. (This may be difficult to do – especially for "strike.")

4. Geologists spend many hours (or years) in the field preparing geologic maps. If you were planning to build a large commercial venture in an area, what valuable information could you get from reading such map?

5. How do the anticlines and synclines demonstrated in the video differ from the hills and valleys you might see in your neighborhood?

6. Name the four basic types of folds, in order of increasing stress.

7. Give a short synopsis of the origin and accumulation of petroleum, from algae and other microscopic organisms floating in the sea, to the final accumulation of oil in a geologic trap.

8. What do petroleum geologists learn from reflection seismographs? According to geologist Terry Crebs, they are working against time to test and develop their property. What is the rush?

9. What special type of faulting is demonstrated in the video? Where, in terms of crustal movement, would you be most likely to encounter this kind of fault?

10. After viewing the video animation and graphics of an angular unconformity, make a sketch of one. Describe to someone the series of geologic events that must have taken place to produce this feature.

You have read the text assignment and seen the video portion of the lesson. This section, made up of a SUMMARY, some CASE STUDIES, QUESTIONS, suggestions for ACTIVITIES, and a SELF-TEST, will help you pull the information together.

Summary

After three or four billion years of tectonic plates colliding, subducting, shearing, or pulling apart, plus an equal period of uplift and erosion, the deformation of rocks at the surface – especially those of the continental crust – is not unexpected. The force required to *bend rocks* and to break them along faults and joints is called *stress* and is of a magnitude that almost defies comprehension. Yet to understand stress, the geologist must be able to measure the amount and type of folding (strain), trace long faults across the continents, unearth and decipher complex unconformities, and relate the structures to tectonic origins over much of Earth history.

Tectonic Forces

In previous lessons, you learned that the forces acting at plate boundaries are either *compressive*, *tensional*, or *shearing* (transform or horizontal sliding). Compression tends to shorten or push together the rock masses; tension pulls them apart; and shear stresses move rocks in opposite directions but parallel to each other (Figures 6.2, 6.3, and 6.4, pages 124-125). The resultant strain may be *plastic*, in which the rocks are *ductile* (capable of considerable *deformation* without breaking) and will bend. This deformation is irreversible because the rocks will not return to their original shape after the stress is removed (Figure 6.1, page 124). If the rock body returns to its original shape, the strain is termed *elastic* – this is reversible or nonpermanent deformation. If the rock formations are brittle, they will tend to *fracture* – a permanent or irreversible reaction; most rocks will fracture under intense stress.

Faulting usually is accompanied by abrupt movement and fracture. Stress builds up slowly as elastic deformation occurs; when the strength of the rock is exceeded, fracturing occurs. Faulting can also result in slow, steady slipping during which ductile deformation may occur at depths of anywhere from 20 or 30 kilometers (12 to 18 miles) to 100 kilometers (60 miles) and where brittle fracture occurs only near the surface.

As you study this chapter, try to visualize the geologic features being discussed in three dimensions, not as two dimensional diagrams on a flat page. The faults and folds may extend many kilometers into the mountains behind the outcrop or road cut you are seeing. Unconformities may be traced great distances as well. The figures that accompany Box 6.1, page 124, regarding the San Andreas fault give a fair indication of the length of this fault.

Why Do Some Rocks Bend?

From your experience, you know that rocks at the surface break into bits when pounded with a hammer (fracture under stress). Under what conditions do they fold or bend? The essential factors that determine whether rocks fracture or undergo ductile or plastic deformation are (1) temperature, (2) confining pressure, (3) time and rate at which movement occurs, and (4) composition.

The higher the temperature, the weaker, less brittle, and more ductile a solid becomes. Deeply buried rocks at higher temperatures and pressures will bend, resulting in folded strata and the distorted *foliation* of certain common metamorphic rocks. If the stress in a rock body builds up slowly and gradually and is maintained for long periods, the atoms in the rock have time to move; the solid material can slowly adjust to the stress and change shape by folding and flowing. If stress is applied suddenly, the rock will break.

Composition of the minerals that make up a rock also affects the response to stress. A mineral that deforms readily, for example, is rock salt, present in the great salt domes of the U.S. Coast along the Gulf of Mexico (Box 21.1, pages 475-476). Other ductile minerals are calcite, dolomite, clay, mica, and talc. The rocks that are ductile are limestone, marble, shale, slate, and schist; those that fracture include sandstone, granite, and the metamorphic rock gneiss.

Geologic Structures: Strike and Dip as Clues to the Past

On an evolving Earth, the forces that moved the plates in the distant past may not be the same as those moving the crustal plates today. A study of the structure of sedimentary or volcanic rocks provides the best clues about what happened to the layered rocks since they were deposited as sediments, lava flows, or ash beds.

The field geologist, in preparing a geologic map, examines exposed rock layers to determine the extent and direction of tilting. These measurements are termed *dip* and *strike* and are described as the relationship between the surface of the inclined bed and an imaginary horizontal plane. (Remember: all sedimentary layers were originally deposited in a horizontal position; see Figure 6.6, page 127.)

If we consider a tilted sedimentary bed as an inclined plane, we know that the intersection of two planes is always a straight line. Visualize a tilted layer of rock entering a lake which has a horizontal surface (Figure 6.8, page 127). The straight line at the intersection has a direction that can be determined with a compass. This direction is called the *strike* and is usually expressed as an angle ranging from 0 to 90 degrees east or west of true north. For example, the strike of a sedimentary layer might be trending 20 degrees east of north and would be recorded on a geologic map as N 20° E (Figure 6.7, page 127).

We need one more measurement to fix the orientation of the tilted rocks. This is the *dip* – the angle between a horizontal plane (such as the imaginary lake mentioned above) and the inclined surface of the rock layer. It is measured down from the horizontal and in a direction always perpendicular to the strike (Figure 6.8). Sometimes it is necessary to take hundreds of dip and strike measurements in the field to determine a large geologic structure. For these measurements, geologists use a Brunton compass, which is an instrument designed just for this purpose (Figures 6.9, 6.10, and 6.11, page 128). Notice the symbol used for dip and strike on the geologic map and note how the direction of dip is indicated.

Folded Rocks

Take a few layers of heavy bath towels and, on a table, slowly compress them or push them together. You will make a series of folds that in sedimentary rock would be called anticlines and synclines (Figure 6.14, page 129). Sometimes the folds can be seen in road cuts or in outcrops (Figures 6.1 and 6.12, pages 124 and 129), but more often the structures are below the surface and must be identified by the changing angle of dip of the exposed layers (Figure 6.15, page 130).

Folding may consist of broad, gentle warping that extends over hundreds of kilometers; or it might be close, tight flexing of microscopic size; or anything in between (Figure 6.20, page 131). An *anticline* is an upfold in sedimentary layers in the form of an arch; a *syncline* is a downfold in the form of a trough. Do not confuse anticlines and synclines with hills and valleys, the latter being erosional features. Synclines can be found on tops of mountains, and anticlines on eroded flatland. If an anticline is deeply eroded, the oldest rock formations will be exposed in the center of the anticline; if a syncline is eroded, the oldest formations will be exposed on the outer limbs. Study Figures 6.16 through 6.19, pages 130-131. Notice the *domes*, the *basins*, and the "bulls-eyes" pattern when the formations are eroded (Figure 6.18, page 131). What is the difference between a dome and an anticline? How would you tell a dome from a basin in the field (two ways!)?

On an anticline, the beds dip *away* from the central hinge line that divides the two *limbs* and dip *toward* the hinge line of a syncline. The folds may be *plunging folds* in which the hinge line is not horizontal. On an eroded surface, the outcropping beds of a plunging anticline or syncline will form a V or a horseshoe pattern. Refer to Figures 6.16 through 6.21, pages 130-132 for excellent representations of folded sedimentary rock layers. Note that all types of folds can be identified or classified in the same way: by the *age* and *pattern* of the beds and the *dip* of the layers.

Fractured Rocks: Joints and Faults

Rocks at the surface are usually brittle and tend to be either *jointed* or *faulted*. Fractures along which little or no movement of rocks has occurred are called *joints*. Joints can be very small, like cracks in a window pane, or large and widely spaced. Sometimes the joints are parallel or are in sets that cross one another, depending on the direction of stress.

Large joints are important in engineering geology, because they weaken the ground surface. Dams and reservoirs located on such a surface might leak (Figures 6.22 and 6.23, page 134).

Faults are fractures with the opposite sides displaced in relation to each other (Figure 6.24, page 136). The movement may be in centimeters or may involve displacement of rocks as great as several hundred kilometers. For evidence of faulting, the field geologist looks for sedimentary layers that have been displaced, fault scarps or cliffs, long straight valleys, uplifted or downdropped land surfaces, dislocated stream beds, and ground up or pulverized rock called *gouge* between the displaced sides.

To describe the movement and the orientation of the fault plane, geologists have adopted two old British coal mining terms used to describe the floors and ceilings of inclined mine shafts. The *hanging wall* is the rock vertically above the fault plane, and the footwall is the surface of the rock below an inclined fault (Figure 6.25, page 136). Classification of faults is based on (1) the inclination of the surface along which movement has occurred (strike and dip of the fault plane) and (2) the direction of relative movement or slippage of rock. A *dip-slip* fault implies vertical movement, and a *strike-slip* fault is horizontal motion parallel to the strike of the fault plane. Most faults will show components of both and are termed *oblique-slip* faults (Figure 6.24, page 136).

Dip-slip faults. Dip-slip faults, based on the direction of movement of the hanging wall in relation to the footwall, may be *normal faults* in which the hanging wall will slip down relative to the footwall, or *reverse faults* in which the hanging wall will move upward over the footwall. Generally, normal faulting is associated with tensional forces that cause the crust to be stretched (Figures 6.26 and 6.27, pages 137 and 138). Sometimes a block bounded by normal faults on either side will drop down, creating a *graben*. If the central block is uplifted, it is called a *horst*. (You learned about grabens and the basin and range fault block mountains in Lessons 6 and 7).

Reverse faults arise from compression that pushes older rocks over younger ones, shortening and thickening the crust. *Thrust faults* are low-angle reverse faults in which the fault plane dips less than 45° and frequently less than 15° from horizontal. Thrusts are common in mountain chains, where compression has moved great slabs of crust many kilometers over the lower block (Figures 6.28 and 6.29, pages 138 and 139).

Strike-slip faults. In most strike-slip faults, the fault plane is essentially vertical, and the movement of opposing blocks is horizontal (Figures 6.24 B and 6.30, pages 136 and 139). *Left-lateral strike-slip faults* involve offsets toward the left when viewed across the fault plane, while *right-lateral strike-slip faults* show displacement to the right. The San Andreas is a right-lateral strike slip and if you were standing on the North American plate looking across the fault, you would see the Pacific plate and part of California moving northward toward your right. If you were standing west of the fault on the Pacific plate, in which direction would

North America appear to be moving? (Again, it would appear to be moving to your right.)

The remarkable strike-slip Great Glen fault slices through the full width of Scotland from southwest to northeast and disappears into the North Sea. The west side moved southwest, and the east side moved northeast. (Is this a right-lateral or left-lateral strike slip?) The trace of the fault is an erosional valley containing a string of lakes, one of which is Loch Ness, home of "Nessie," the famous Loch Ness Monster! The fault was active in the Paleozoic Era, over 250 million years ago; it cut through an ancient granite mass and displaced rock on either side horizontally about 100 kilometers (60 miles).

Displacement on faults is not always easy to measure, and there is uncertainty about the actual horizontal displacement along the well-studied San Andreas fault. Survey points on the ground are frequently checked along the San Andreas to determine direction and rate of movement.

The Great Glen fault apparently is no longer active, but the San Andreas is one of the most active faults on Earth. (Refer to Box 6.3, pages 140-141, noting the photographs and diagrams in the box). Active faults are classified by some experts as those that have moved in recorded time; others say within the last 10,000 years, although some experts extend the time limit to 100,000 years. Careful field work can generally determine when ancient faults last moved. As most faults are inactive, only the most active and those in urban areas are the objects of intense study (Box 6.2, page 135).

Unconformities

An *unconformity* is a break in the geologic record that records the sequence of geologic events in a region. It tells of deposition of sedimentary materials, plate movements causing deformation of the sedimentary strata, uplift, and a period of erosion followed by deposition of new sedimentary layers. The unconformity records the long period of erosion and informs the geologist that part of the rock record is missing. The unconformity may also result from a change in environmental conditions that caused deposition to cease for a period of time. Unconformities are classified into three types: disconformities, angular unconformities, and nonconformities.

Disconformities can be difficult to see. The rocks above and below the break are parallel to each other, and the disconformity may appear as a bedding plane between sedimentary layers. This type of unconformity is defined as an erosional surface that was covered by later deposition with no tilting of the lower formations. How does the geologist determine there was a break in the geologic record? It requires a careful analysis of the fossils contained in the beds above and below the break to reveal the missing geologic time (Figure 6.31, page 142).

The most obvious type of unconformity is the *angular unconformity* in which the older strata were deformed, tilted, and acted upon by erosion before the younger horizontal layers were deposited (Figures 6.32 and 6.33, page 143).

The third class of unconformity is the *nonconformity*, which indicates uplift, complete erosion of overlying sedimentary strata followed by long-continued erosion of metamorphic or igneous rock below. This cycle is followed by subsidence and deposition of younger sedimentary layers on the ancient erosional surface (Figures 6.34 and 6.35, pages 143-144).

A study of faults, folds, and unconformities brings out the relationships between crustal movements and the forces of erosion and deposition. Today, you are walking on a potential surface of unconformity, as streams and glaciers, wind and waves erode the uplifted and exposed rocks of the land. The eroded sediments are being carried away and deposited elsewhere, perhaps on some old erosional surface. Again we see the interactions between the internal and external processes that have been going on throughout Earth's long history.

CASE STUDIES

Your text has some excellent case studies related to this lesson. Read Box 6.1, page 133, "Is There Oil Beneath My Property?" Box 6.2, page 135, "Faults and Nuclear Power Plants," and Box 6.3, pages 140-141, "California's Greatest Fault – The San Andreas." Study the diagrams showing the San Andreas and the location of Los Angeles and San Francisco in relation to the fault.

Questions

1. What is meant by structural geology?

Structural geology is the study of rock deformation – including folds and faults, unconformities, shapes, arrangement, and relationships of bed rock units. Geologic structures are studied to determine causative forces and to relate them to plate movements of the past. Episodes of uplift and deposition, of tilting, folding, and displacement can be read and displayed on a geologic map.

2. What is the difference between stress and strain?

Stress is a *force* acting on a rock unit; it results from movement or other forces generated within Earth. *Strain* is the deformation of rock in response to the stress placed upon it. As result of stress, the rock may fracture, fold, or become jointed or faulted or otherwise adjust to the force applied.

3. What are the three types of strain observed in rock bodies?

The three types of strain are:

Plastic strain. Occurs when the deformed rock does not return to its original shape after stress is released. Plastic strain is related to temperature, confining pressure (or the weight of overlying rock), the length of time stress is applied, and the presence of ductile minerals. Rocks become less brittle and more susceptible to folding at higher temperatures and pressures when stress is applied slowly and the rocks are at some depth in the crust. Certain minerals such as rock salt are ductile and will flow easily rather than break under stress.

Elastic strain. Describes rock that returns to its original shape after stress is released. Most solid rock is somewhat elastic as you learned in the discussion of seismic waves passing through the body of Earth.

Fracture. If a rock body breaks in its response to stress, it is said to *fracture*. Most of the rocks on the surface are brittle and will break when hit. But rocks will fracture after prolonged exposure to intense stress, even if their original response was plastic or elastic strain.

4. **How can folds and faults be used to interpret the geologic past?**

Sedimentary rocks, normally layered and deposited in a horizontal position, offer the most easily interpreted indications of past geologic history. Marine rocks on mountain tops indicate huge uplifts usually due to compression during plate collision. If the layers are essentially horizontal as in the Grand Canyon, the forces were equally applied over a vast area that was uplifted as if by a giant piston. Tilted and folded rocks indicate unequal forces applied at depth within the crust over long periods of time before uplift and erosion exposed them on the surface. Thus we can see that plate tectonics, folding and faulting, and mountain building are all genetically related.

Faults also are a clue to plate movements; the elastic limits of the rock are passed and the rock breaks. The strike-slip faulting on the San Andreas is definitive evidence of the northward movement of the Pacific plate against the edge of the North American plate. Reverse and thrust faulting accompany compression, while normal faulting is an indication of tension or pulling apart of the crust. Small scale folding adjacent to a fault plane can be a clue to relative movement of blocks on either side of the fault. In old inactive faults, it is sometimes difficult to decide whether normal or reverse faulting occurred.

5. **What kind of information can a geologist gather from a geologic map?**

A good geologic map provides a wealth of information. The various surface formations (distinct rock units) will be displayed in different colors, coded to a key that gives the name of the formation, the age, and the kind of rock. The kind of rock would include lava flows, marine sandstones, shales or limestones, metamorphic rocks, igneous rocks that solidified at depth but are now exposed on the surface, ancient non-marine rocks, as well as ore bodies, surface deposits such as uplifted marine terraces or glacial debris, and even old landslides. Superimposed on the map of the geographic extent of the formations will be symbols showing the dip and strike with direction and angle of dip. The traces of faults will be marked, and sometimes the unconformities and the contacts between adjacent formations will be indicated. In mountainous areas, the geologic map may be an object of extreme complexity, with faulting and folding of many rock units, to which might be added the alien formations of the exotic terranes.

There are excellent geologic maps of many localities. The work involved in preparing a geologic map is not inconsequential; it may take years of field work, sometimes under difficult living conditions. The cost to the map buyer is minimal, however, because the work is usually supported by state or federal agencies. A geologic map of your area will provide many hours of enjoyment and will help you understand the geology of the area where you live.

6. How is strike shown on a geologic map?

Strike is shown on a geologic map by a line marked with a compass direction that shows the orientation of a tilted bed as determined in the field. Since the intersection of two planes is always a straight line, a special compass that can create an artificial horizontal plane can determine the direction of the intersection of the bedding plane with the horizon. The direction is usually expressed as degrees east or west of true north.

7. How is dip different from strike?

Dip – the angle between a tilted bed and a horizontal plane – is measured at right angles to the strike to get the maximum angle of tilt and is shown as degrees from horizontal on a geologic map. The direction of dip is also shown because the strike alone does not indicate in which of the two possible directions the bed is tilted (Figure 6.8, page 127). Strike and dip are important in discovering the geologic structure of folds that cover large areas or that have been deeply eroded so their structure is not clearly apparent.

8. What is the difference between a joint and a fault?

A joint is a break or fracture in which little or no movement has occurred along the plane of the fracture. A fault is also a fracture, but because there has been movement on the fault plane, the opposite sides have been displaced relative to each other.

9. What is the difference between a dip-slip fault and a strike-slip fault?

A dip-slip fault, as the name suggests, moves parallel, either up or down, to the dip of the fault plane. The dip of the fault plane is measured down from the horizontal at right angles to the strike, as if it were a tilted sedimentary layer. Normal faults indicate tensional forces because the upper block above the fault plane (the hanging wall) slips down the fault plane. Reverse and thrust faults are the result of compressional forces pushing the hanging wall upward over the footwall.

Strike-slip faults move horizontally along a vertical fault-plane strike, which has a compass direction. The strike of the San Andreas fault, for example, is generally northwest-southeast, although there are a few bends in its 1,000 kilometer (600 mile) path through western California.

10. At one time, the West Coast seemed ideally situated for construction of nuclear power plants. What caused the change in plans?

The West Coast, close to the shore, has a very low population density, which would be good in case of a radiation leak. In addition, the coast is close enough to urban settlements to keep the cost of transmitting electricity to a minimum. The availability of cooling water from the ocean is a major benefit for removing waste heat. California, however, is on a major plate boundary, is extremely active seismically, and is crisscrossed with large and small faults. Most of the faults are considered inactive, and the location of major active faults has been carefully mapped. Determination of active versus inactive faults is not an easy task, however, and some "inactive" faults may cause strong earthquakes after a century or two of quiescence. Because of public concern about potential damages nuclear power plants and the uncertainty about active versus inactive faults, plants already in operation will continue to produce electricity, but few new plants will be built.

11. What type of geologic structures do geologists seek as possible traps for petroleum?

One of the economic benefits of studying geology is knowing where to find oil. Petroleum – aside from water – is one of the very few natural substances that exists as a liquid on this planet. The source materials of oil and gas are marine organisms, mostly *plankton*, myriads of microscopic floating plants and animals that fall to the sea floor and are buried by sediments before being eaten or decomposed. After deep burial and long heating of these plants and animals, oil droplets form and will migrate upward, following fractures and interconnecting pore spaces in the rock. The oil sometimes follows the cracks to the surface, where it oozes out on land or under the sea as an oil or tar seep.

If an impermeable layer of rock, such as a hard dense shale, blocks the path of migration, the oil may accumulate in a porous formation, which becomes a *reservoir rock*. There also must be a geologic structure to hold or *trap* the oil and prevent further migration. A favored structure is an *anticlinal trap*, where gas fills the upper part of the arch, oil accumulates below, and water fills the lower part of the trap. (Remember: oil floats on water!) (Refer to Figure 1 in Box 6.1, page 133.) There are other types of structural traps, such as accumulation against faults, or under unconformities, or where the rock changes from permeable to impermeable because of changes in the conditions of sedimentation (Figure 2 in Box 6.1, page 133 and Figure 21.1, page 472).

A perfect anticline and a potential reservoir do not guarantee commercial quantities of oil, however, and the ultimate test – at huge expense – is to drill a well. Geologists use many means to try to "see" underground before drilling, but even with all the techniques available, only about one well in ten will produce enough oil to pay for itself.

Activities

Look through the diagrams and photographs that show faults, folds, or unconformities in other chapters in your text. For example:

1. Look at Box 21.1, Figure 2, page 475 showing oil and gas trapped against a salt dome. What kind of faults are shown? (The half-arrow shows the direction of movement.) Why is the oil trapped against the salt, and why doesn't the salt dissolve away? What is the general structure of the formations invaded by the ductile salt? What geophysical methods are used to detect the salt dome? (Did you know that salt, which is easily soluble in water, does not dissolve in oil?)

2. Look in Chapter 8 in your text, and follow Figures 8.3 through 8.12, page 178-181 to learn how a geologist interprets past events through unconformities. Also, locate the angular unconformities and the nonconformity in the Colorado plateau in Figure 8.14, page 181.

3. In Figure 16.46, page 373, what term is used for the right-hand geologic structure shown in yellow in the diagram? What geologic structures are shown in Figure 16.47, page 374? Turn to Chapter 19, and look at Figures 19.10 and 19.11, pages 439-440. These valleys were formed in the same way as that in Figure 16.46, page 373. Death Valley, California, Figure 19.9, page 439, is a perfect example of this structure. (Compare with Figure 6.26 D, page 137 and decide if this is a result of tension or compression in the crust.)

4. In Figure 6.20, page 131, notice the *isoclinal* or *hairpin folds*, the overturned folds and the recumbent folds. In which case would the stress or force be the greatest? Compare with photographs in Figure 6.21, page 132.

5. Obtain a geologic map of your area or an area of interest to you. If there is a county, state, or federal geologic or mining office in your area, call the workers there and ask if they sell geologic maps of the area. If they do not, they may direct you to another agency that has maps. Private map dealers may have what you want.

You can also write to the United States Geological Survey, Map Distribution Office, P.O. Box 25286, Federal Center, Denver, Colorado, 80225, or to United States Geological Survey, Map Distribution, 507 National Center, Reston, Virginia, 22092. People there can give you lists and prices of maps of your area. Most of their maps cost about $3.60, with the most expensive going for about $13.

6. Some natural history museums or planetariums have exhibits, and their libraries or gift shops may have guide books on local geology. Call them.

1. If you were doing field work near a plate boundary that you believed had undergone collision, which of the following structures would you be least likely to find?

    a. Anticlines
    b. Reverse faults
    c. Grabens
    d. Thrust faults

2. In your field work, you notice that the rock formations are folded rather than broken or fractured. You can assume that the deformation occurred

    a. on the surface or at very shallow depths.
    b. in rocks that lacked ductility.
    c. as a relief of pressure from overlying formations.
    d. over long periods of time at depth in the crust.

3. Of the following types of strain that result from stress on rock bodies, only one is reversible. It is

    a. plastic.
    b. elastic.
    c. fracture.
    d. flowage.

4. If the oldest rocks in a folded and eroded sedimentary structure are found nearest the hinge line, you know you have discovered a(n)

    a. anticline.
    b. recumbent fold.
    c. angular unconformity.
    d. syncline.

5. Faulting is sometimes hard to discern in the field. All of the following would be clues except

    a. long straight valleys.
    b. offset sedimentary layers.
    c. tilted sedimentary rocks overlain by horizontal layers.
    d. displaced stream channels.

6. Many people in Los Angeles know they are slowly creeping northward along the San Andreas fault. To the geologist, this means the San Andreas is a

   a. right-lateral strike-slip fault.
   b. normal fault.
   c. reverse fault.
   d. left-lateral strike slip fault.

7. If you were a petroleum geologist logging a drilling well and you drilled through a sequence of rocks that were younger at depth than those above, you would know you drilled

   a. into the center of a syncline.
   b. through a disconformity.
   c. through a strike-slip fault.
   d. through a thrust fault.

8. If you, as a paleontologist (a specialist in the study of fossils), discovered that a whole succession of fossil forms that occurred elsewhere was missing from a sedimentary rock sequence, you would infer that you had encountered a(n)

   a. disconformity.
   b. normal fault.
   c. isoclinal fold.
   d. nonconformity.

9. From a series of aerial photographs, you notice that the patterns of eroded exposed strata form a series of Vs or horseshoes. From this you realize that you are looking at

   a. an angular unconformity.
   b. plunging folds.
   c. grabens.
   d. joints in sedimentary rocks.

10. When taking measurements of tilted sedimentary rocks, you are always careful to measure

    a. dip at right angles to the strike.
    b. strike at right angles to the horizon.
    c. dip parallel to the strike.
    d. strike in a downward direction from the horizon.

11. If you are observing folded sedimentary strata in the field and the plane that includes the hinge line is almost horizontal, you are seeing

    a. a syncline.
    b. elastic strain.
    c. a recumbent fold.
    d. low energy stress.

12. You are looking at the dips and strikes recorded on a geologic map. You notice a strange structure in which the dips are pointed away in all directions from the eroded center. You determine

    a. this is a hill or mountain.
    b. this is a structural dome.
    c. this is a volcanic cone.
    d. the center exposes the youngest rocks in the area.

13. The sequence of events that can be "read" in angular unconformities is as follows:

    a. erosion, uplift and tilting, deposition, and renewed erosion.
    b. deposition, erosion, tilting, and renewed deposition.
    c. uplift, deposition, erosion, and renewed uplift and tilting.
    d. deposition, uplift and tilting, erosion, and renewed deposition.

14. A nonconformity always has a cycle of

    a. tilting of sedimentary layers.
    b. lava flows above an erosional surface.
    c. uplift and erosion of igneous or metamorphic rocks
    d. an erosional surface between parallel layers of sedimentary strata.

15. All of the following structures may serve as "traps" for petroleum except

    a. anticlines.
    b. synclines.
    c. faults.
    d. unconformities.

9

EARTHQUAKES

This lesson will help you understand the nature and consequences of earthquakes.

GOAL

After reading the textbook assignments, completing the exercises in this study guide, and viewing the lesson's video portion, you will be able to:

LEARNING OBJECTIVES

1. Describe the factors that contribute to the occurrence of earthquakes.

2. Recognize the relationship between faults and earthquakes.

3. Describe the characteristics of P and S waves and contrast them with surface wave motion.

4. Discuss how the location of an earthquake is determined.

5. Explain the difference between the intensity and the magnitude of an earthquake and how each is determined.

6. Compare earthquakes that occur in the eastern United States to earthquakes that occur in the western United States.

7. Describe the effects of earthquakes.

8. Explain the origin of tsunamis.

9. Trace major concentrations of earthquakes on a map of the world.

10. Relate the concept of plate boundaries to the distribution of earthquakes.

EARTHQUAKE – the tectonic reminder that we are living on the moving lithosphere of a restless planet! As the great plates slowly crush together, tear apart, or grind past each other, stress builds in the solid crust, and rocks suddenly break, move, and set off vibrations that ring through the whole body of Earth. These tremors – of which over one hundred thou-

INTRODUCING THE LESSON

sand are felt each year and several million others are recorded on sensitive *seismographs* – are *earthquakes*, a major geologic hazard but also a primary source of information about the unseen interior of our planet.

Even during a *quiet* period, Earth is continuously shivering and trembling, producing tiny wiggles – *microseisms* – on the seismograph. These little unfelt quakes originate from surface phenomena such as tides, strong winds, the pounding of heavy surf, and human and industrial activities such as traffic vibrations near the site. The events that cause the seismograph needle to swing wildly, even jump off the paper, are results of tectonic forces and movement along faults. (In California and Nevada alone, about 5,000 noticeable earthquakes occur each year.) As Californians say, "Living on the San Andreas fault is a very moving experience!"

Earthquakes, however, are not just a California story. The three largest earthquakes in the United States occurred on the Missouri New Madrid fault in 1811 and 1812 and were felt over a large part of the United States. Earthquakes occur world-wide and are particularly destructive in countries where building standards are poor and the unreinforced structures are generally inadequate to survive the quakes.

In this lesson, you will learn how tectonic stress causes solid rock to break, suddenly releasing vast amounts of stored energy that moves blocks of crust and causes the seismic waves we sense as earthquakes. You will learn how an earthquake is recorded on *seismograms* around the world and how, from this information, the locations of the *focus* and *epicenter* are determined. *Magnitude* and *intensity* will be compared, and earthquakes' effects on human life and structures will be evaluated.

You will become familiar with how earthquakes are measured and why they generally occur within narrow belts (something you can probably deduce from your experience with plate boundaries, mountain building, and geologic structure). Prediction of earthquakes is advancing slowly, but preparedness is now in the public eye (especially in the Pacific Rim of Fire, where most earthquakes occur). This preparedness should go a long way toward mitigating the effects of a major shake.

LESSON ASSIGNMENT

Completing the following eight steps will help you master the lesson objectives and achieve the goal for this lesson:

Step 1: Read the TEXT ASSIGNMENT, Chapter 7, "Earthquakes."

Step 2: Study the KEY TERMS AND CONCEPTS noted in the study guide.

Step 3: Watch the VIDEO portion of the lesson using the VIEWING GUIDE in the study guide.

Step 4: Read the PUTTING IT ALL TOGETHER section in the study guide, which will help you summarize and integrate all the information in this lesson.

Step 5: Complete any assigned lab exercises.

Step 6: Complete any assigned ACTIVITIES found in the study guide.

Step 7: Review the material for the lesson and complete the SELF-TEST found in the study guide.

Step 8: Go back to the LEARNING OBJECTIVES and make sure you have learned each one.

1. Read Chapter 7, pages 147-172, making sure you include the chapter's "Introduction" and "Summary." This is a long chapter, but you should be familiar with many of the concepts and key terms from previous lessons.

2. Try to answer the "Questions for Review" and the "Questions for Thought."

3. The diagrams and photographs are important for visual understanding of this subject. Study them carefully.

4. Review Lesson 3 and the part of Chapter 2, "Evidence from Seismic Waves," pages 26-27, that refers to seismic waves.

The key terms listed under "Terms to Remember," page 172 in your text, will aid your understanding of the lesson's material. Make sure you look up unfamiliar words in the glossary or in the chapter.

Once you have read the text material for the lesson and studied the terms, you are ready to view the lesson's video portion. Review the questions that appear in this section to help you watch for important points and to prepare you for what you will see in the video.

Before viewing, answer the following questions:

1. What are the two categories of seismic waves?

2. Describe the characteristics of P waves and S waves.

3. Explain the difference between intensity and magnitude in describing the strength or size of earthquakes.

After viewing, answer the following questions:

1. Visualize Earth without the effects of the internal heat engine. According to the video, what would be different, besides an absence of earthquakes?

2. What is the principal source of stress in Earth's crust?

3. When an earthquake occurs, how is the energy expended at the moment of slippage?

4. You learned about the P and S waves in a previous lesson. Knowing that the different seismic waves travel at different velocities, how would you determine how far away the epicenter is? And how would you find the actual location of the first slippage?

5. The video showed some buildings swaying dangerously during an earthquake, while others seem to remain still and stable. What are the conditions that cause these different responses to the same quake?

6. Parkfield, California has become the site of much seismic instrumentation, monitoring, and long term studies. Why is there all this activity at Parkfield, and what are the goals of the seismologists working there?

7. Describe the workings and purpose of some of the sophisticated instruments shown in the video that are being used at Parkfield.

8. After viewing the video, do you feel scientists are now are able to predict earthquakes successfully over both the long and short range? If you lived in San Francisco, and the U.S. Geological Survey predicted a major quake next Friday, what would you do? (Don't forget: everyone else in town will also know of the prediction.)

PUTTING IT ALL TOGETHER

You have read the text assignment and seen the video portion of the lesson. This section, made up of a SUMMARY, two CASE STUDIES, QUESTIONS, suggestions for ACTIVITIES, and a SELF-TEST, will help you pull the information together.

Summary

Plate movements and their accompanying earthquakes have been geologic phenomena during most of Earth history, but never before have human populations been so large and so spread over the globe. It has been estimated that over the last 4,000 years about 15 million people have died as a result of earthquakes. Because earthquakes are a major geologic hazard, modern well-equipped earthquake observatories have been established in many areas, including in deep abandoned mine shafts below the surface microseisms that mask small Earth movements. There is now a network of about 130 observatories distributed in 60 countries; these stations quickly locate and measure global earthquakes and have greatly increased our understanding of Earth movements and of the fine structure inside Earth.

Causes of Earthquakes

Most of the major earthquakes, as you already have learned, are tectonic in origin and are usually located at or near a plate boundary. Earthquakes can also occur during violent volcanic eruptions, but these quakes are rare compared to those on known active faults. Faults are not continuously active, however; in fact, most are inactive.

Earthquakes also occur on previously unknown faults that show no surface displacement. Under Los Angeles, for example, there seem to be buried, active, low-angle thrust faults with no surface evidence, another source of disquiet in earthquake-prone California (refer to Box 7.3, pages 170-171). The earthquakes of the eastern United States are generally small, but even the very strong quakes are not always associated with surface faults that can be monitored (Box 7.1, "Earthquakes in the Eastern United States – Rare but Occasionally Strong," page 156).

Faulting, as you learned in Lesson 8, is a response to stress from plate movements. Stress builds in the crust until the rock undergoes *elastic deformation*. Think of a steel spring that is being compressed. Strain builds slowly; the rocks bend since friction between the two sides of the fault prevents smooth slippage. The stress in the deformed rock will finally exceed the breaking point of the rock, and suddenly the rock cracks and energy is released. The rocks on opposite sides of the fault will move in opposite directions, releasing the stored energy and returning to their original shape through *elastic rebound* (Figure 7.1, page 149). The released energy will be spent moving and fracturing rocks as they are forced against other rocks, generating heat through friction and transmitting vibrations (seismic waves) through Earth (Figure 7.2, page 149). Later adjustments, called *aftershocks*, are generally of lesser magnitude but may persist for months (Figure 7.16, page 160). The stress will be relieved for a time but, as the plates continue their endless march, it will build again and eventually cause another earthquake.

Most earthquakes take place at relatively shallow depths where the crust is brittle and cool. The maximum depth for the occurrence of earthquakes is about 670 kilometers (400 miles), with the greatest number (85 percent) having a shallow focus of 0 to 70 kilometers (0 to 42 miles) (Table 7.1, page 153). As you learned in your study of Earth structures, hot rock under pressure tends to flow and fold rather than suddenly break.

Earth Shakes

The vibrations or seismic waves start at the *focus*, a point within Earth on a fault where the movement or first release of energy takes place. Once a rupture on a fault starts, it does not remain at the focus but propagates along the fault at several kilometers per second. The total length of the rupture associated with the 1906 San Francisco earthquakes, for example, was about 432 kilometers (260 miles). The *epicenter* is the point vertically above the focus (Figure 7.2, page 149), where *body waves* first reach the surface.

Earthquakes produce two main types of seismic waves that differ markedly in speed of transmission and other properties. The waves that travel through the interior of Earth are called *body waves*; others that travel at or close to the surface, not surprisingly, are called *surface waves*.

Body waves radiate with a spherical wave front from the point of focus, travel through the interior of Earth, and emerge at the surface where they are detected by seismographs. The energy is transmitted by solid rock, which moves slightly as the wave passes and then, being elastic, returns to its original shape. The body waves are of two types – P and S waves. The fastest are termed *P waves* for *Primary* or *Push-Pull*; they travel about 6 to 7 kilometers (3.6 to 4.2 miles) per second and are compressional and extensional, similar in motion to sound waves. (See Figure 7.3 A, page 150, noting the wave propagated by the "slinky.") The rock particles vibrate back and forth with a pulsing motion, in the same direction as the waves are propagated. From your experience, you know that sound and push-pull waves are transmitted through air, through liquids (SONAR), and through solids.

The second type of body wave is the *S wave* for "Secondary," "Slower," "Shear," or "Shake," travelling at about 2 to 5 kilometers (1 to 3 miles) per second. S waves cause the rock particles to shake from side to side at right angles to the direction of propagation of the energy (Figure 7.3 B, page 150). From your study of Earth's interior, (Lesson 3) you learned an important property of S waves: they are transmitted by solids only and are not propagated by liquids or gases because fluids have no elasticity or restoring force. From this property, seismologists were able to deduce from the *S-wave shadow zone* that part of Earth's core was molten or liquid or acted as a liquid (Figure 2.9, page 32).

Surface waves, originating from body waves that reach the surface, radiate from the epicenter, spreading out like ripples in a pond. These are the slowest of the seismic waves but are the most damaging to human structures because they produce the greatest ground motion. One type of surface wave creates side to side horizontal movement while the other has a strong vertical component. Both can be especially disruptive to poorly reinforced buildings. Much information about the shallow crust has been gleaned from the study of surface waves.

Detecting Earthquakes

The study of earthquakes is *seismology*, from the Greek word for earthquakes, *seismos*. The instrument used to detect the shocks and vibrations caused by earthquakes is a *seismometer*. The *seismograph* is a recording device and the paper record it produces of Earth's vibrations is a *seismogram*.

The problem of detecting and measuring vibrations was not easy to solve, since the ground, building, floor, and all the instruments shake together. There is no stable, fixed frame of reference from which to make the measurements. The earliest record of a seismometer is from ancient China, but modern seismology was founded by an English mining engineer, John Milne, in 1880, after a trip to earthquake-torn Japan.

The seismometer was transformed from an interesting curiosity into a precision instrument.

Seismographs make use of the property of inertia, which is the resistance of a large mass to sudden movement. The mass, such as a block of iron, is suspended from a spring or thin filament and tends to remain almost stationary, essentially floating in space, unaffected by the movement of the ground and the frame. During an earthquake, the spring expands and contracts, and the scale or recording sheet attached to the frame moves, but the needle or stylus connected to the weight remains still (Figures 7.4 and 7.5, pages 150-151). Modern seismograms are recorded on a drum that turns at precisely four revolutions per hour. As the drum turns, it slowly spirals ahead. Minutes and seconds are marked off so that the exact time of the onset of motion can be easily read.

Seismology has come a long way in the last 100 years. There are instruments to record strong and weak vibrations, long and short seismic waves, and waves approaching from different directions. Other seismographs are so sensitive they can detect vibrations as tiny as one hundred millionth of a centimeter! Portable seismographs were taken to the moon on the Apollo missions and relayed to Earth the recordings of "moonquakes" over a long period of time.

Locating the Epicenter

The P, S, and surface waves all originate at the point of focus at the same time. The surface waves are created when these body waves arrive at the surface. Because they all travel at different speeds, they reach the seismographs in a definite order, with the P waves arriving first. The further the epicenter is from the seismograph, the greater the time difference between the arrival of the faster P waves and the slower S waves. The last to arrive will be the surface waves (refer to Figures 7.6 A and B and 7.7, page 152, noting the travel-time curve, which plots the seismic-wave arrival time against distance). What would the travel-time curve look like if it were recorded in Colorado from an earthquake in China? Earthquakes around the globe are recorded at seismic stations, and from their travel-time curves, the distance to the epicenter can be calculated.

But this does not determine the *location* of the epicenter; a single station cannot determine direction, only distance. Each station must draw a circle, the radius of which is the distance to the epicenter. At least three such circles from widely spaced seismographs must be used, and the earthquake's epicenter will be where they intersect (Figure 7.8, page 153). In practice, the three circles rarely meet at a single point because earthquakes do not start at a point but at an elongate rupture along a fault line. Deep focus earthquakes also produce ambiguous records – they often lack surface waves because these waves lose much of their energy as they travel great distance to the surface. The P waves seem to arrive earlier than they should because the deeper, denser rock transmits the body waves more rapidly. Using computers, the depth of focus can also be calculated from the arrival times recorded at many different seismograph stations (Figure 7.6, page 152).

Measuring the "Feel" or Strength of an Earthquake

Trying to estimate the strength of an earthquake while your house is shaking and the dishes are rattling is difficult but is actually one method of finding the earthquake *intensity* – defined as a measure of the effect on people and buildings. It is expressed in Roman numerals ranging from I to XII on the modified *Mercalli scale* (Table 7.2, p. 153). There are certain drawbacks to the Mercalli scale, primarily because damage and intensity diminish farther from the epicenter. Also, damage is dependent on the structure of buildings and on the substrate – whether it is solid rock (which has the least shaking from the energy being transmitted through the rock) or unconsolidated sediments (which have the most shaking because the earthquake energy is absorbed). Human response is based on taking thousands of surveys, which together provide an accurate assessment of the intensity. (Refer to Figures 7.9 and 7.10 A and B, pages 154-155. Notice the distribution of high intensity earthquakes, especially in southeastern Missouri and in the northeastern United States.)

The second method of calculating the strength of an earthquake is to estimate the total energy released by measuring the *amplitudes* (heights from crest to trough divided by 2) of the S waves recorded on a special seismograph. The signal is adjusted for distance from the epicenter and is assigned a number called the *magnitude*. The *Richter scale* (Table 7.3, p. 154) is a numerical scale starting at 1 and increasing upward. Each unit increase in magnitude corresponds to a tenfold increase in the amplitude of the seismic wave signal. For example, a magnitude 2 has an amplitude that is ten times larger than a magnitude 1, and a magnitude 3 is a hundred times larger. The energy corresponding to a Richter scale increase of one is more than a 30-fold increase (31.5 to be precise). Thus, the energy difference between a magnitude 1 and a magnitude 3 would be 31.5 x 31.5 or about 992 times greater.

The largest recorded earthquakes to date had Richter magnitudes of about 8.6. They released as much energy as 10,000 atom bombs of the kind that decimated Hiroshima at the end of World War II – as much energy as produced by detonation of about 100 million metric tons of TNT!

What are the Effects of Earthquakes?

Major earthquakes can be catastrophic not only because of the direct damage from ground motion and faulting but because of the indirect damage set in motion by the quake. Some of the effects are as follows:

1. *Ground motion*, the result of seismic waves, especially the slow-moving surface waves passing through surface-rock layers and the loose soil cover. Sometimes the ground motion is visible as rolling waves,

which can bring on a form of seasickness for susceptible people. The shaking may tear buildings loose from their foundations, collapse bridges and freeways, and cause injuries from falling debris (Figure 7.12, page 157).

2. Where faults break the ground surface, any structures that cross the fault, such as fences, buildings, roads, or pipelines, can be broken apart. The trace of the fault may be a small cliff or *scarp* (Figure 7.15, page 159). Sudden changes in elevation of the land may lift shore lines many meters, exposing living coral reefs, for example, to the lethal effects of drying air and baking sun or may cause subsidence, subsequent flooding, and destruction by sea water (Figure 7.17, page 160).

3. An indirect effect that is often more serious than the original shock is *fire*, brought on by ruptured gas and water mains and fallen electrical wires. In the disastrous earthquakes of San Francisco in 1906, and Tokyo and Yokohama in 1923, more than 90 percent of the damage to buildings was caused by fire (Figure 7.13, page 158).

4. In regions of steep slopes and abundant loose soil, the shaking can cause *landslides* and other rapid mass movements. This is particularly true in Alaska, parts of southern California, China, Iran, and Turkey. In 1970, an earthquake in Peru set off a series of landslides that buried more than 17,000 people. (Refer to Box 13.1, page 286-287, and Figure 7.14 A, page 158.)

5. The sudden shaking and disturbance of water-saturated sediment and soil can turn seemingly solid material to a liquid-like mass such as quicksand. This is called *liquefaction* and was one of the major causes of damage in the Anchorage, Alaska, earthquake of 1964. In Nigata, Japan, also in 1964, shaking caused the ground to act like quicksand; earthquake-resistant buildings toppled over intact as the ground beneath them failed (Figure 7.14 B, page 158).

6. Seismic sea waves, called *tsunamis* occur following violent movement of the sea floor after strong coastal earthquakes, submarine landslides, and large volcanic explosions (Figures 7.18 and 7.19, page 162). The term *tidal waves* is a misnomer since ocean tides are not involved here. These great waves are extremely destructive, especially within the Pacific basin, and in some cases have caused a greater loss of life than the accompanying earthquake.

Plate Boundaries and Earthquakes

Each type of plate boundary – diverging, transform, and converging – has a characteristic pattern of earthquake distribution and motion (Figure 7.26, page 166). At *diverging* boundaries, as in the mid-oceanic ridges, where plates are moving away from each other, the earthquakes are shallow and restricted to a narrow band. The quakes are usually along the sides of the rift valley, where the horizontal tension is tearing apart the sea floor. Divergent boundaries within a continent also produce rift valleys and shallow-focus earthquakes 0 to 70 kilometers (0 to 42 miles) deep (Figure 7.27, page 166).

At *transform* boundaries, where two plates move past each other, the earthquakes are also shallow and are produced by horizontal movement along strike-slip faults. The famous San Andreas fault was long considered to be a single fault but is now believed to consist of a broad system of parallel faults that form the plate boundary (Figure 7.28 A and B, page 167). A surprising discovery was made from the records of the October 17, 1989, Loma Prieta earthquake. That quake occurred about 80 kilometers (48 miles) south of San Francisco in central California and had a magnitude of 7.1. In addition to about 1 to 2 meters (3 to 6 feet) of right-lateral strike-slip movements near the focus, about 1 meter (3 feet) of reverse faulting or vertical movement occurred, but there was no surface rupture. A portion of the Pacific plate west of the San Andreas fault zone was pushed up and over the North American plate to the northeast. Seismologists agree that at least this part of the San Andreas is a far more complex fault zone than had previously been appreciated. (Find the San Andreas fault in Figure 7.20, page 163.)

Converging boundaries may involve continent-continent, continent-ocean, or ocean-ocean convergence, each producing its own pattern of earthquakes. During continental collision, broad zones of shallow earthquakes will occur as one continent overrides part of another. The seismic zone is complex, earthquakes are frequent, and the seismic belt may be several thousand kilometers wide. (Refer to Figure 7.20, noting the earthquake zone between India and Asia, and to Figure 7.29 A, page 167, showing collision of two continents.)

If the convergence is between an ocean and a continent, the sea floor, being denser, will bend and subduct in a deep sea marginal trench at the edge of an ocean basin. As the oceanic plate starts to descend, it will stretch, and normal faults will occur as a result of this tension. Below the trench, the subducting plate is under compression as it underthrusts beneath the continental plate (Figure 7.29 B, page 167). Near the trench, there are shallow focus quakes. At greater depths, the earthquakes are confined to a thin zone, about 20 to 30 kilometers (12 to 18 miles) thick, located within the descending lithosphere (Figure 7.30 A, page 168). The depth of focus increases landward from the trench and defines the Benioff zone. (Review Lesson 6.) Deep focus earthquakes occur at depths ranging from 350 to 700 kilometers (210 to 420 miles). From a study of these earthquakes, it seems the quakes are caused by compression and underthrusting, although it may be that the denser plate is being pulled down and is under tension (Figure 7.30 B, page 168). Evidence isn't clear which is operating. There seem to be conflicting opinions.

Ocean-ocean convergence, where the older, denser sea floor subducts beneath younger sea floor, also is marked by deep sea trenches, subduction of cold sea floor, and earthquakes increasing in depth in the descending Benioff zone. The convergence is the site of island arc andesite volcanoes found in the northern and western rim of the Pacific Basin.

Progress in the prediction of earthquakes is moving slowly, although there have been some outstanding successes, such as in China in 1975 when a destructive 7.1 magnitude quake was predicted 5 hours before it happened. Of course, most earthquakes do not send clear signals before disaster strikes, as in 1976, also in China, when a 7.6 quake struck with no warning and killed an estimated 240,000 people. The San Andreas is heavily monitored to detect earthquake precursors (Box 7.3, "Waiting for the Big One," pages 170-171). Perhaps science will catch up with nature, and there will be adequate warnings before the next "big one."

Tsunamis remain a true and constant threat to the people of Hawaii and the Pacific basin. What is a tsunami and what causes it? It is a series of low, broad, waves in the deep ocean, most commonly caused by violent movement of the sea floor. Submarine faulting that causes a block of ocean floor to be thrust upward or to suddenly drop will start the disturbance in the water. Such fault movements are accompanied by earthquakes, but the earthquakes do not cause the tsunami; they both result from the same movement on a fault.

The second most common cause of tsunamis is landslide, either coastal or entirely underwater. The highest tsunami waves ever reported were produced by a landslide at Lituya Bay in Alaska. Here, on July 9, 1958, a massive rockslide at the head of the bay produced a tsunami wave that surged up to a high-water mark 530 meters (1,740 feet) above the shoreline!

The third cause of tsunamis is volcanic activity, a result of nearshore or underwater volcanoes. In 1883, the violent explosion of the famous island volcano, Krakatoa, in the Indian Ocean, sent tsunami waves as high as 40 meters (130 feet) crashing ashore in Java and Sumatra. In all, more than 36,000 people perished as the waters surged over the low-lying islands of southeast Asia.

Tsunamis are unlike wind waves that normally disturb the sea surface. They have extraordinarily long wave lengths of about 160 kilometers (96 miles) from crest to crest, while wind waves seldom exceed 300 meters (1,000 feet) in wave length. The most astonishing features of tsunamis is their speed; they move through the deep sea at a speed of about 720 kilometers per hour (430 miles per hour) – the speed of jet airliners! In comparison, wind waves travel about 16 to 32 kilometers per hour (10 to 20 miles per hour), rarely reaching 80 kilometers per hour (about 50 miles per hour). Yet in deep water of the open ocean, the height of the tsunami is less than 1 or 2 meters (3 to 7 feet). The highest wind wave ever recorded was 36 meters (120 feet) in the open sea, with heights of 2 to 3 meters (7 to 10 feet) being more usual. Because of the long wave length, the whole water column down to the deepest parts of the oceans will move with the passing tsunami. Travelers on a cruise ship will be unaware of the energy roiling the sea beneath

CASE STUDY I:
Tsunami:
The Great Wave

them, but a soft-bodied sea cucumber lying in the mud 4,545 meters (15,000 feet) below will feel the tug of the tsunami.

As the waves approach the shore, the water becomes shallow, and the topography of the local sea floor has a strong effect on how the wave will behave. A gently sloping shelf around the larger islands, submarine ridges extending offshore, and funnel-shaped bays or inlets can force the tsunamis to great heights. When the waves "feel bottom", they will start to move more slowly, dropping to around 65 kilometers (40 miles) per hour. As the wave is confined, the tremendous energy will be spent lifting the water, peaking at 15 to 30 meters (50 to 100 feet), before moving onto the shore. Sometimes, as a wave approaches, the water level will lift slightly, not enough to be really noticed; this will be followed by the trough of the tsunami, which will drain the water seaward and will be very much noticed as it exposes the floor of bays and harbors. The flopping fish and crawling lobsters have lured people to exposed sea floor, which was inundated when a roaring wave came in a few minutes later. The drop in water level is now understood as a warning to run for high ground.

One popular misconception is that there is just one giant wave. Actually, a tsunami may consist of ten or more giant waves in what is called a *tsunami wave train*. The individual waves will crest between 5 and 90 minutes apart, and the first may not be the largest. Sometimes the third and fourth waves are the highest and most destructive. The waves most often arrive like a very rapid high tide that will continue to rise for five or ten minutes because of the long wave length, causing extensive flooding and inundation of coastal installations on long sections of the coast. Even the withdrawal of the tsunami can cause significant damage. As the water is drawn back to the sea, it can scour out sediments, remove entire sandy beaches, and undermine the foundations of buildings. The outflow of water can be rapid and turbulent, and observers have reported its loud hissing, roaring, and rattling noises.

The havoc wrought in terms of destruction and loss of life has made tsunamis a major hazard around the Pacific Rim. Once the wave has been generated, the wave form will travel across the ocean to break on any distant shore in its path. Because Hawaii is in the center of the Pacific basin, no matter where a tsunami starts, the wave is likely to strike Hawaiian islands. (Refer again to Figure 7.20, page 163). The Tsunami Warning Service was established in Honolulu in the late 1940s, and a "Tsunami Watch" is automatically issued for all earthquakes greater than Richter magnitude 7 occurring in the area of the Aleutians and for all quakes greater than 7.5 occurring elsewhere in the Pacific basin. The Warning Service is a fine example of practical geology in action.

Earthquakes are understandably most frequent near plate boundaries, but what accounts for earthquakes in the stable continental interior? In fact, what is a stable continental interior? These are the bed rock *shields* or *cratons* – the ancient hearts of the continents – some of which contain rock more than 3 billion years old, and the platforms of sediment-covered bed rock that surround the shields. These are supposed to be the seismically quietest parts of continents, far from mountain building and volcanoes. Some tectonic activity can extend far from the actual plate boundaries, however. Where an oceanic plate is subducting beneath a continent, volcanic activity and mountain building may take place on the overriding continent in a band hundreds of kilometers wide, as it does in the Andes. In the Himalayas, the folded, uplifted, and faulted region is several thousand kilometers wide.

Most of the faults responsible for stable continental earthquakes are deeply buried under layers of sediment that cover much of the old crust. Accurate measurements indicate that even the stable crust is being slowly deformed. Geophysicists are now measuring tiny changes in distance between stations thousands of kilometers apart as the crust is being stretched or compressed. The stresses from compression, for example, are incredibly small – about one millimeter over thousands of kilometers. But the continuing compression within a continent seems to reactivate old faults, causing them to slip and to generate earthquakes.

Some areas that have undergone extension – such as ancient passive margins or failed rifts (*aulacogens*) – are areas where stresses continue to build slowly. A failed rift underlies the site of the New Madrid earthquakes (1811-1812). The New Madrid fault spawned three quakes in excess of magnitude 8.0, which formed the large Reelfoot Lake, altered the path of the mighty Mississippi River, and rang church bells in Boston. A repeat of the 1811-1812 earthquakes would be devastating, because the epicenter region affects seven states, with a current total population of over 33 million. Few structures, including buildings, bridges, highways, pipelines, and so on, meet present standards of earthquake-resistant design. Yet some earthquake experts believe there is a good chance of the New Madrid fault unleashing shock waves in the range of magnitude 8.0 by the year 2030.

The earthquake that severely damaged most of Charleston, South Carolina in 1886 was several times larger than the event that struck San Francisco in October 1989, yet occurred on a geologically quiet zone that has not had any plate boundary activity since the opening of the Atlantic some 180 million years ago. The Charleston area is located on an ancient fault system created during the breakup of Pangaea; it intersects faults from the Appalachian mountain building period. These deeply-buried faults became reactivated by the continual stress of drifting continents. The city itself is also on an extended passive margin, a structure now known to experience more earthquakes of various sizes than the interiors of the ancient shields.

For sites already identified by an historical earthquake on a visible fault, it may be possible to determine the frequency of large events – a step toward forecasting seismic hazards. For earthquakes on deeply buried faults with no surface identification, the problem of prediction is difficult. The earthquakes are infrequent, and the intervals of recurrence may be thousands of years apart. When they do occur, they may be far stronger than those on the better known faults and will be far more unexpected. As can be seen, even in the "quiet" interiors of the continents or on "passive" continental margins, the term "stable" is only relative.

Questions

1. What really is an earthquake? Why will the effects of the same quake vary so at different places?

An earthquake is a shaking, jolting, rumbling, rolling, rattling, trembling, unsettling motion of the ground caused by seismic waves traveling within Earth as well as on the surface. Earthquakes happen suddenly and will vary in intensity and motion depending upon the distance from the epicenter. The seismic waves produced from Earth movements are emitted over a wide range of frequencies from low frequencies that cause a rolling motion to high frequencies that cause sudden jolts. The ground absorbs the energy of the moving shock waves and causes them to die out eventually. The higher frequency waves which cause the sudden jolts die out first; the low frequency waves roll along and can be felt at a greater distance from the epicenter. P waves have a higher frequency and will be the first to arrive at a seismograph and also the first to die out. If you feel a sudden jolt, you may be sensing the arrival of the P waves, and you may be close to the action. If you feel only a roll or rumble, you are far from the epicenter, but the rolling may persist for many seconds.

2. What are the basic causes of earthquakes?

Earthquakes are caused primarily by movements of Earth's great tectonic plates as they either converge upon each other, pull apart, or move horizontally past each other on long strike-slip faults. Earthquakes, in fact, generally define the plate boundaries which indicate where the great quakes are most likely to happen. Earthquakes also occur in areas of active volcanism, which are likewise determined by tectonic activity.

3. What starts an earthquake?

The energy released during an earthquake is a result of the moving plates setting up stress in the crust and lithosphere, causing rock to bend. Rock is elastic to some extent and resists deformation. As the plates continue to move, the elastic limits of the bending rock will be reached, and suddenly the rock will break. The break is a fault or plane along which the rock masses will move in directions opposite from each other. At the point of initial movement, called the *focus*, the blocks of rock will shift against other rocks, fracture, generate some heat by friction, and emit the vibrations felt around the globe.

4. What are the special characteristics of the different kinds of seismic waves?

There are two general types of seismic waves: body waves, consisting of the primary and secondary (P and S) waves, and the surface waves. The P waves are "push-pull" or compression and extension waves moving in the direction of propagation of the wave. These are the fastest waves, the first to reach a seismograph, and they have the ability to be transmitted by solids, liquids, and gases. The S waves, which have the greatest amplitude on seismograms, are slower, lag behind the arrival time of the P waves at the seismographs, and are shake or shear waves that can be transmitted only through elastic solids. Because they shake from side to side, S waves can be very damaging to buildings, which are more susceptible to horizontal than to vertical motion. The surface waves are derived from the emergence of the body waves at the epicenter. They are slower but have the greatest amplitude and longest duration. Surface waves can be the most destructive to buildings and other structures since they cause the greatest ground motion and shake in a complex manner.

5. When seismologists in California say an earthquake took place 20 miles off the coast of Peru and that it had a magnitude of 6.3, how did they make these determinations?

The distance to the epicenter of an earthquake can be calculated by using time-travel charts. The speed of the P and S waves is known, and by noting the time of arrival of the P waves and the lag time before arrival of the S waves, observers can determine the distance to the epicenter. The greater the time interval between the arrival of the P waves and then the S waves, the greater the distance to the source of the waves. These measurements give distance but not exact location. When the distances to the epicenter from at least three seismograph stations are sent to a central station, observers there draw circles with their radii proportional to the calculated distances. The common intersection of these circles reveals the location or near-location of the epicenter.

The magnitude is a measure of the energy released by the earthquake. The magnitude is computed from actual marks on a seismogram, in which the amplitude (the height of the S-wave from crest to trough divided by 2) is carefully measured. The distance to the epicenter must be included in the calculations because the amplitude of an S wave of a very distant shake is small. From these figures, the actual energy release will be determined. This number should be approximately the same regardless of where it is measured.

These are simplified basic concepts of how distance and magnitude are calculated. There are many variables, however, that influence the final result. The speed of the waves through Earth and around the surface will vary, depending on the solidity or elasticity of the rock. As they pass through Earth, body waves may be slowed, speeded up, or not received at all on certain seismographs. Magnitudes may also be unreliable because of the variations in the crust and lithosphere. And finally, the epicenter may not be a point but may be a section on a rupturing fault. It is only after observers at many seismograph stations report their data that the results can be refined.

6. Earthquakes occur in many of the states of the U.S. but seem to be concentrated in the western states including Alaska. What is the difference between the origin of the earthquakes in the eastern states and those in the west?

Earthquakes in the western states are the result of proximity to the active plate boundary between the Pacific and North American plates. Some boundaries are not a thin zone but a broad area in which the effects are felt at great distances from the actual suture or subduction zone. Because of the frequency of earthquakes in the west, the major faults along which earthquakes occur repeatedly and which demonstrate surface displacement are well studied and monitored. The Parkfield experiment in central California shown in the video is a good example.

The eastern borders of the United States are on a passive continental margin, half an Atlantic Ocean from an active spreading center. The earthquakes there are less frequent and usually of a lower magnitude. They are not generally associated with surface displacements and seem to have deeper points of focus than most western earthquakes. Strong quakes do occur, however, on deeply buried faults, in failed rifts, and in ancient passive margins under extensional stress. The maps of seismic risk are interesting yet puzzling because our knowledge of causes of earthquakes in the continental interior and on the East Coast is still incomplete (Figure 7.11, page 156).

7. What was the largest earthquake ever recorded?

The largest earthquake recorded was the 1960 plate-boundary event in Chile, in which the energy released was equivalent to that of an average hurricane but released in one or two minutes rather than over 10 days!

The largest earthquakes on a stable continental interior (and in the United States) were centered in New Madrid, Missouri during the winter of 1811-12. The New Madrid events were felt over a wider area than any other earthquake in history. The strong rock crust of plate interiors transmits seismic waves far more efficiently than the faulted crust near plate boundaries. The New Madrid quakes were centered about 1,000 miles inland but damaged masonry as far away as the East Coast and collapsed the scaffolding erected around the U.S. Capitol Building.

8. How are earthquakes related to plate boundaries?

Earthquakes actually define the location of the plate boundaries. The belts of earthquakes around the globe shown in Figure 7.20, page 163 delineate with remarkable clarity the edges of the great plates, the mid-oceanic ridges, the smaller plates of the Pacific basin, even the strangely shaped Australia-India plate. The plate boundaries are sites of strongest interaction between moving plates, and earthquakes result from stress upon the rocks and their elastic rebound after fracture.

9. Do different types of plate boundaries cause different types of earthquakes?

Different plate boundaries, because of the relation of the plates to one another, produce different stresses, which will be expressed in the type of earthquake, the depth to focus, the range of magnitude, and the width of the seismic zone. At divergent plate boundaries, occurring at the mid-oceanic ridges and rises and at rift zones within the continents, the earthquakes are shallow and restricted to a rather narrow belt. At transform boundaries, such as along the San Andreas fault or other strike-slip faults where the plates are moving horizontally, the focus is shallow, but the magnitude of some of the earthquakes can be high. The effects of these faults may extend laterally and affect wide areas in the crust. At convergent boundaries, especially where two continents are colliding, earthquakes may be frequent and strongly felt. The seismic zone, as in the Himalayan belt, may have a width of thousands of kilometers. The *Alpine zone* is a great convergent boundary, extending from the Alps and the Mediterranean region through the Middle East and across the Himalayas. The devastating quakes of Turkey, Armenia, Iran, India, and China are related to this zone of active collision.

At an ocean-continent convergent boundary, the earthquakes near the deep sea trench are shallow where the sea floor is beginning to subduct beneath a continent. As the sea floor descends and slopes landward, the focus deepens within the Benioff zone (see Lesson 4). Earthquakes at this boundary – as in the Andes – are among the strongest recorded and, with the accompanying landslides from the high mountain slopes, have caused untold destruction and loss of life.

The Benioff zone also lies beneath the curved lines of islands in the island-arc systems, where ocean-ocean convergence is occurring. The quakes in the Benioff zone are also frequent and of high magnitude. The strong quakes felt in Japan, the Philippines, and the Aleutians occur in zones of ocean-ocean convergence. Essentially all of the world's intermediate and deep focus earthquakes occur in the Benioff zones that surround the Pacific basin.

10. Why are tsunamis most frequent in the Pacific Ocean and almost unheard of in the Atlantic? Have tsunamis ever occurred in the Indian Ocean?

Tsunamis are directly related to the plate boundaries characteristic of the Pacific Rim. Tsunamis that accompany earthquakes in South America and the Aleutian-Alaskan region, however, have posed a greater hazard than those generated locally on the West Coast of the United States. The 1960 Chilean earthquake produced wave heights of 12 to 16 feet at Crescent City in northern California and did considerable damage across the Pacific in Japan. The 1964 Alaskan tsunami generated waves of more than 20 feet at Crescent City, where it caused $7.5 million in damage (1960 dollars) and 11 deaths. In contrast, the 1906 San Francisco earthquakes produced local tsunami waves of only about 2 inches.

The Atlantic, however, is surrounded by passive continental margins and lacks extensive vertical fault displacement. While earthquakes do occur, they are rare, and historically no tsunamis have been generated on the East Coast. No tsunami occurred during the Charleston, South Carolina, earthquake of 1886, for example, one of the largest earthquakes ever in the United States.

The Indian Ocean has an active subduction zone parallel to the western margin of the islands of Java and Sumatra, where one of the greatest tsunamis in history was generated. The cause was volcanic in this case, the major eruption of Krakatoa in 1886, which caused the giant waves that spread through the Indian Ocean and into the western Pacific. An estimated 36,000 people perished on the low-lying islands of the region. The tsunami was the real catastrophe of Krakatoa even though the explosion was heard in Australia, 4,800 kilometers (2,880 miles) away. (A small but poorly documented rise in water level was even reported in the British Isles!)

Activities

1. Look at Figure 7.20 A, page 163. It will show you that earthquakes occur predominantly in belts located for the most part around the Pacific basin. This is the famous *circum-Pacific belt* which is the site of about 80 percent of the shallow-focus quakes, about 90 percent of the intermediate focus quakes, and essentially 100 percent of the deep-focus quakes (Figure 7.21, page 164)! This belt marks the boundaries of the Pacific tectonic plate.

2. Study Figure 7.20 B page 163, comparing the plate boundaries and subduction zones with the belts of earthquakes shown in Figure 7.20 A.

3. Study Figure 7.10, page 155. If you wanted to move to an earthquake-free state, what state(s) would be a good candidate? Now turn to Figure 7.11, page 156. Are these figures more reassuring?

4. Call a nearby university or government agency that has an operating seismograph and ask if it is open to the public. You might say you are a geology student interested in earthquakes. The scientists there might even give you a copy of a seismogram.

5. If there is a natural history museum or a planetarium nearby, ask if it has an operating seismograph. If no seismograph is available, write to your local state university and ask for a copy of a seismogram. You'll find it very interesting to study.

6. If you live in a seismically active region, call the geology department in a nearby college or university, the local State Bureau of Mines and Geology, or an office of the U.S. Geological Survey and ask where you can see major faults or where there was visible quake damage from the last earthquake. You will discover that public agencies are usually very helpful to geologists and geology students.

7. If there is a large public library nearby that has old newspapers on file, ask to see articles from October 17 to 20, 1989, on the Loma Prieta Earthquake, in Santa Cruz and San Francisco, California.

Self-Test

1. Maps showing the global locations of earthquakes reveal that

    a. earthquakes are evenly distributed around the world.
    b. most earthquakes follow the borders of the Atlantic Ocean.
    c. most earthquakes occur in the interior of continents.
    d. most earthquakes surround the Pacific basin.

2. Few earthquakes occur below about 700 kilometers (420 miles) because

    a. plate movements do not extend that deep.
    b. the rocks are hot and flow plastically below that depth.
    c. rocks are hard and compressed below that depth and do not break.
    d. faults large enough to cause earthquakes are surface phenomena.

3. The energy released during an earthquake is a result of

    a. earthquake motion.
    b. seismic waves.
    c. stress from moving tectonic plates.
    d. strain from elastic rebound.

4. The forces that cause an earthquake are least likely to result in

    a. plastic flow.
    b. movement on faults.

c. seismic vibrations.

d. fracturing of rocks.

5. The point within Earth where seismic waves originate is the

a. epicenter.

b. seismocenter.

c. Mercalli point.

d. focus.

6. One difference between plastic and elastic deformation is in the

a. elastic limit.

b. permanence of strain.

c. amount of stress.

d. rate of applied stress.

7. A map showing geographic locations of earthquakes indicates that the epicenters

a. locate all of the known faults on the continents.

b. occur the boundaries of lithospheric plates.

c. are rarely on a line of surface displacement.

d. are useful in determining the thickness of oceanic crust.

8. The time-travel curve is used to measure the

a. direction from a seismograph station to an epicenter.

b. speed of the P and S waves.

c. distance between the seismograph station and the epicenter.

d. direction and distance between two seismograph stations.

9. The seismic waves that are transmitted by solids, liquids, and gases are also

a. high frequency and the first to arrive at a seismograph.

b. fastest moving and most damaging as they shake the ground from side to side.

c. the slowest of the waves that have a simple up and down motion.

d. compression and extension waves with the highest amplitude.

10. The physical principle that enables the seismometer to detect and record earthquake waves is based on

a. a steel spring that can be elastically deformed.

b. the inertia of a heavy suspended weight.

c. a revolving paper-covered drum that turns at a specific rate.

d. a weight that swings while the frame and drum remain still.

11. In terms of total quake energy, about 85 percent of the energy released by earthquakes occurs in

a. subduction zones as deep focus quakes.

b. continental interiors on reactivated faults.

c. the 1 or 2 major earthquakes that take place each year.

d. thousands of shallow focus earthquakes.

12. If you want to report the damage that an earthquake caused to structures, you would use the

    a. calculated distance to the epicenter.

    b. scale showing depth to the focus.

    c. Mercalli scale of intensity.

    d. Richter scale of magnitude.

13. An earthquake with a magnitude of 7 releases about _____ more energy than an earthquakes a magnitude of 4.

    a. 30 times

    b. 1,000 times

    c. 15,000 times

    d. 30,000 times

14. The largest earthquake recorded in North America was in

    a. Missouri on the New Madrid fault.

    b. San Francisco on the San Andreas fault.

    c. Charleston, South Carolina.

    d. Mexico City.

15. Observers comment that they have heard a deep rumble during an earthquake. This may be related to the

    a. sound of structures within buildings moving against each other.

    b. noise of rock moving against rock at a deep point of focus.

    c. seismic waves reaching the surface and setting the air molecules in motion.

    d. sound of seismic waves being reflected from deep layers within Earth.

MODULE III

GEOLOGIC TIME AND LIFE

10

GEOLOGIC TIME

This lesson will help you develop a sense of the vast amounts of time over which geologic processes have been at work.

GOAL

After reading the textbook assignments, completing the exercises in this study guide, and viewing the lesson's video portion, you will be able to:

LEARNING OBJECTIVES

1. Discuss the importance of the concept of geologic time.

2. Recognize the contributions of Hutton to our understanding of geologic time.

3. Discuss the concept of uniformitarianism.

4. Differentiate between the relative geologic time scale and the absolute geologic time scale.

5. Explain how the principles of original horizontality, superposition, and cross-cutting relationships are applied by geologists to solve geologic problems.

6. Describe how rock units in separate areas are correlated through the methods of physical continuity, stratigraphic position, similarity of rocks, and comparison of fossils.

7. Explain the evidence for the age of Earth and why the oldest expected rocks will probably never be found.

Welcome to the Fourth Dimension, time, *geologic time*, to be exact. In the last hundred years or so, humans have had to expand their horizons to include an Earth that is 4.6 billion years old rather than a few thousand as decreed by early philosophers. Yet it is possible that before the turn of the century, scientists may have new data that will add a few more million or hundreds of million of years to the age of the planet.

INTRODUCING THE LESSON

The concept of the immensity of geologic time is one of the most significant contributions to human thought made by the science of geology. An understanding of the enormous scope of geologic time is basic to our acceptance of plate tectonics, and mountain building, as well as the origin of great, thick sedimentary deposits, and erosion that can reduce the highest peaks to almost flat plains. The span of geologic time is needed for life to originate in the early seas, to develop into the increasingly complex plants and animals that slowly appear and disappear throughout the fossil record, and finally to culminate in the vast assemblage of living things on Earth today.

The modern sense of time began in rural Scotland late in the 18th century through the work of a gentleman scholar, James Hutton. Hutton reasoned that the ancient rock of the Scottish landscape originally was formed by processes we see operating in the world today. This became known as the principle of *uniformitarianism* – a keystone of modern geology – and is summed up by the saying "The present is the key to the past."

Geologists began to find ways to determine both the *relative* age of rock units and, later, their *absolute* age. They used unconformities of all kinds plus the principles of original horizontality, superposition and cross-cutting relationships, and assemblages of fossils to decipher the sequence of events in a landscape. It wasn't until early in the 20th century that the natural radioactive decay of certain elements was suggested as a way to obtain absolute ages for rocks. A geologic time scale was developed that is the worldwide basis for correlation of geologic events, rock formations, and fossils.

In this lesson, you will learn the story of geologic time. You will see how the geologist uses a whole battery of principles, laws, reasoning, common sense, plus highly skilled technical ability to work out the geologic history of Earth. And with that history, the geologist is able to reaffirm that this is indeed a very old planet with a mobile lithosphere that has been moving around and changing its face for a long, long time.

LESSON ASSIGNMENT

Completing the following eight steps will help you master the lesson objectives and achieve the goal for this lesson:

Step 1: Read the TEXT ASSIGNMENT, Chapter 8, "Time and Geology."

Step 2: Study the KEY TERMS AND CONCEPTS as noted in the study guide.

Step 3: Watch the VIDEO using the VIEWING GUIDE in the study guide.

Step 4: Read the PUTTING IT ALL TOGETHER section in the study guide, which will help you summarize and integrate all of the information in this lesson.

Step 5: Complete any assigned lab exercises.

Step 6: Complete any assigned ACTIVITIES listed in the study guide.

Step 7: Review the material in this lesson and complete the study guide's SELF-TEST.

Step 8: Go back to the LEARNING OBJECTIVES and make sure you have learned each one.

1. Read Chapter 8, pages 175-191, making sure to include the "Introduction" and the "Summary" in your reading.

2. Carefully study the diagrams in the text assignment, especially those illustrating cross-cutting relationships.

3. Refer to Lesson 3, and re-read the sections on the birth of the solar system and the evidence for the age of Earth.

4. The "Questions for Review" will help you prepare for the SELF-TEST and other examinations.

TEXT ASSIGNMENT

Using your text and the glossary, learn the "Terms to Remember," page 191 in your text. Make a list of the *eras*, putting the oldest at the bottom. Using Table 8.2, page 184, enter the *periods* opposite the proper era. Enter the absolute ages at the beginning of each era from Table 8.4, page 189. Spend some time learning this list, which summarizes the major periods of geologic time.

KEY TERMS AND CONCEPTS

Once you have read the text material for the lesson and studied the terms, you are ready to view the lesson's video portion. Review the questions that appear in this section to help you watch for important points and to prepare you for what you will see in the video.

VIEWING GUIDE

Before viewing, answer the following questions:

1. Explain the saying, "The present is the key to the past."

2. What is the difference between relative age and absolute time?

3. Why do you think the oldest rocks formed on Earth will probably never be found?

After viewing, answer the following questions:

1. According to Dr. Jim Sadd's calendar of geologic time, about when did oxygen first appear in the atmosphere? If Earth is about 4.6 billion years old, and life appeared at the end of March on his calendar, about how long was Earth barren and lifeless? When did *Homo sapiens*, modern man, appear?

2. What were the contributions made by the Scottish naturalist, James Hutton, to the field of geology? What is the significance of his statement that "the present is the key to the past"?

3. What is the law of superposition and why is it important? Describe the Law of Original Horizontality and tell why it is very useful in geology.

4. Why is the law of cross-cutting relationships fundamental to the interpretation of Earth history?

5. How did Dr. Dee Trent and the host explain how rock formations can be correlated over vast distances?

6. What is a fossil, according to Dr. Webb? Describe the different kinds of fossils.

7. How is relative age of rock formations determined? What was the basis for dividing geologic time scale into eras and periods? Briefly describe the dominant life forms of the Precambrian, the Paleozoic, the Mesozoic, and the Cenozoic.

8. How was radioactivity discovered?

9. Pay attention to Dr. Jason Saleeby's techniques for radioactive dating of rocks. What information does radioactive dating give that fossil relationships does not?

PUTTING IT ALL TOGETHER

You have read the text assignment and seen the video portion of the lesson. This section, made up of a SUMMARY, a CASE STUDY, QUESTIONS, suggestions for ACTIVITIES, and a SELF-TEST, will help you pull the information together.

Summary

At one time, it was assumed that Earth was no more than a few thousand years old and appeared as we see it today from the time of its birth. Your study of plate tectonics and the shifting of continents and opening and closing of ocean basins over long spans of time should have already firmly dispelled these simple notions.

Uniformitarianism: The Birth of Modern Geology

The first step away from attributing aspects of Earth to catastrophes and divine intervention came with the writings of James Hutton, who proposed that features seen on Earth could be explained by natural processes that have always been in operation. This principle, *uniformitarianism*, assures us that physical and chemical laws were acting in the past as they are today. It must be noted, however, that certain *rates* of geologic activity have changed over the eons, such as volcanic activity and sedimentation. In addition, sites of geologic activity have changed (for example, the sites of plate collisions and mountain building). It is true that earthquakes, violent eruptions, hurricanes, and meteorites are sudden and catastrophic, but what is implied in the principle is that the laws of physics and chemistry, and not magical or supernatural powers, determine features of Earth.

The other great contribution of geology to human thought involved the concept of the immensity of *geologic time*. Once Hutton's writings became accepted, it was soon realized that lengthy time was required to accumulate the great piles of sedimentary rocks and sculpt the landscape by erosion. Earth was far older than ever imagined in the years before Hutton's conclusions. (Refer to Chapter 1, "Geologic Time," pages 20-22, and re-read the section on uniformitarianism.)

Uniformitarianism, a big word for what seems an obvious and simple concept, was controversial, even revolutionary in its day. It is such a basic premise today that it is rarely used in technical discussions, and we tend to forget that it forms the cornerstone of modern Earth Science.

Relative Time

With the acceptance of the great age of Earth came the need to know *how long ago* and *when* events took place in relation to other geologic happenings. *Relative time* deals with the sequence in which events occurred. For example, which rocks are oldest in a sedimentary pile? When did the intrusion of igneous rock take place relative to deposition of the sedimentary rocks? When were the layers folded? And so on.

Geologists use only a few principles, applied one at a time, to unravel the relative time sequence of an area. Each one, however, represents a discovery made by an individual only after painstaking observation. The first, the principle, the law of *original horizontality*, states that beds of sediment (sands or muds) deposited in water (oceans or lakes) will form horizontal or nearly horizontal layers. The layers of tilted sedimentary rocks seen standing on end or in tight folds in mountain ranges remind us that they were once horizontal and that it took enormous forces to uplift and re-arrange the old sea floor.

The second principle or law, known as *superposition*, simply states that within a sequence of undisturbed sedimentary rocks, the oldest layers will be on the bottom and that the rocks will get younger going toward the top. The laws of superposition and original horizontality were established by Nicolaus Steno (1638-1686), a Danish physician and theologian who arrived at his conclusions in 1669.

The third principle, known as the principle of *cross-cutting relationships*, states that any pattern of rock, specifically a pile of sedimentary layers, is older than any process that disrupts it, such as a fault, an igneous intrusion such as a dike or a vein, or the erosion of a valley. In other words, the layers were there first and were acted upon later by younger events.

Another principle is that a sequence of marine rock layers may be consistent and traced over hundreds of square kilometers. When you think of the size of an ocean floor where the sediments were first deposited, it is an understandable phenomenon and is called the principle of *lateral continuity*. But eventually, the layers thin out at the edge of the basin or are either faulted away, truncated, or cut off by erosion. Any activity that disrupts the lateral extent is younger than or happened after the deposition of the original sedimentary layers.

Correlation of Rock Units

While the relative ages and events can be worked out for a single area, it is not easy to correlate rocks from one region to another or from one continent to another. Part of the evidence for the existence of the supercontinent Pangaea, for example, was based on correlation of a distinctive rock sequence in Africa with a similar sequence in South America.

Physical continuity. One of the best methods of correlation is to compare the physical features of the rocks themselves. If a long sequence of identifiable rock layers can be traced or matched with a similar sequence elsewhere, correlation is fairly assured. But trying to correlate individual rock types from one region to another is less certain. A black shale layer (or white sandstone) in one area does not necessarily mean it is the same age or formation as that in a distant area.

Fossils. Fossils are defined as any remains or traces of prehistoric life found preserved in rock. The word *fossilus* is from the Latin, and means "something dug up." A fossil may be a footprint of a dinosaur, an impression of a leaf, a bit of bone, shells, tracks, trails, burrows, or an insect in amber (Figure 8.16, page 182). They are common in marine sedimentary rocks and have proved to be of great value in correlating strata and in determining the relative age of the layers in which they are found. Fossils are especially useful in correlations between widely separated areas, even different continents.

Life has changed continuously in form and kind throughout geologic time. Fossil species found in older rocks at the bottom of a thick sedimentary pile are most unlike forms living today; generally, they are all extinct. As the formations become younger, more of the fossils will resemble living forms. Most groups changed with time and fossil species succeed one another in a definite and recognizable order, usually from primitive to more complex forms. This is known as *faunal succession*. The value of fossils as a means of correlating rock formations is based on the concept that certain assemblages of organisms lived only at a specific time; this concept provides a method for determining the relative age of the strata in which they occur (Figure 8.15, page 182).

William Smith (1769-1839), a practical canal-builder in Britain, was the first to notice that there was a relationship between stratified rocks and the fossils they contained. He made the following observations: (1) sequences of rock always occurred in the same order; that is, chalk beds were always found above coal beds (physical continuity); (2) different layers contained assemblages of distinctive fossils (correlation); and (3) the distribution of sedimentary rocks followed a pattern that could be represented on a map. The geologic map of England and Wales that Smith published in 1815 was the world's first. (Interestingly, Smith was not aware of the evolutionary significance of the fossils he collected so patiently over the years. To him they were little more than uniquely shaped objects, but their utility as indicators of relative age was too practical to deny.)

The Standard Geologic Time Scale

Fossils proved so successful in correlating formations and determining their relative age that by the middle of the 19th century, a system of dividing up geologic time according to changes seen in the fossil record was established and is still in use today. The standard scale divides geologic time into three large segments: the *eras*, the oldest being the *Pleozoic* or "old life," followed by the *Mesozoic*, the time of "middle life," and the *Cenozoic*, or "new life." The boundaries between the eras represent major changes in the fossil record, such as mass extinctions and the appearance of many new forms. For example, the beginning of the Paleozoic is marked by a sudden flourishing of abundant and complex life. Before the Paleozoic, in the very long Precambrian time that lasted almost 4 billion years, organisms lacked shells or other hard parts and were rarely preserved; therefore the Precambrian fossil record is very sparse.

The long eras in the time scale are divided into shorter *periods*, which in turn are divided into smaller units of time called *epochs*. Each of these subdivisions is based on changes in the fossil assemblages that generally are recognized worldwide (Table 8.2, page 184). Where do we humans fit into the geologic time scale? We live in the Recent or Holocene Epoch, of the Quaternary Period of the Cenozoic Era. When did the dinosaurs flourish? In the Mesozoic Era, and they died out at the end of the Cretaceous Period. What group of animals expanded after the demise of the great reptiles? The mammals, in the Cenozoic Era. (See Figure 1 in Box 8.4, page 190.)

Geologists have problems and questions about the age of rock formations when they try to make long-distance correlations. Rock sequences do not match everywhere, as the conditions of erosion and deposition changed over time and from place to place. Life itself changed as a response to changing biologic environments. Some groups of plants and animals evolved into more complex creatures, while others seem to have withstood their predators and environmental changes and remained essentially unchanged over the eras. Still others, such as all the ancient trilobites, all the species of dinosaurs, even such recent mammals as the sabre-toothed tiger and the woolly mammoth, became extinct.

Absolute Age

The geologic time scale was continually refined and expanded after the days of Hutton, but it was still a scale of *relative* ages. Until the last few decades, there was no way of assigning *absolute* ages to rock units or to the divisions in the time scale. The discovery of radioactivity in 1895 by

Henri Becquerel, a French physicist, was followed by the work of Ernest Rutherford who, in 1905, suggested that certain radioactive elements could be used to date rocks.

The general concept is not difficult to understand. Certain forms (isotopes) of elements such as uranium, thorium, potassium, and rubidium spontaneously undergo natural nuclear disintegration or decay, throwing off sub-atomic particles such as protons and neutrons. When a proton is emitted during decay, the atom becomes a different element. With each decay event, a portion of the original radioactive element becomes a new *daughter product*, a stable non-radioactive element that will not change any further and will accumulate in the mineral in which it occurs (Figures 8.17 and 8.18, page 186). The rate of change from the radioactive element to the daughter product is fixed, predictable, measurable, and is described in terms of the element's *half-life*. This is the time needed for one half of the original radioactive atoms in a mineral to convert to the stable form. At the end of the first half-life, only one half of the original radioactive atoms remain. After the second half-life, only one quarter remains, and so on. Thus, to determine the length of time the element has been undergoing nuclear decay and the time since the mineral and rock formed, it is necessary to compare the remaining amount of radioactive elements with the amount of stable decay products (Figure 8.19, page 187).

While the concept is rather simple, obtaining reliable dates from rocks and minerals is actually very complicated. First, it is necessary to use radioactive elements that have a very long half-life. The element rubidium-87 decays into another element, strontium-87 and has a half-life of 49 billion years! This and other elements with long half-lives enable geochemists to date the very oldest materials on Earth. (Refer to Table 8.3, page 186.)

Radioactive dating also requires elaborate laboratory procedures to precisely measure extremely small amounts of the radioactive element and its daughter products. The atomic clock starts when a body of molten magma, for example, solidifies and the radioactive elements are frozen into certain minerals. The principal source of error in *reading the clock*, however, may be built into the rocks themselves. If a mineral is to produce a reliable date, it must have remained a closed system since its origin. This means that no parent or daughter materials have entered or left the mineral during its entire existence. Thus it is important to use only fresh unweathered rock materials that have not been chemically altered.

The rock must have contained sufficient radioactive material originally that it can be accurately measured. In addition, enough time must have elapsed that measurable quantities of the daughter product have accumulated. When geologists use the decay of potassium-40 to argon-40, the possibility always exists that argon, which is a gas, has diffused away from the host rock, giving a much younger reading for the age of the rock. Because of these uncertainties, usually more than one radioactive test must be used on the same sample to verify the dates.

Dating methods are most successfully applied to igneous rocks, since it is unlikely they will be contaminated by other materials after solidification. Sedimentary rocks are not used, since igneous and metamorphic fragments included in the layers would yield the age of the source rocks but not the time of deposition. Metamorphic rocks also yield unreliable dates since the process of metamorphism involves intense heat, which can drive out the daughter product and even the original radioactive parent and thus reset the atomic clock. Rocks younger than about 250,000 to 500,000 years cannot be successfully dated by these methods. This is because the half-life of radioactive minerals is so long that very little daughter product forms in this short period of time.

Relative Ages Now Combine with Absolute Ages

Through the application of radiometric methods during the last 50 years – especially the last 20 years – geologists have been able to refine the geologic time scale, which was previously based on fossil evidence. Fortunately, radiometric dates support quite well the time scale created with relative dates. It can be said with some certainty that the rocks at the start of the Paleozoic – the beginning of a time of enormous change in abundance and complexity of life – were formed 570 million years ago and that these dates can be verified worldwide. (See Table 8.4, page 189.) When did the dinosaurs die out? About 65-66 million years ago. (Refer to Box 8.1, page 183, on this controversial topic.) Did early people have to fend off the predatory great reptiles (as they still do in old monster movies)? Not at all: early humans were Pleistocene creatures whose ancestry goes back probably less than 2 million years. (Read Box 8.4, "The Longest Movie Never Shown – The Earth's Story," page 190, to better understand events measured against geologic time.)

One other method of radioactive dating should be mentioned, even though it usually is not used in connection with the geologic time scale – carbon-14 dating. Carbon-14 decays to nitrogen-14 with a half-life of only 5,730 years. It can be useful in dating back to about 50,000 years and is valuable in dating archaeologic artifacts, organic materials of recent origin, and recent geologic events, such as lava flows and features of the later phases of the Glacial age. But for dating the life cycles of trilobites or the break up of Pangaea, other elements with longer half-lives must be used. (Read Box 8.3, page 189, "Radiocarbon: Dating 'Young' Events.")

As you can see from the geologic time scale (Table 8.4) the eras, periods, and epochs become shorter and shorter, a reflection of the fact that there is more data available in the younger rocks in terms of both relative and absolute dating. The older rocks have been destroyed, either by weathering and erosion, or subduction and mountain building, and the fossil record is accordingly incomplete. It also appears that the rate of change in the biological world is increasing, especially since the end of the lengthy Precambrian. Certainly the rate of extinction of species is being hastened in recent times, which some biologists blame on the arrival of one species in particular . . . well known to all of us.

The question "How old is Earth?" has been around for about 2,500 years. As far as we know, Xenophanes of Colophon (570-470 B.C.) was the first of the early philosophers to focus on this problem. He recognized that fossils were remnants of former life that had lived on the sea floor. He also concluded that such rocks and fossils must be of great age. Unfortunately for Xenophanes, his brilliant intuitive perception won him no converts, since the prevailing thought among the learned Greek establishment at that time was that there was no beginning and no end to Earth.

Around 450 B.C., Herodotus, the great Greek historian, travelled through the lower Nile River valley. He was the first to apply the term *delta* to the great triangular deposit he observed at the mouth of the Nile. He reasoned that the delta must have been constructed layer by layer from a series of floods. If a single flood could lay down only a thin veneer of sediment, it must have taken many thousands of years to build up the Nile Delta.

Aristotle and other Greek and Roman philosopher-naturalists strengthened and expanded this rather scientific approach to the problem with observations combined with deduction. But this kind of thinking was temporarily lost in late medieval time when a solution to the problem of the antiquity of Earth and all things on it was found in one book: *Genesis*. The literal interpretation of this book of the Bible led to the pronouncement by Archbishop Ussher of Ireland in 1664 that Earth was created at 9 A.M., October 26, 4004 B.C.! Based on this, other dates could be and were calculated for Noah's flood and other Biblical events.

In 1760, Comte de Buffon, using rates of cooling of an iron ball (which he correctly likened to the interior of Earth), suggested that Earth was far older than Ussher's proclamation and that it was at least 75,000 years old. Toward the end of the 18th century, James Hutton and his colleagues, recognized that Earth must be very much older because geological processes to operate slowly under the principle of uniformitarianism. Hutton wrote that in viewing Earth on these terms, he could envision "no vestige of a beginning, no prospect of an end."

Time moves on: in 1854, Herman von Helmholtz, one of the founders of the science of thermodynamics, using gravitational attraction within the sun as the source of solar energy, came up with estimates of 20 to 40 million years for the age of the sun and Earth. Meanwhile, the theories of Charles Darwin, written in 1859, required as much time as possible for the evolution of the amazing array of life forms on Earth today, including at least one million years for the development of human beings alone.

Lord Kelvin, a great physicist, investigated von Helmholtz's estimate, and finally agreed that the most probable age was indeed 20 to 40 million years. This figure was based on the rate of cooling of an Earth that was presumably molten at the time of origin. His arguments seemed flawless, for no other terrestrial or solar heat sources were known at that time. Geologists, however, were reluctant to accept such a modest age but were incapable of launching a strong counteroffensive. As late as the turn of the century, Lord Kelvin, at an important scientific meeting, reiterated his faith in his calculations on the age of Earth. At the same meeting, the American geologist T. C. Chamberlin, challenged Kelvin's assumption that Earth had begun as a molten body. He proposed that the planets formed by accretions of cold, solid chunks of matter, an idea still accepted today. (Review Lesson 2, on the origin of our solar system.) Some other process, he reasoned (not knowing of radioactivity), must have caused internal heating, independent of solar heat.

Another approach was taken by an Irish scientist, Joly, in the 1890s. Using the then-known average rate of delivery of salt to the sea by rivers, he found it would have taken about 100 million years to develop the present salinity of our oceans. Rates of sedimentation were also estimated, based on assumed rates of deposition, and multiplied by the known thicknesses of strata, giving an age of at least 75 million years.

These early attempts to date Earth showed ingenuity and progress in one direction, toward older and older estimates for our planet's birthdate. By the middle of the last century, most geologists were fairly certain that the total age must be on the order of several hundred million years, but they had no numerical methods to prove this hunch.

The last great breakthrough occurred in 1896 in the Paris laboratory of physicist Henri Becquerel, where he and his assistants Marie and Pierre Curie discovered the phenomenon of radioactivity. Further knowledge came rapidly. After the turn of the century, the British physicist Ernest Rutherford reasoned that natural emission of radioactive atoms occurs in certain elements in some regular fashion through time and that the rate of nuclear decay could be expressed mathematically.

In 1905, a Yale chemist, Bertram Boltwood, suggested that lead, a stable element, was among the disintegration products of uranium. (This was an interesting reversal from the goals of the old alchemists: to turn lead into gold. Uranium at that time was a rare element worth far more than gold, and it regularly turned to lead!) Based on their lead-uranium ratios, Boltwood calculated the age of minerals and found his results ranged from an astounding 410 million to 2.22 billion years!

During the middle 20th century, *absolute* dating of minerals changed tremendously – the use of isotopes of radioactive elements has become a science in itself. Based on the age of rocks brought back from the moon and from meteorites, and assuming the entire solar system formed at the same time, the best estimate for the age of Earth (and the planets) is 4.6 billion years. The original rocks of Earth's surface most likely will never be found because plate motions, subduction, and erosion have removed the ancient original surface. And while controversies will still rage in the scientific literature, the science of geology has forever changed the way we view the place of humans in geologic history and how we regard geologic time.

Questions

1. Why is the concept of geologic time considered by some to be a revolution in science?

For a long time, almost into the 20th century, the age of Earth was guided by medieval theological concepts. The present scientific evidence of the enormous age of Earth provides enough time for tectonic plate movements to occur and for life forms to develop, change, and become extinct while new organisms take their place. Many geologic processes operate slowly, and to understand that mountains can slowly be uplifted from the sea, folded, faulted, and equally slowly eroded away, requires information not available to the philosophers of old. Besides, many of them accepted the idea that Earth was formed intact as it exists today, complete with fossils, mountains, and life forms. The new methods for dating both relative and absolute time have given scientists dependable numbers to work with, a great contribution to modern thought that has influenced almost every field of human endeavor.

2. Why are dinosaurs considered ancient when on the geologic time scale the time of their demise was not that long ago?

The human perception of passage of time is based on the daily rotation of Earth, the monthly transits of the moon, and Earth's yearly revolution around the sun. Geologic time is greater by many magnitudes and is a difficult concept to grasp. To better understand the enormity of geologic time, visualize a long strip of paper representing the geologic time scale spread out on a football field, so that 4.6 billion years equals 100 yards. The first 70 yards or so (almost three-quarters of the length of geologic time) would consist of Precambrian rocks that are poorly exposed on the surface of Earth, difficult to understand, and almost barren of multi-cellular life. On the paper strip, where would we see the last of the dinosaurs? They became extinct about 65 million years ago, only 5 yards from the end of the field. The appearance of humans would be in the last few inches; of course human civilizations, inventions, and discoveries are a hair line from the 100-yard line. "Ancient," then, is a relative term based on human observation.

3. What was the great contribution to human thought made by James Hutton?

Hutton saw Earth as a body created and developed by natural laws of chemistry and physics that could be observed in operation today. He removed the origin of Earth from the realm of the mystical, theological, supernatural, and catastrophic, instead believing that geological features of the landscape could be interpreted by natural phenomena. That is essentially what he meant in his saying, "The present is the key to the past." The geological processes that can be observed today can be used to explain the origin of geologic features of the past. Of course, rates of change, plate positions, and so on are different now than in the past; but the concept of natural forces at work is a correct and profound observation.

4. What is the importance of original horizontality and superposition in determining relative time?

To interpret the sequence of events from rock outcrops or exposures, the geologist must take into consideration that all water-laid sedimentary rocks form originally as horizontal strata. Because of plate activities and stresses within the lithosphere, few of the originally horizontal layers remain in that position. (Some do, of course. See the diagrams (Figure 8.12 and 8.14, page 181) on the horizontal sedimentary beds in the Grand Canyon.) By measuring the present dip and strike of rocks in the field, the geologist is able to calculate the original position of the layers, and from the principles of superposition, determine which were the first to be deposited, and hence are oldest in the sedimentary pile.

5. Not infrequently, especially in regions of intense deformation, the sedimentary pile will be tilted into a vertical position or completely overturned. Obviously the principle of superposition would not be valid here. How do geologists recognize overturned strata, and how do they determine which end was originally up?

First, geologists will study the sequence of formations of about the same age in areas that were not subjected to such intense stress. If a rather well-defined sequence can be found, the formations in the suspected overturn can be compared. Also, geologic structures such as recumbent anticlines may provide clues to overturning (see Lesson 8). And finally, a careful analysis of the fossil succession and comparison with a section that obviously was not overturned may indicate the correct superposition, with the oldest on the bottom of the rock sequence.

6. What is the value of the relative geologic time scale if it doesn't tell how many years ago events took place?

The relative time scale is first of all based on the fossil record. The fossil record is important because it can be used to determine relative ages of the rocks in which the fossils occur. Because life changes with time in a more or less orderly way, any rock sequence can be correlated with any other rock sequence having the same fossil assemblage, and it can be assumed they were deposited at the same time. For example, the fossils in the lowermost horizontal layers of the Grand Canyon are comparable to those collected in Wales, Great Britain, and other localities. It can be said that the rocks in which they occur – although widely separated – all formed at about the same time. This is a powerful tool for correlating rock units worldwide. The fossil record is never totally complete, but enough matching units exist to construct a reliable relative time scale.

7. What is an index fossil, and how does it differ from other fossils?

Many species of organisms survived over long spans of time, and their remains are found in rocks covering lengthy geologic periods. These forms are not particularly useful in limiting the age-range of the rocks in which they occur.

Certain other fossil species called *index fossils* are especially well suited for purposes of correlation and limiting time intervals. These species are easy to identify because they differ considerably from other fossil groups and have their own characteristic markings. They are generally abundant, geographically widespread, and can be used to correlate rocks over large areas. And finally, they are restricted to narrow time intervals, allowing for precise correlation. In other words, they appeared, flourished, spread widely in the seas of that time, and abruptly became extinct.

Few index fossils exhibit all of these traits. One group of planktonic (floating) microscopic fossils related to the amoebas but possessing a shell are called foraminifera and have been successfully to bracket rather narrow ranges of time. The various species are easily identified under the microscope by *micropaleontologists*. Since the *forams* tended to float across large areas of the sea, they settled to the sea floor in many different sedimentary environments and have been useful in correlating late Mesozoic and Cenozoic sediments. Another group of microfossils called *radiolarians* have recently been extracted from sedimentary formations and are being used to trace the movements and collisions of exotic terranes (refer to Lesson 7).

8. How are rocks from the Precambrian correlated if abundant fossils do not appear until after the end of the Precambrian and the beginning of the Paleozoic? Are there any fossils at all in the oldest Precambrian rocks?

Before the use of radiometric dating, scientists found it exceedingly difficult to establish even relative ages for Precambrian rocks. They used the principle of superposition and decided, after long study and extensive field work, that a particularly distinctive association of rocks in the Precambrian (Canadian) Shield seemed to represent the oldest Precambrian interval, which became known as the *Archean Eon*. Correlation of these truly ancient rocks was done mostly by similarities in the deeply metamorphosed and distorted formations exposed at the surface and from periods of mountain building. The accuracy and resolution of time of events is still much less precise in the Precambrian, where the rock record is discontinuous, than it is in the later eras, where there is an almost continuous record of sedimentary deposition. The Archean interval of Earth history is now defined as extending from the time of Earth's origin to approximately 2.5 billion years ago, based on radiometric dating.

Life does appear in the Archean time. Primitive blue-green algae formed structures called *stromatolites*; other forms of algae and bacteria are represented in Archean rocks by fossil cells that lack cell nuclei and chromosomes. But these fossils are so small, so scarce, and so undistinguished that they cannot be used as index fossils for correlation. Radiometric dating is used today, but that technique also has problems since these ancient rocks have been severely metamorphosed, contorted, and eroded through the long passage of time.

Activities

1. Your text has an excellent example of how geologic history of a region is interpreted using the basic principles discussed in this chapter. Start with Figure 8.1, page 177. Look it over carefully, noting the names given to the various formations. In Figures 8.2 and 8.3, page 178, note the *original horizontality* of the sedimentary formations. In Figure 8.4, page 178, that strange new body injected into the bottom of the pile is a granitic igneous intrusion. From your text, read the article on *cross-cutting relationships*, pages 177 and 178, and follow the development of the region through the diagrams to Figure 8.11, page 180. This is the method geologists use to unravel the history and relative ages of formations such as those seen in the Grand Canyon and other regions.

2. In Figure 8.14, page 181, the cross section through the Colorado plateau, note the Navajo Sandstone in the diagram. Do you think this formation ever occurred in the Grand Canyon? What happened to *physical continuity* here? What kind of an unconformity is exposed at the base of Bright Angel Shale, shown at the left-hand side of the diagram? What kind of an unconformity is shown to the immediate right side of Bright Angel Shale?

3. Read Box 8.1, page 183, on "Demise of the Dinosaurs – Was it Extraterrestrial?" This topic has generated much discussion and many scientific papers. But after several years of field studies all over the world, the question is still not settled.

Self-Test

1. The greatest significance of Hutton's statement, "The present is the key to the past," lies in its explanation of

    a. the importance of relative age dating of formations.
    b. the laws of original horizontality.
    c. using natural laws observed today to explain geologic features of the past.
    d. the value of fossils in showing evolution of the past forms.

2. The current conceptions of the length of geologic time support all of the following theories except

   a. the movements of tectonic plates.
   b. rate of terrestrial erosion.
   c. rates of evolution of life on this planet.
   d. dates of events as described in theologic texts.

3. The law of superposition implies that

   a. the oldest rocks in a sedimentary sequence will be on the bottom.
   b. in any sedimentary sequence, the rocks on the bottom will always be Precambrian in age.
   c. the youngest rocks in any rock series will always be Cenozoic in age.
   d. the rocks containing dinosaur bones will be at the bottom.

4. Any sequence of rocks containing a layer of sandstone overlying a layer of shale

   a. must be the same absolute age as any other similar sequence.
   b. must be the same relative age as any other similar sequence.
   c. must be studied further before correlation can be made
   d. would most likely contain the same fossil succession as any other similar sequence.

5. Correlation of distant sedimentary rock units, such as on separate continents, can best be accomplished through the use of

   a. similarity of rock types.
   b. uniformitarianism.
   c. cross-cutting relationships.
   d. index fossils.

6. In any pile of sedimentary rocks, if the youngest fossils are in the lowest layers, you are probably dealing with

   a. overturned strata.
   b. reversed evolution.
   c. intrusion and cross-cutting.
   d. a normal, expected sequence of deposition.

7. The standard geologic time scale divides geologic time by

   a. correlation of rock units.
   b. sequence of fossil assemblages.
   c. sequence of igneous intrusions.
   d. carbon-14 dating methods.

8. Most of geologic time occurred during the

   a. Paleozoic Era.
   b. Cretaceous Period.
   c. Holocene Epoch.
   d. Precambrian Eon.

9. If the remaining amount of a radioactive element left in a mineral is only one eighth of the original amount, how many half-lives have gone by?

   a. 2
   b. 3
   c. 4
   d. 5

10. Which of the following statements would refer to cross-cutting relationships?

    a. All rock formations are originally horizontal.
    b. The oldest rocks in any sequence are Paleozoic in age.
    c. Any igneous intrusion is younger than the rocks it intrudes.
    d. Faulting is always older than the disrupted rock.

11. Radioactive dating methods are best used to

    a. date individual fossils.
    b. date individual sedimentary layers.
    c. determine the sequence of geologic events.
    d. determine how many years ago geologic events took place.

12. Index fossils have all of the following characteristics except

    a. wide time range in the fossil record.
    b. wide geographic distribution.
    c. easily recognizable characteristics.
    d. narrow time range in the fossil record.

13. The technique of using uranium-lead dating is to

    a. measure the amount of radioactive lead found in igneous rocks.
    b. compare the decay rate of uranium with that of potassium.
    c. compare the amount of uranium with that of lead in a mineral.
    d. determine how much lead has decayed into uranium in a mineral.

14. Human life appears in the fossil record in each of the following time divisions except the

    a. Holocene epoch.
    b. Cenozoic Era.
    c. Mesozoic Eon.
    d. Quaternary Period.

15. The best estimate of the age of Earth is based on

    a. radioactive dating of Archean rocks in the Precambrian Shield.
    b. the rate of sedimentation multiplied by the thickness of sedimentary layers.
    c. the computed loss of heat from the early molten planet.
    d. radioactive dating of moon rocks and meteorites.

11

EVOLUTION THROUGH TIME

This lesson will help you describe the development of life on Earth.

GOAL

After reading this study guide, the textbook assignments, completing the exercises, and viewing the lesson's video portion, you will be able to:

LEARNING
OBJECTIVES

1. Discuss current theories about why life on Earth changes through geologic time.

2. Describe the major types of life that dominated the Precambrian, Paleozoic, Mesozoic, and Cenozoic Eras.

3. Describe the processes that can either destroy the remains of an organism or cause them to become fossilized.

4. Explain why stromatolites are important.

5. State the various explanations for mass extinctions, such as the extinction of the dinosaurs.

6. Discuss radiometric dating and its application to geologic time.

7. Describe the basis for developing a geologic time scale.

This lesson is a special addition to *Earth Revealed*. Because there is not a specific corresponding chapter in your text, this study guide chapter is longer than the other chapters of the study guide. The evolution of life is an important aspect of Earth history, and this overview will introduce you to the fossil record, to the mechanisms of change in the living world, and to the sequence in which plants and animals came to populate Earth.

INTRODUCING
THE LESSON

"The present is the key to the past" is a catchy phrase you learned in Lesson 10. In this lesson, you will learn that "the past can be the key to the present." To try to follow the great parade of evolving life on Earth, we will examine the fossil remains found in rocks. You will discover that fossils do not form a perfect record, however, partly because some organisms were more suitable for preservation than others. There are countless remains of shelled *brachiopods, trilobites, crinoids*, and bones and teeth of *vertebrate animals* but few jellyfish and other soft-bodied creatures (although their numbers must have been enormous). From the scattered remains, we can observe the most primitive algae and bacteria of the Precambrian, the explosion of life in the early Paleozoic, the amazing diversity and drama of the great reptiles that flourished during the Mesozoic, and finally, the more familiar mammals of the Cenozoic.

The geologist and the biologist understand that evolving species responded to the changing face of the globe and to new environments provided by shifting tectonic plates. You will read about repeated *mass extinctions*, including the great unsolved puzzle about the demise of the extremely successful dinosaurs which, after dominating life on Earth for millions of years, perished in a relatively short time. Although the great reptiles are no longer on Earth, did they leave any descendants? Equally puzzling is why the inconspicuous little mammals, no larger than cats, survived the Mesozoic mass extinction to become the leading lights of the Cenozoic Period. Thus, past life offers clues to present life and tells the story of biologic evolution, including catastrophes, extinctions, variations, adaptations, and survivals on a dynamic planet.

LESSON ASSIGNMENT

Completing the following eight steps will help you master the lesson objectives and achieve the goal for this lesson. Note that for this lesson you are to read a major portion of the study guide *before* continuing with the next step in the LESSON ASSIGNMENT.

Step 1: Read the study guide lesson's PUTTING IT ALL TOGETHER section as far as the ACTIVITIES. Review study guide Lesson 10, particularly the sections dealing with fossils and relative age. Also, in Lesson 2, re-read the "Case Study: Some Ideas about the Origin of Life."

Step 2: Read the TEXT ASSIGNMENT, consisting of scattered references in your text.

Step 3: Study the KEY TERMS AND CONCEPTS noted in the study guide.

Step 4: Watch the VIDEO using the VIEWING GUIDE in the study guide.

Step 5: Complete any assigned lab exercises.

Step 6: Complete any assigned ACTIVITIES found in the study guide.

Step 7: Review the material in this lesson and complete the study guide's SELF-TEST.

Step 8: Go back to the LEARNING OBJECTIVES and make sure you have learned each one.

1. In Chapter 1, review "Geologic Time," pages 20-22, and carefully study Table 1.1, page 22, showing important ages of life on Earth.

2. In Chapter 14, study the photographs that show how fossils occur in rock, pages 303-326.

3. In Chapter 8:

   a. Review "Correlation by Fossils," pages 181-182. See Figure 8.16, page 182, showing trilobites.
   b. Read Box 8.1: "Demise of the Dinosaurs," page 183.
   c. Read Box 8.4, page 190, and study Figure 1.
   d. Review the "Terms to Remember," page 191. They are important in this lesson, too.

TEXT ASSIGNMENT

Many of the terms in this lesson will not be found in your text or glossary. If you have access to a recent text in biology or a good dictionary, you will find the words and concepts discussed there.

Adaptation: Changes in an organism that enable it to take advantage of new or changing environmental conditions. Adaptations are specialized features, either body parts or behavior patterns, that allow animals and plants to perform one or more functions that are useful to them and permit their survival.

Algae (al-jee): A group of single-celled plants and seaweeds that contain chlorophyll and are photosynthetic. There are seven or eight distinct groups of algae, of which five are predominantly marine. All except the blue-green algae contain cell nuclei.

Blue-green algae or procaryotes: Primitive single-celled plants that lack a cell nucleus and organized chromosomes. Blue-green algae are found today almost everywhere light and water are available; despite at least 3 billion years of existence, they have changed little and apparently function today in the same way as the earliest forms.

Stromatolites: Knobby mounds, pillars, or dome-shaped organic structures that live in the intertidal zone and are produced by thread-like primitive blue-green algae that form sticky mats. The algae trap carbonate muds then grow through the mud to form another mat. The accumulation of many layers of mud and mats forms the stromatolite mound. These are among the most primitive groups of organisms in the world since they extend back through more than 2.5 billion years of geologic time. Stromatolites similar to the Precambrian forms are found living

KEY TERMS AND CONCEPTS

today in the intertidal zone of Western Australia and other regions bordered by warm seas.

Phylum: One of the primary divisions of the animal and plant kingdoms; a group of closely related classes of animals or plants (plural: phyla).

Genus: A group of closely related species of plants or animals.

Species: A group of individuals – either plants or animals – that have substantially the same body structure, habits, geographic and geologic range, and can successfully interbreed. For example, the genus of humans is *Homo*, and the species is *sapiens*. (The genus name is always capitalized, and the species name is lower case.)

Brachiopod: A marine animal with two unequal shells, each of which is bilaterally symmetrical. They are also called *lamp shells* and have a superficial resemblance to clams, although they are unrelated. During the Paleozoic Era, brachiopods were extremely successful, and over 30,000 extinct species have been described. Today, only about 300 species exist, some of which are almost unchanged from their Paleozoic antecedents; most are relatively rare. Brachiopods are an interesting example of a group that survived the repeated mass extinctions that eliminated more spectacular animals from the fossil record.

Trilobite: An extinct Paleozoic *crustacean* (a class that includes lobsters, crabs, barnacles, and various *planktonic* species) that was especially abundant in the early Cambrian Period. The name is descriptive of the body plan, which is divided into three longitudinal grooves forming *three lobes*. Although the trilobites were a very successful group of *arthropods* (the joint-legged animals), none survived the late Paleozoic mass extinctions.

Mass extinction: An episode of biologic extinction in which large numbers of species disappear in a few million years or less. Although the mass extinction at the end of the Mesozoic is probably the most famous because it marked the end of the dinosaurs, at least five others have been equally devastating. The fossil record reveals that, at certain times, between 50 and 90 percent of all the species on Earth died during very brief geologic intervals! In fact, of all the species that have existed in the course of Earth's history, only a small fraction remain alive today.

Ecologic niche: A position of an organism in its environment or "what the organism does for a living." The niche provides certain kinds of food, physical and chemical conditions, and opportunities for interaction with other species. Organisms occupy their niche as predators, carnivores, herbivores, parasites, crawlers, flyers, scavengers, and so on.

Once you have read the study guide and text material for the lesson and studied the terms, you are ready to view the lesson's video portion. Review the questions that appear in this section to help you watch for important points and to prepare you for what you will see in the video.

Before viewing, answer the following questions:

1. Why are some organisms more likely to become preserved as fossils than others?

2. Describe life in Precambrian time and discuss why the fossil record is so skimpy.

3. Briefly, what are the two most promising theories about the extinction of the dinosaurs?

After viewing, answer the following questions:

1. Where do blue-green algae fit in the great parade of life on earth?

2. What are some of the different ways fossils can be formed?

3. Why are paleontologists so interested in stromatolites and how do they form? Are there any examples of these very ancient structures living today?

4. Why was the advent of sexual reproduction a major step in the evolution of life?

5. Describe the Paleozoic "Cambrian Explosion."

6. What was the great contribution made by early algae toward making Earth a habitable planet?

7. In terms of plate tectonics, what was happening during the late Paleozoic?

8. Based on the interviews with Jamie Webb and Dr. Stanley Awramik, how would you describe a mass extinction? Do you think one is going on right now? What is the evidence?

9. How have new discoveries changed our old ideas about dinosaurs?

10. Why did the discovery of a world-wide iridium layer cause such interest among geologists and paleontologists? Are there any other theories that attempt to explain the same phenomena but through other events?

11. Were any dinosaur bones discovered in the California La Brea Tar Pits? What can be learned from detailed study of bones and microfossils recovered from the Tar Pits? During what time period did the Rancho La Brea animals live in this are?

12. What influence, if any, do you believe early humans had on the mammals of the Tar Pits and other Pleistocene localities?

**PUTTING IT
ALL TOGETHER**

Here are a SUMMARY, a CASE STUDY, QUESTIONS, suggestions for ACTIVITIES, and a SELF-TEST to help you integrate your learning of the material in this lesson. You should read all parts except the ACTIVITIES and SELF-TEST before you view the video.

Summary

Just as land areas of Earth evolved from scattered small land masses into huge continents, so life evolved from microscopic cells in Precambrian seas to the large complex organisms that populate Earth today. This thin film of life clinging to a moving lithosphere is unique in the solar system and is a fascinating story worthy of our attention.

What is a Fossil?

Before we investigate "Evolution Through Time," which is based primarily upon the fossil record, let us decide – what *is* a fossil? Fossils may be actual remains or tangible signs of ancient life preserved in rock.

It is difficult to decide exactly when something becomes a fossil. Last year's clam shell on the beach is not considered a fossil. On the other hand, an insect found preserved in an glacial age deposit at the Tar Pits in Southern California would be considered a fossil even though the same species is still living. By definition, only organisms that lived before historic times are considered fossils.

The most readily preserved features of animals are their *hard parts*: teeth and bones of vertebrate animals and solid structures, such as shells and spines, of *invertebrate animals*. Although plants do not have skeletal structures, their cells have rigid walls of cellulose. Woody tissues, leaves, seeds, and other parts are often well preserved. Microscopic plants called *diatoms* develop structures of insoluble silica that are suitable for preservation.

Invertebrate animals that lack skeletons and are completely soft-bodied have left a poor fossil record or, probably in most cases, no fossil record at all. Sometimes soft or semi-hard organisms will leave tracks, trails, or burrows or will be a carbon impression on a slab of flat rock. Fossils resulting from entrapment of living animals are the best-preserved, such as the dire wolf or sabre-toothed cat found in the Tar Pits of Southern California or the famous woolly mammoth of Siberia and the baby woolly mammoth trapped in ice and preserved in the Alaskan permafrost. These actual remains, however, are very young, perhaps only about 18,000 years old.

It is important to recognize, however, that most species of animals and plants that lived on Earth have never been discovered. Uncommon species and those that lacked skeletons are unlikely to be found in fossilized form. Even most species with skeletons have left no permanent record because of the vagaries of sedimentary deposition and fossilization.

Processes that Destroy or Limit the Fossil Record

A variety of processes in nature act to destroy organic remains. Once an organism dies – whether by predation, disease, accident, old age, or catastrophe – there is usually intense competition among other organisms for the nutrients stored in its body. What is not immediately consumed may be decomposed by exposure to sun and rain, or the remains may be moved and scattered. If they are buried where they lie, they have a better chance to become lithified or mineralized and preserved as fossils.

Because of *the way* they live (habits) or *where* they live (habitats), some organisms are more likely to become fossils than others. Marine or other aquatic organisms are more frequently preserved than those that live in eroding uplands or other places where their remains have little chance for permanent burial. Organisms that live in burrows or have roots in the substrate have a head start on fossilization because they are already *pre-buried*. But even after burial and lithification (the replacement of organic material with rock, frequently calcium carbonate, accomplished by circulating ground waters), metamorphism or melting may obliterate any signs of life.

The Mechanisms of Change: Adaptations

As you remember from Lesson 10, the early geologists, attempting to correlate sedimentary strata, did not recognize that fossils are remains of once-living animals. Today, it is understood that, just as Earth has been changing and evolving since the beginning of geologic time, biologic change has also been occurring. Biologic evolution is the continuing change in populations of organisms; such evolution has occurred in the geologic past and is occurring today as a result of changing environments.

When we examine the vast numbers of living forms that inhabit our planet, it is impressive to consider how well each species functions in its own habitat. The teeth of horses, for example, are well suited to cropping and grinding harsh grasses that contain small bits of hard silica. Whales, dolphins, and fish have wonderfully streamlined bodies for rapid movement through water. Plants – especially those in difficult environments such as the desert – have fine hairs or a waxy coating on the leaves, or tiny leaves or spines to hold water and prevent evaporation. Each organism has many *adaptations* that enable it to live, find food, avoid predation, and reproduce successfully.

Before the middle of the 19th century, however, the nature of adaptations was not well understood. Scientists know today that adaptations are usually small changes from what is already present. Evolution only works by *remodeling* and *not by designing* each new organism from scratch.

Common Ancestry

Each group of living things has certain features in common with other groups, suggesting common origins or common ancestry. An interesting example is the common origin of the toes of land-dwelling mammals and the wings of bats. Bat wings are actually formed of four toes, the appearance and bone structure of which resemble those of walking mammals. Another example of common ancestry is found in *vestigial organs* that have no present use but are functional in other creatures. The presence in modern whales of small bones that resemble the functional pelvic bones of other mammals reflects a biological past in which legs were present. All organisms carry with them a veritable museum of useless reactions, leftovers from some very ancient ancestor. Can you wiggle your ears? What does your appendix do for you? When you are cold and get *goose bumps*, does the hair on your arms fluff up? The list could go on and on.

Charles Darwin and Natural Selection

By 1859, when Charles Darwin published his great work, *On the Origin of Species by Natural Selection*, fossils were known to be organic remains, but few biologists gave much thought to the concepts of evolution. In 1831, when 22-year-old Darwin set sail as an unpaid naturalist aboard the *Beagle*, he was not planning to develop a new theory or start a revolution in science. But many of his observations — some of which were purely geological — convinced him of the validity of the theory of uniformitarianism and led to his conclusion that natural processes transform life in the same way that they transform the physical features of Earth. (Refer to Lesson 10.) Darwin's observations in the Galápagos Islands, South America, and other lands led him to believe that all organisms evolved from ancestral forms but changed to be able to function, survive, and reproduce in changing environments.

Because Darwin lived before the advent of modern genetics, he could not explain how an organism could pass along a favorable genetic trait to its offspring. It was difficult to convince others that natural selection could effectively produce change and new species.

Darwin proposed that the process of *natural selection* operates in nature in much the same way as animal breeders use artificial selection to improve livestock. Darwin reasoned that certain individuals would survive longer than others if they were better equipped to find food, avoid predators, resist disease, or deal with various environmental conditions. These individuals, because they lived longer, would produce more offspring than other species. They would also pass along their traits to a large number of individuals of the next generation, and the next, becoming a large successful population in that environment.

Mutations

With the discovery of genes came the concept that genes could be altered. Genes are chemical structures that undergo spontaneous chemical changes or *mutations* that provide variability upon which natural selection operates. Some mutations are lethal, or inhibit the success of an organism. Others enhance the ability to survive, and these gene changes are passed along to the offspring, which also have a better chance to survive and reproduce.

This is a very brief glance at a fascinating topic: the origin of species, the influence of changing landscapes, and the interaction between the biologic and geologic world. Now let us turn to the record of evolving life on Earth.

Early Records of Life

To the uninitiated, *dinosaurs* seem to be among the earliest living creatures, but dinosaurs first appeared less than 250 million years ago. From the fossil record, we know that life is at least as old as the oldest rocks discovered – about 3.5 billion years!

Going back to beginnings, we have to examine the records from the Precambrian Eon, which includes almost 90 percent of geologic time, starting with the birth of Earth 4.6 billion years ago. This is the longest and should be the most interesting chapter in geologic history because here we should be able to see the origin of life on Earth. But the difficulties of dating these rocks (Lesson 10), the repeated deformation, metamorphism, melting, and fragmentation caused by long erosion make this study one of the most difficult and demanding aspects of geology.

The oldest biologic structures discovered are *stromatolites*, large mounds formed by blue-green algae and discovered in Australia in rocks 3.4 to 3.5 billion old. Actual fossils of blue-green algae and bacteria have also been tentatively identified in Archean age rocks (older than 2.5 billion years), some from western Australia and others from southern Africa. They are *prokaryotic*, consisting of cells that do not have a nucleus and whose DNA is not clustered into discrete chromosomes. They also lack other internal organelles (organ-like structures) that are present in more advanced forms of life. After about 1.5 billion years of evolution, it seems that Earth was still populated almost exclusively by primitive single-celled prokaryotic forms of life. The blue-green algae were photosynthetic and released free oxygen that eventually accumulated in the atmosphere, profoundly changing the global ecosystem. Despite their incredible antiquity, the Archean algae and bacteria very much resemble similar cells living today.

The Proterozoic: Late Precambrian

The Proterozoic followed the Archean, beginning about 2.5 billion years ago and ending about 570 million years ago, at the beginning of the

Paleozoic Era. In the course of Proterozoic time, there was a great expansion and change from the simple cells of the Archean. Stromatolites became abundant in the fossil record about 2.3 to 2.2 billion years ago and did not start to decline until early in the Paleozoic Era. The biggest advance during this period was the appearance of cells with organized nuclei (*eukaryotes*) about 1.4 billion years ago. All cells today, with the exception of the blue-green algae, are eukaryotes – that is, cells with chromosomes, nuclei, and other advanced internal structures. It is not known exactly when these cells evolved but probably not before the late Proterozoic.

The early Precambrian seas were a huge thin vegetable soup. In the late Proterozoic, however, about 670 million years ago, abundant multicellular plants and animals appeared. The Ediacara fauna of South Australia is the most famous of late Precambrian *soft-bodied* fauna and reveals that off the sandy shore of the bare continent, soft coral fronds lifted from the ocean floor. Some primitive arthropods (joint-legged animals) roamed the shallow seas, jellyfish undulated in the currents, and marine worms plowed their way through the soft sediment. Some of the animals were so unusual that their relationship to later forms has not yet been determined.

Oxygen continued to enter the sea water and atmosphere from the photosynthetic activities of marine algae and probably had increased to about 10 percent of its present amount. Organisms, which until then were generally microscopic, became larger. Animals that used oxygen had more energy for food gathering, reproduction, and growth; natural selection provided for the success of their species. All of the ancestral forms that appeared in the early Paleozoic must have evolved during the very late Precambrian, even though we have recovered relatively little evidence of their soft bodies from the ancient rocks.

Some paleontologists believe that with the appearance of multicellular organisms true evolution actually began, because it is here that sexual reproduction began. Until that time, single cells simply split into two and produced daughter cells with the same genetic material as the parent. With the advent of sexual reproduction and the combination of genetic material from two different parents, great diversity became possible, increasing the capability to adapt to new or changing environments. Sexual reproduction is one of the most important developments in the history of life.

The Cambrian Explosion of New Life Forms

About one million years later, 570 million years ago, the fossil record improved immensely because the seas teemed with the first appearance of the major groups of animals. This was also the first appearance of animals with shells, spines, hard plates, tubes, or scales, which some biologists think was a response to the first appearance of fierce predators: an early arms race. The predators are not well represented in the fossil record because many were soft-bodied and are not preserved.

Those that are preserved (for example, in the Burgess shale, a mid-Cambrian marine community found high in the Canadian Rockies of British Columbia) were a lethal, aggressive cast of characters, some as large as a half-meter long. Even the well-armored arthropods, the trilobites, showed bite-sized chunks missing from their sides.

Many of the Cambrian phyla have survived until today; other forms are so unusual or bizarre that they cannot be assigned to any known phylum. The Cambrian saw a higher percentage of experimental groups of animals than any other interval in the history of Earth, but some were short-lived. Many creatures evolved during the Cambrian to fill ecological niches that had never before been filled. This was a time when supercontinents were breaking up; because extensive tropical shorelines and abundant marine food supplies developed, many new niches or ways of making a living became available. The most astonishing aspect of the Cambrian fauna is that so many radically different types of animals appeared in such a short time. Whatever their origin, the earliest Cambrian innovations – such as shelled organisms, predators, and deep burrowers – went on to colonize the world.

Life in the Early Paleozoic

During the latter part of Cambrian time, several minor mass extinctions eliminated many of the trilobite species not only in North America but in other regions of the world as well. The last of the Cambrian mass extinctions, at the very end of the period, eliminated large numbers of molluscs and still other trilobite groups. A few trilobites are found in younger Ordovician strata, but they never again reached the abundance and diversity seen in some of the Cambrian limestones. Trilobites are the principal index fossils of the Cambrian strata. (See the information in Lesson 10 on index fossils.)

In the Ordovician Period that followed, the seas were populated by many groups of animals that continued to flourish in later Paleozoic periods. Brachiopods, sedentary shelled animals, are the most conspicuous group of well-preserved fossils in Ordovician rocks and in all younger Paleozoic systems as well. Corals and the beautiful waving crinoids (sea lilies) were common. Jawless fish left a few scattered remains in the rock record, the first evidence of vertebrate life. The most active predators on larger animals were the invertebrate starfish and the *nautiloids*, related to the colorful living *chambered nautilus*, squids, and octopuses. Coral reefs were also established during this period. During the Cambrian and Ordovician, a widespread distribution of stromatolites carried over from Precambrian time. But the end of the Ordovician Period was also marked by a great mass extinction, and large stromatolites became rare. On a global scale, over 100 families of Ordovician marine animals failed to survive into the following Silurian Period.

Middle Paleozoic Time

The oceans of the world stood high during most of the middle Paleozoic Era, which included the Silurian and Devonian Periods, leaving a widespread marine sedimentary and fossil record on every continent. The shallow seas that inundated the continents teemed with life after recovery from the mass extinction at the end of the Ordovician. In the tropical zones, huge complex reefs were populated by new varieties of corals. Predators were larger and more advanced, and the first jawed fish – some of which were the size of large sharks – roamed the great seas. Plants colonized the barren areas first in marshy environments and later formed large forests that looked nothing like modern forests. The first known insects appeared and, near the end of the Devonian Period, the first terrestrial vertebrate animals, *amphibians*, crawled onto the land. The fins of their ancestors – through mutations and natural selection – had been transformed into legs. More than 80 million years elapsed between the arrival of land plants and the arrival of land-living amphibians. This is a reasonable sequence since animals cannot live on land without an adequate supply of edible vegetation. Another mass extinction near the end of the Devonian swept away large numbers of marine forms, including some of the primitive fish, and the once rich fauna of the seas was reduced again to a few hardy survivors.

The Late Paleozoic World

The late Paleozoic Era included the Mississippian, Pennsylvanian (together known as the *Carboniferous*), and Permian Periods, lasting from about 360 million to about 248 million years ago. This was a time when the changes taking place on land were more remarkable than events happening in the sea. The name *Carboniferous* is given to rock formations with an abundance of plant fossils and to huge coal deposits (carbon) made of leaves, stems, and trunks of ancient trees that accumulated in lowland swamps. The forests of the Carboniferous were made up of spore-bearing tree-like ferns and seed ferns, reaching heights of about 30 meters (100 feet). One of the most famous of the seed ferns was *Glossopteris*, whose fossil imprints were confirming evidence of the existence of Gondwanaland (Figure 4.3, page 67). This was the first appearance of *gymnosperms* or naked-seed plants, which include such living conifers as pines, redwoods, spruces, and their relatives. By the end of the Paleozoic, most of the exotic tree ferns declined, but the gymnosperms expanded and flourished during the following Mesozoic Era.

Insects, some of them enormous, flourished during the late Carboniferous. This was the time of the giant dragonflies and the supercockroach (whose descendants are the scourge of civilization). The large amphibians, pioneers of the land, were replaced by early reptiles as the dominant vertebrates. Animals adapted rapidly to the many environments, and many new species appeared. In the seas, the skeletal debris from marine planktonic organisms accumulated to form widespread limestones, evidence of the tremendous biologic productivity of the late Paleozoic time.

The end of the Permian was marked by the greatest mass extinction in all of *Phanerozoic* time (since the end of the Precambrian Eon). From the evidence found in the rocks and fossils, we know the late Paleozoic world underwent major climatic changes. Glaciers, for example, spread over the south polar region of Gondwanaland. Sea level rose and fell several times with the growth and melting of the glaciers. During Permian time, there was a general drying of the climate at low latitudes, leading to a decrease in the coal swamps and the extinction of many spore-bearing plants and the large amphibians, both of which require moist conditions. (Living amphibians are small and include frogs, toads, and salamanders.)

In addition to the great mass extinction, the end of the Paleozoic saw Gondwanaland collide with Eurasia and start a great cycle of mountain building in Europe and eastern North America. Some of the most biologically productive continental shelf and shallow sea habitats were obliterated by the continental collision. By the time this major suturing event was completed, almost all of the supercontinent Pangaea was in place.

The Mesozoic Era: The Interval of Middle Life

It was during the period of transition from the Paleozoic to the Mesozoic that Pangaea took its final form and included almost every segment of Earth's continental crust. Pangaea was so large that much of its land area lay far from any ocean, and widespread aridity prevailed. At the end of the Triassic, the earliest period of the Mesozoic, another mass extinction occurred. During the middle period of the Mesozoic, the Jurassic, starting about 213 million years ago, sea level rose, and marine waters spread rapidly over the land, leaving a record of shallow marine deposition. About this time, Pangaea began to fragment, and once again Gondwanaland separated from the northern land masses.

Life in the early Mesozoic differed greatly from that of the Paleozoic. As you have learned, after every great mass extinction, an explosion of new species repopulates the environment, filling the newly vacated ecologic niches. The *molluscs*, for example, recovered from their decline and expanded to become even more diverse than in the Paleozoic, a success that has continued to the present. (Molluscs include clams, oysters, mussels, snails, squids, octopuses, abalone, the chambered nautilus, *nudibranchs*, and other forms.) The marine ecosystem saw the appearance of modern reef-building corals, and swimming reptiles and fish became the dominant predators of the seas. Flying reptiles and birds appeared. The most dramatic event in the terrestrial habitat was the emergence and diversification of the dinosaurs. *Mammals* also evolved during the early Mesozoic, but they remained small and inconspicuous, the largest being the size of a house cat.

The Mesozoic is known as the Age of the Dinosaurs for good reason; they dominated the terrestrial environment by their enormous diversity, filling many niches, and by the incredible size of certain species. The oldest dinosaur giants are found in Australia in lower Jurassic rocks. The most spectacular Jurassic assemblage of fossil dinosaurs in the world, however, is found in the Upper Jurassic Morrison formation, which extends from Montana to New Mexico.

The Cretaceous Period and the End of the Mesozoic

Widespread shallow marine deposits on the continents, together with some deep sea sediments and non-marine materials, indicate that the Cretaceous Period, from 144 million to 65 million years ago, was a time of transition and instability. At the start of the Cretaceous, the continents were still clustered together, but by the end of the Cretaceous, the Atlantic Ocean had widened and Gondwanaland had fragmented into most of its present continental masses.

Life in the Cretaceous was also a mixture of modern and archaic forms. Modern-appearing clams, snails, and other invertebrates populated the seas along with a variety of *ammonoids* (ancestors of the nautilus) and reptilian marine predators, both of which became extinct. All the living families of sharks, as well as most of the modern fish, had evolved by the end of the Cretaceous. Flowering plants appeared and expanded, while the older gymnosperms declined. Snakes, crocodiles, turtles, and salamanders appeared, although dinosaurs continued to dominate the terrestrial ecosystem.

In the sea, two groups of planktonic (microscopic floating) organisms, a single-celled group of animals called *foraminifers* and tiny plants called *coccolithophores*, both with shells of calcium carbonate, contributed vast quantities of sediment in oceanic areas. In fact, the coccoliths proliferated in such volume that their microscopic platelets formed thick fine-grained limestone called *chalk* on the sea floor. (Cretaceous chalk is widely used for writing on blackboards.)

The chalks are famous as the White Cliffs of southeastern England and western Europe but also occur in Kansas and in the Gulf Coast region of the United States. Coccoliths are still living, but never before or since have they contributed such vast deposits of limy sediments, and it is still not known what accounts for their amazing productivity during the Cretaceous.

Mass Extinction at the End of the Cretaceous

As you have seen, mass extinctions occurred all through geologic time, marking the end of one period and the start of another. But none have so captured the interest and imagination of scientists and lay persons as the mass extinction at the end of the Mesozoic, which marks the end of the dinosaurs. The dinosaurs, of course, were not alone. In the sea, the *ammonites*, the beautiful coiled *cephalopod molluscs*, and the large marine reptiles also died out. Many snails, clams, corals, and bryozoans also disappeared. Some groups of marine plankton – which had been so widely dispersed in the Cretaceous – were decimated.

The cause of the mass extinction remains a mystery. Over the years, many hypotheses have been proposed, such as those involving cosmic radiation from a supernova, a sudden lowering of sea level, bursts of volcanic activity that temporarily screened Earth from the sun, a sudden change in Earth's magnetic field, or a spill of cold fresh water into the Atlantic Ocean from the previously isolated Arctic basin. Climatic changes such as a marked cooling at the end of the Cretaceous have been proposed, but there is no evidence of any factors that might have caused this widespread refrigeration. The two most widely researched theories today involve a great meteor impact or a long interval of volcanic activity. Both theories have strong points and weaknesses. You'll read more about this fascinating mass extinction in the "Case Study."

The Cenozoic Era: The Age of Mammals or the Time of Modern Life

The Mesozoic ended about 65 million years ago, and the Cenozoic Era – the Age of Mammals – commenced. The first five epochs of the Cenozoic are assigned to the *Tertiary* Period (see Table 8.4, page 189), while the Pleistocene (glacial age) and the Recent (Holocene) are placed in the *Quaternary* Period. The mammals – having inherited Earth from the reptiles – underwent a remarkably rapid evolution and spread into many ecologic niches during the early Cenozoic. The first mammals were small creatures that resembled modern rodents. By mid-Paleocene, the first period of the Cenozoic Era, true mammalian carnivores had emerged. By the end of the Paleocene, about 58 million years ago, the earliest members of the horse family had evolved, but they were the size of small dogs.

During the Eocene – about 58 to 37 million years ago – modern varieties of hoofed herbivores began to appear, including the first members of the elephant order. In the seas, the most distinctive marine organisms were the whales, which evolved from carnivorous land mammals and achieved great success as large marine predators. The top marine carnivores were large sharks descended from the sharks that lived during Cretaceous time. The Eocene was a time when moist tropical and subtropical forests covered much of North America and Eurasia. A minor mass extinction marked the end of the Eocene and may have been related to the cooling climate.

It is interesting to note that the extremely arid desert of Egypt – the land of the pyramids and the Sphinx – was once an Eocene sea teeming with giant foraminifera. (Refer to the origin of "chalk" discussed above.) The pyramids were built of light-colored Eocene limestone, and the shells of the forams can be easily seen in these giant blocks quarried near Cairo.

It was during the following Oligocene – about 37 to 24 million years ago – that the largest mammal ever to walk Earth appeared, a member of the rhinoceros family. It stood about 6 meters (19 feet) at the shoulder, the height of a good sized modern giraffe. Other Oligocene mammals were large saber-toothed cats, bear-like dogs, and a group that resembled modern wolves. The horse family had disappeared from Eurasia during the Eocene but survived in North America. This was also the time of first appearance of monkeys and ape-like primates. Although they were small, the monkeys were important in the terrestrial habitat. The Oligocene was a time of cooling, and the subtropical and tropical forests retreated to low latitudes, where they remain to this day.

The Late Tertiary

The late Tertiary is particularly interesting because during this time the modern world took shape, including all life forms and topographic features we see today. The major changes were the great diversification of the mammals and the spread of grasses and weedy plants. Snakes, songbirds, frogs, rats, and mice expanded dramatically, and modern reef- building corals recovered from the terminal Cretaceous extinction and spread throughout the tropical seas. Youthful mountains such as the Alps, Himalayas, and the Rockies had late Cenozoic origins (see Lesson 7).

About 6 million years ago, the Mediterranean Sea dried up then rapidly refilled. (What is the evidence that the Mediterranean dried up? It consists of remarkable cores taken by research drilling ships from the Mediterranean sea floor. The cores contained *evaporite minerals* such as salt and gypsum left behind when the seawater evaporated!). The most widespread physical change on Earth was continental glaciation, which began in the Pliocene but is best known for effects on the Northern Hemisphere continents during the Pleistocene.

Miocene-Pliocene Epochs

The Miocene saw further expansion of whale species including toothed whales, such as the huge sperm whale and the better known *killer whales*, and the *baleen* whales with plates of feathery baleen in their mouths that allowed them to feed by straining plankton from the sea. Dolphins, which are also small toothed whales, became abundant in Cenozoic seas.

On land, the deer family, cattle, antelopes, sheep, and goats all expanded, as did the giraffe family and the pig family. Many types of elephants experienced great success during the Miocene and Pliocene intervals but later declined. Today, there are only two elephant species: the large-eared African elephant and the smaller, more docile Indian elephant commonly trained to perform in circuses.

The monkeys expanded, forming the *Old World Monkeys*, which now live in Africa and Eurasia, and the *New World Monkeys* in South America. The New World Monkeys differ from their Old World relatives in possessing a prehensile (grasping) tail.

The Quaternary: Pleistocene Continental Glaciation

Evidence that rapid cooling had started about 3 million years ago is found in the record left by certain cold-water species of microplankton, coarse ice-rafted sediments in North America, and glacial deposits above dated lava flows in Iceland. The beginning of the Pleistocene Epoch is placed at 1.8 million years ago. From the geologic evidence, it is believed that four major glacial intervals occurred, separated by warm interglacial ages. About 18 minor glacial expansions also occurred when glaciers advanced even during interglacial times. During major expansions of the ice, sea level dropped by as much as 100 meters (330 feet) because the water was contained as ice on the land. Glaciers now are small, having receded about 10,000 years ago, but there is no reason to think they won't expand again. (You will learn more about glaciers in Lesson 23. You might look ahead to Chapter 18 in your text, for a preview of this interesting subject.)

The Pleistocene life of North America was dominated by large mammals such as mastodons, mammoths, ground sloths, sabre-toothed cats, bears, and giant beavers, as well as horses and camels. Horses and camels invaded the Old World from the Americas, across a Bering Sea land-bridge. Horses became extinct in the Americas, but camels survived in South America as llamas and their relatives. Most of these animals died out in North America about 8,000 years ago, when the last continental glaciers retreated. No one knows why this mass extinction occurred; it was not the climatic change, because these species had lived through at least four major glacial advances and retreats. The arrival of humans in North America and the impact of their hunting has been suggested as the cause of the large mammal extinctions.

Evolution of the Human Family

The development of humans – the most successful of mammals – is rather poorly understood. Fossil humans are quite rare; the number of individuals existing during the Pleistocene was small, and the habitat in which they lived – the eroding uplands – was not conducive to fossilization of their remains. In addition, any graves that might have yielded human remains were shallow, and most were destroyed by marauding scavengers.

Australopithecus, the oldest known genus of the human family, has been found in African deposits ranging age from about 1.3 to 4.0 million years. The oldest of the *Australopithecus* species includes the famous skeleton known as *Lucy*. The brain capacity was larger than that of a chimpanzee but much smaller than that of the modern human brain. This was a small species, about 4 feet in height. The pelvis was remarkably like that of humans, indicating an upright posture. The skull, however, retained many ape-like features including a low forehead and a projecting, muzzle-like mouth.

The genus to which humans belong is *Homo*, the species of man. *Homo erectus* (upright Man) was alive in Africa about 1.6 million years ago and apparently was a great traveler. Remains have been found not only in Africa and Europe but also in China, where they were referred to as *Peking man*, and in Java, where they have been called *Java man*. Fossils of Homo erectus represent a long interval of time, extending from about 1.6 million ago to perhaps 400,000 years ago. Homo erectus was closer to modern humans in size, but the maximum brain size measured from their skulls only reached about the average for modern humans. Homo erectus was also a toolmaker and produced some magnificent stone hand axes.

Olduvai Gorge, Tanzania, is a unique site in which a continuous stratigraphic record going back two million years is exposed. The interval from about 100,000 to 400,000 years ago has produced very few fossils; unfortunately, this was the time when species ancestral to our own species, *Homo sapiens* (Man the Wise), were appearing. Starting at about 100,000 years ago, the record improves, and abundant well-preserved specimens of *Neanderthal* man (Homo sapiens neanderthalensis) have been recovered. The stone culture developed by this group included excellent knives, scrapers, and spear or projectile heads. Neanderthal was also quite a traveler, and records extend from Spain to central Asia, ranging in time up to about 35,000 years ago. Radiocarbon dating tells us that the Neanderthals vanished from eastern Europe about 40,000 years ago and from western Europe about 35,000 years ago. At about this time, the *Cro-Magnon* people, who were undisputed members of Homo sapiens and were anatomically almost identical to modern humans, spread throughout Europe. The culture of Cro-Magnon used elements from the Neanderthals but improved upon them, and a sophisticated variety of specialized tools was invented. The magnificent cave paintings found in France and Spain – primarily of animals and done in glowing colors with great skill and sensitivity – fill us with wonder and admiration for these ancient peoples.

Many intriguing questions about human paleontology still remain. What happened to the Neanderthals that caused them to vanish without a trace? And where did the populations of Cro-Magnon originate? It is likely that all later humans in Europe were descendants of Cro-Magnon, but were these the same peoples that reached the Americas? The oldest human fossils found in the New World are about 11,000 to 13,000 years old, on the Palouse River in Washington. Humans must have reached the New World over the Bering Sea land-bridge between Siberia and Alaska perhaps 30,000 years ago, although no fossils of this age have been found in the Western Hemisphere.

The evolution of life, of the crustal plates, and of the planet continues. Our species is not an end point of evolution. The amazing diversity of life results from the processes of evolution you learned about in this lesson; this is a fascinating story that started over 3 billion years ago with single-celled green algae.

It is interesting to contemplate that of all these species, only humans can sense the continuity from the first simple cells in 3 billion year old rocks to the enormous complexity and diversity of life today. And only humans can appreciate the relationships among the multitude of modern life forms with each other and with life of the past. This insight is both a gift and a responsibility since we must become the guardians of life on Earth.

About 65 million years ago, at the end of the Mesozoic, some event of great importance wiped out the dinosaurs – the undisputed rulers of the animal kingdom for tens of millions of years – but spared the humble mammals to inherit Earth. The fossil record changed drastically, for not only did the dinosaurs become extinct but so did more than half the species of plants and terrestrial and marine animals. Earth had become a difficult if not impossible place to make a living. The survivors, especially the mammals, moved into the ecologic niches vacated by the giant reptiles and eventually became the dominant large animals.

The clues to what happened are the dinosaur bones (relatively rare), abundant fossils of microscopic foraminifera and other marine life, and the thickness and chemical composition of sedimentary or igneous deposits laid down at this critical interval. The problem today involves data interpretation, which has led to the great debate: What caused the mass extinction? Two leading theories have been proposed. In the layers of sedimentary rock found in the Apennine Mountains of Italy and elsewhere, some scientists see traces of a huge impact from space. Other scientists see evidence of a world shaken and polluted by enormous volcanic eruptions that formed the widespread lava flows called the *Deccan Traps* in India.

The *Extraterrestrial Impact* theory holds that some 65 million years ago a giant asteroid or comet plunged out of the sky, striking Earth at a velocity of more than 10 kilometers (6 miles) per second. The enormous energy liberated by the impact touched off a series of environmental disasters, including storms, tsunamis, cold, darkness, greenhouse warming, acid rains, and global fires. When quiet returned at last, half the flora and fauna had become extinct. The extinction was apparently abrupt in geologic terms, but we do not know how many years it actually took.

Discovered in Italy, a clay layer about one centimeter thick marked the Cretaceous-Tertiary (earliest period of the Cenozoic) boundary (called the *KT* boundary). A study of the amount of the element *iridium* – which reaches Earth through cosmic dust fallout – was used to determine the rate of accumulation of the clay layer. To everyone's astonishment and confusion, the boundary clay contained far more iridium than could be deposited from the usual sprinkling of cosmic dust. Iridium is relatively abundant in certain meteorites, in far greater concentration than in Earth's crust. The solution to this anomaly was that a large comet or asteroid about 10 kilometers (6 miles) in diameter had struck Earth and dumped an enormous quantity of iridium into the atmosphere. Further research showed high levels of iridium at the KT boundary at 95 sites throughout the world.

The effects of such an impact would include dust filling the air, blocking sunlight needed for photosynthesis, and causing food chains to collapse. Also, an *impact winter* would occur, bringing on extremely cold climates. Water lofted into the air would bring about greenhouse warming after the impact winter. Survivors of the cold winter might have perished in the subsequent period of extreme heat. Acid rain might have formed from the combination of nitrogen and oxygen in the air, which would explain the extinction of plants and animals whose calcium carbonate shells are soluble in acidic water. Some of the KT boundary clay layers contained sizeable amounts of soot, suggesting the heat of impact caused fires around the globe.

But the site of impact has never been discovered. The crater might be hidden under Antarctic ice, or it might have occurred on the sea floor, later to be subducted down some deep-sea trench.

The *Volcanic Eruption* theory also presumes clouds of dust and chemical changes in the atmosphere and oceans that created an ecological domino effect eradicating large numbers of animal and plant families. The impact theory assumes that the mass extinctions all occurred at about the same time, whereas the volcanic eruption theory implies that the mass extinctions occurred over tens or even hundreds of thousands of years. This would correspond to an episode of violent volcanic eruptions in India that occurred at the time of the mass extinction. Other extinction events in geologic history also appear to have occurred roughly at the same time as periods of major volcanic activity.

Volcanoes can inject into the atmosphere vast amounts of carbon dioxide that would trigger abrupt climate changes and alter ocean chemistry. The extensive volcanism in India produced lava flows known as the *Deccan Traps*. (*Deccan* means "southern" in Sanskrit, and *trap* means "staircase" in Dutch). The size of the Deccan Traps is enormous. Individual lava flows extend well over 10,000 square kilometers (3,600 square miles) and have a volume exceeding 10,000 cubic kilometers (2,180 cubic miles). In places, the accumulation of flows is 2,400 meters (7,900 feet) thick, although the original deposit, before 65 million years of erosion, was certainly much greater.

Using new dating techniques, the age of Deccan lavas was determined to be between 64 and 68 million years. Magnetic and fossil studies together have determined that the duration of Deccan volcanism was about 500,000 years. (Review magnetic reversals in Lesson 5.) It seemed that since this great volcanic event occurred just at the KT boundary, some link must exist between the Deccan Traps and the mass extinction.

The next question was how to account for the iridium enrichment of the clay layers at the KT boundary. It was known that particles emitted from Kilauea volcano in Hawaii, for example, were enriched in iridium. Another study of the clay layer at the KT boundary indicates it consists primarily of a mineral that is altered volcanic ash. Either theory would account for the world-wide occurrence of the clay layer, because the ash and dust in each case would reach the upper atmosphere and be widely distributed.

The timing of the extinction seems to be the major difference between the two theories. Some of the extinctions began 300,000 years before the KT iridium event, and another took place about 50,000 years after the boundary. Environmental changes began at least 200,000 years before the KT event, more consistent with a longer period of volcanic activity.

In any case, catastrophes plus concepts of uniformitarianism have broadened the view of geologists and paleontologists regarding the course of evolution. If an impact or volcanic episode wiped out half the life at the end of the Cretaceous, then *survival of the fittest* is not the only factor that drives evolution. Mass extinctions, whatever the cause, also opened up ecological niches and permitted new organisms to develop. What appeared to be catastrophes may have been agents essential in the evolution of life. For this, all living forms owe their presence on Earth to the successful but ultimately unlucky creatures of the late Mesozoic, including the dinosaurs.

NOTE: For further reading, see *Scientific American*, "What Caused the Mass Extinction?" October 1990, Vol. 263, Number 4, pages 76-92, the article from which this "Case Study" has been excerpted.

1. What is a fossil?

A fossil may be any actual remains or tangible sign of ancient life that has been naturally preserved in rock. Recent shells on a beach or cow bones in an old pasture are not considered fossils, as any remains deposited since the beginning of historic time are excluded from the definition.

2. What factors determine whether an organism will be preserved as a fossil?

Many factors are involved in the process of preservation. Organisms with hard parts have a much greater chance of being preserved than completely soft-bodied creatures. This fact tends to distort our vision of prehistoric life since we know so little of the soft-bodied fauna and tend to assume their numbers were small. Rapid burial is important, and marine forms, living in basins where sedimentary deposition is taking place, are more likely to be preserved than those living in areas of active erosion. In the natural world, nothing that could be considered food will go to waste. Scavengers, from tiny insects through many species of the animal kingdom, even birds, will quickly consume the nutrients of an animal that perished. While this does not permit many organisms to enter the fossil record, it *does* keep the landscape clean of organic debris.

3. What are adaptations, and what is their value to an organism?

Adaptations are body parts that have evolved to enable an organism to survive in a specific environment or ecologic niche. There are over 1.5 million living species that have been named and described and probably twice that number that have not yet been discovered and named, all with adaptations to their specific environment. Birds, for example, have adaptations to flight such as hollow bones, a light skeleton, excellent distance vision, feathers, and, of course, wings, but no teeth. Darwin noted the relationship between the shape of the bill and the diet of the finches in the Galápagos. Behavior may also be a worthwhile adaptation, as seen in certain coral reef fish, some of which feed by day and some by night.

4. What are the principles of evolutionary change proposed by Darwin from evidence he found in the fossil record?

The evidence Darwin found led to the following concepts:

a. He felt that present-day organisms had descended by gradual changes from ancestors different from themselves, just as geologic features changed gradually by natural processes. Fossil families show slight changes in structure over long periods of time (faunal succession).

b. The structural similarities between living forms indicated a common ancestry. For example, there is a similarity between the human arm, the forelimbs of a porpoise, and the forelimbs of a penguin. There have been modifications, of course, but the basic structures are amazingly similar.

c. In any population, a certain amount of variation normally exists between individuals. Today we understand that such variation expresses genetic differences resulting from mutations (spontaneous changes in the genes) and sexual reproduction. Domestic animal breeders have capitalized on these variations to breed for certain desired traits. Oriental goldfish breeders have used natural variations for over ten centuries to develop characteristics that were not obvious in the first generation but were much admired in succeeding generations.

d. As there were natural variations in any wild population, natural selection would favor plants and animals that had a better chance to survive and reproduce. The offspring would inherit characteristics that, in turn, would provide an improved ability to compete and survive. This improved *fitness* may result from an increased resistance to disease, starvation, or climatic variations, or it may be simply a capacity to reproduce faster. The term *survival of the fittest* is rather broad, and since the rules and conditions for survival change continuously on dynamic Earth, organisms seldom evolve to fit their total environment perfectly. The tiny fraction of survivors on Earth today are temporary winners, especially in view of the enormous changes being wrought on Earth by human activities.

5. What were the earliest forms of life in the Precambrian seas?

The earliest life forms were, to the best of our knowledge, bacteria and single-celled small plants called *blue-green algae*. They were primitive in that they lacked a cell nucleus and DNA within organized chromosomes. The blue-green algae formed mounds called *stromatolites*, the earliest biologic structures known, some of which still form on beaches of unusually warm salty bays.

Towards the end of the Precambrian, after almost 4 billion years of evolution, life forms became more complex: they were multicellular, larger, and included many experimental types that did not survive into the Paleozoic. The ancestors of the myriad animals that suddenly appeared in the Cambrian must have been present in the late Precambrian, but because they lacked shells, spines, or other hard parts, were poorly preserved, and little is known about them. The vertebrate animals of later periods had not yet evolved and would not appear in the fossil record for another 100 million years.

6. What was life like in the Paleozoic, the time of Old Life?

The early Paleozoic fossils were invertebrates but cannot be considered *simple*. The trilobites, index fossils of the Cambrian Period, belong to the phylum Arthropoda, a group of highly developed, complex organisms that includes all the joint-legged animals such as spiders, insects, crabs, and lobsters. The trilobites radiated into many niches and were all sizes from planktonic species to large forms that crawled the sea floor. Jawed fish came on the scene during the Devonian, about 400 million years ago, and some were the size of large modern-day sharks. Perhaps these predators contributed to the final demise of the rather defenseless trilobites. Near the end of the Devonian, about 350 million years ago, the first vertebrate animals crawled onto the land. Thus, while the Paleozoic is the time of *Old Life*, it was a time of abundant life, of great variety, with plants and animals filling many niches in the sea and even invading the previously barren surfaces of the continents.

7. When did the amphibians and reptiles first appear?

The amphibians, which are the most primitive four-legged vertebrates, first appeared in the latest Devonian Period and apparently derived from the lobe-finned fish. They became abundant in the Carboniferous (which has been called the *Age of Amphibians*) and developed into many species. Some resembled alligators, others were small and snakelike, and a few were large lumbering plant eaters. Some measured 6 meters (about 20 feet) from the end of the snout to the tip of the tail and showed little resemblance to their modern relatives, the frogs, toads, and salamanders, the only members of their class that survive in the modern world.

The earliest reptile fossils were found in sediments of the Carboniferous Period but were widespread during the Permian, the last period of the Paleozoic Era. The tremendous mass extinction at the end of the Paleozoic, the greatest in all of Earth history, caused extinction of many of these large amphibians, of the early reptiles, and of much of the life in the sea.

8. What were the dinosaurs really like?

As researchers bring increasing knowledge to bear on their studies of dinosaur remains, many earlier notions about how these creatures looked and behaved have been drastically changed. First, regarding their size: they were not all huge; most were small animals, less than 1 meter (3 feet) long. Not all were the slow hulking beasts usually portrayed; some were agile and moved at great speed, adapted for two-legged running similar to that of the ostrich and other flightless birds. The legs and feet were positioned more directly under the body rather than sprawled out to the sides as in the amphibians, and this gave them greater mobility. Also, from their tracks, the long strides suggested they moved rapidly. Their tracks also showed they traveled together in herds and cared for their eggs and young. A question much debated was how the largest dinosaurs, which had relatively small jaws, could chew enough food to live on. These giants were herbivores, subsisting entirely on vegetation, and had *gizzard stones* like those of birds (of course, much larger) that helped them grind their food after it had been swallowed. Even the giant dinosaurs existed on dry land and walked on all fours, contrary to the old depictions of dinosaurs half submerged in water to buoy up their enormous weight. And they didn't drag their tails.

The skies were populated by flying reptiles, the most spectacular of Cretaceous animals. The largest known species is estimated to have had a wingspan of 15.5 meters (50 feet)!

The most debated question is their metabolism: were they warm- or cold-blooded? Since dinosaurs are classified as reptiles, it was assumed they were slow-moving cold blooded like living crocodiles, lizards, and snakes. Most dinosaurs were very active animals, however. And in order to compete with early warm-blooded mammals, dinosaurs had to sustain high rates of activity for long periods of time – activity which the cold-blooded reptiles were not equipped to do. On this basis many scientists believe that at least some dinosaurs were warm-blooded, a matter still under debate.

9. Were the dinosaurs ancestral to the mammals?

During the Permian Period, late in the Paleozoic, a group known as the *mammal-like reptiles* appeared, but they were not dinosaurs. They were called *mammal-like* partly on the basis of their teeth.

Reptilian teeth are all the same type from front to back, usually sharp pegs. Mammalian teeth have flat frontal incisors for nipping, side fangs for tearing, and back molars for shearing and grinding. The mammal-like reptiles had teeth somewhat like those of a dog. The body structure differed from the amphibians and early reptiles in that the legs were positioned more vertically beneath the body and allowed more rapid movement. They may have been warm-blooded and might even have had hair.

Based on the upright posture, the complex teeth and jaws, and the ability for greater activity, these animals approached the mammalian level of evolution in anatomy and behavior. Most of the mammal-like reptiles died out near the end of the Permian Period, but those that survived the mass extinction expanded in the early Triassic, only to decline again in the Jurassic Period. They left an important legacy, however, in the form of the true mammals, which evolved from them near the end of the Triassic Period.

10. When did the birds appear?

A most remarkable discovery took place in 1861 in Germany in a fine-grained Jurassic limestone, about 140 million years old. It was an unmistakable feather. A few months later, the entire skeleton of the species to which the feather belonged was discovered. This feathered animal was given the name *Archaeopteryx* which means *ancient wing*. The skeleton was so much like that of a dinosaur that it would have been regarded as such except for the feathery plumage. This was the link between birds and their dinosaur ancestors. The reptilian teeth, large tail, and clawed forelimbs – all lacking in advanced birds – strongly reflect the dinosaur ancestry of the Archaeopteryx. Modern birds still carry vestiges of their reptilian ancestry. Take a close look at birds' legs and feet: they still retain the scales! (If you can't catch a bird, a chicken in the supermarket will do; it has the same scales on its legs.)

To answer an earlier question: did the dinosaurs leave any descendants? The answer is yes, but we call them *birds*.

11. Why is it so difficult to determine the lines of human ancestry?

There are several reasons why the line of descent leading to modern man is uncertain. The fossil record is very poor. In terms of numbers of individuals, the ancestral forms were very few compared to the multitudes of fossils of other species we have studied. Sediments representing the interval in which the ancestral groups evolved are rare, scattered, or thin, and were deposited on land or in river basins where erosion recurred. Thus, remains were carried away, weathered by sun and rain, or otherwise naturally destroyed.

The question of where the biologically modern populations of Homo sapiens known as Cro-Magnon originated is still not completely solved. One possibility is that they evolved in Europe from a population of Neanderthals. Another is they migrated north from

Africa about 40,000 years ago as descendants of Homo erectus, who are believed to be direct ancestors of our species. The poor fossil record does not dispel these uncertainties.

The next question is what happened to the Neanderthals? From the abundance and wide distribution of their remains, we know their populations were large. Did Cro-Magnon exterminate them during periods of warfare? Did Cro-Magnon have a greater intellect than Neanderthal that somehow caused them to disappear? Did they interbreed so that the Neanderthal traits were absorbed into the general population? These are some of the gaps in human paleontology that time and further research may fill.

Activities

1. Turn to Table 8.4, page 189 in your text. Review the sequence of geologic eras and periods. Include the dates for the beginning of the Paleozoic, Mesozoic, and Cenozoic Eras. This is basic information needed for an understanding of geology.

2. In Box 8.4, Figure 1, start with the Archean and slowly wind your way through geologic time. What kind of life would you see during the Archean and Proterozoic Eons? What arthropod group should be shown as dominant in the Cambrian Period? What kind of forests are shown for the Carboniferous Periods? Note in the Permian Period the reptile with the tall fins on its back; this is *Dimetrodon*, a famous form that, surprisingly, is a *mammal-like reptile*. This was a predator, over 2 meters (6 feet) in length. Since it doesn't look much like any known mammal, why is it considered *mammal-like*?

   On the spiral diagram, when did humans appear? Do you think that the appearance of humans was in some way related to the disappearance of the large Pleistocene mammals? Here's a clue: What is the tiny human Box 8.4, Figure 1 holding in its hand?

3. Be sure to review in Chapter 8 the sections on determining relative age and absolute age, including the technique of radiometric dating. What method is used to date human fossils? See Box 8.3, "Radiocarbon: Dating 'Young' Events," page 189.

4. Now is the time to go to your local natural history museum. They probably have exhibits of dinosaurs or even older fossils to acquaint you with life of the past. Some exhibits show mammals of other continents. A natural history museum can be a treasure trove that will enhance your understanding and enjoyment of this lesson in *Earth Revealed*.

5. Now is also the time to go to your local zoo. As you see the wonderful animals, keep in mind the principles of mutations, adaptations, common ancestry, natural selection, and survival of the fittest. Also remember that repeated mass extinctions occurred over geologic time and that the animals of today are descendants of hardy survi-

vors of those great biologic catastrophes. As you look at the various species, try to figure out what ecologic niche they occupy in their natural environment and what adaptations made them successful. A mass extinction is going on right now. What might be some of the causes?

6. If there is an aquarium near you, it is worth a visit to see the many adaptations of aquatic plants and animals, including fish, porpoises, and birds.

7. If you have a pet, such as a dog or cat or even something more exotic, look at its teeth. You will see the *differentiated teeth* characteristic of mammals. If your pet is an alligator, will it too have differentiated teeth? (No: it is a reptile and will have a set of peg-like teeth similar to each other from front to back.) If you do not have a pet, stand in front of a mirror, smile widely, and examine your own *dentition*.

8. If you live in an area where domestic animals have been bred for certain traits, what are some of the characteristics that breeders emphasized?

Self-Test

Note that this section has 20 items – more than the typical 15 found in other lessons. This is because the text does not devote considerable space to the topic of this lesson and having more "Self-Test" items may help you master the material better.

1. In general, the organisms most likely to be missing from the fossil record are

   a. animals with shells.
   b. animals without backbones.
   c. soft-bodied animals lacking hard parts.
   d. small mammals that lived on the land.

2. The pretty colored shells you found on the beach last summer are

   a. all examples of fossils.
   b. not fossils because they are shells and not bones.
   c. not fossils because the species is still living.
   d. not fossils because they are not from prehistoric times.

3. Fossilization is least likely to occur in

   a. uplands on continents.
   b. shallow seas and areas of rapid burial.
   c. swamps and marshes.
   d. lagoons and deltas.

4. Spontaneous changes that take place in the genes are called

   a. natural selection.
   b. survival of the fittest.

   c. mutations.

   d. adaptations.

5. The best explanation for the tiny pelvic bones within the body of whales is that the pelvic bones

   a. are recent mutations.

   b. indicate evolution from an ancestor that had legs.

   c. help the whales survive in their present habitat.

   d. indicate whales are starting to evolve into animals with legs.

6. The concept of evolution contains all of the following ideas except

   a. natural selection.

   b. slow changes within a population as the environment changes.

   c. sudden changes that create completely new animals and plants.

   d. survival of the better adapted plants and animals.

7. If you could study life in a late Archean sea, you would most likely see

   a. trilobites crawling on the sea floor.

   b. corals waving in the currents.

   c. small fish with jaws.

   d. microscopic blue-green algae.

8. Some scientists feel that evolution really started when multi-cellular animals and plants developed because

   a. they were bigger and better able to defend themselves against predators.

   b. there is greater variation for natural selection to work on when genes are contributed from two parents.

   c. multicellular animals were always the *fittest* in any environment.

   d. multicellular animals were better able to survive the mass extinctions.

9. Stromatolites are interesting to paleontologists because they

   a. are the only Precambrian plants ever discovered.

   b. are the first of the great predators.

   c. all perished and left no descendants.

   d. are the earliest biologic structures ever discovered.

10. The fossil record of the Cambrian Period is much better preserved than that of the late Precambrian, mostly because

   a. of the appearance of animals with shells and other hard parts.

   b. of the first appearance of photosynthetic plants that provided oxygen.

   c. the Precambrian organisms all died and left vacant niches.

   d. this was a time of great stability of the continental and oceanic plates.

11. The Paleozoic Era saw the appearance of all the following except the

    a. first insects.
    b. great coal-producing swamps and marshes.
    c. first corals.
    d. first birds.

12. The end of the Paleozoic Era was a time when

    a. there was a great expansion of all life forms.
    b. the supercontinent of Pangaea was assembled.
    c. many productive environments opened on shallow continental shelves.
    d. there was a world-wide warming, and tropic conditions prevailed everywhere.

13. The best place for a dinosaur fossil hunter to look is in

    a. old Carboniferous coal swamps.
    b. Jurassic and Cretaceous rocks.
    c. Cenozoic marine formations.
    d. glacial debris from the Pleistocene glacial age.

14. It was long assumed that all dinosaurs were cold-blooded because

    a. all large animals of the size of dinosaurs are cold-blooded.
    b. they all lived in warm climates and had no need of temperature regulation.
    c. they were all slow moving because of slow chemical reactions.
    d. they were all reptiles, and living reptiles are cold-blooded.

15. The principal difference between the result of comet impact and of extensive volcanism as a cause of the Mesozoic mass extinction concerns the

    a. huge dust cloud that was raised.
    b. occurrence of the iridium layer.
    c. length of time involved in the mass extinction.
    d. breakdown of food chains caused by shading from the sun.

16. The importance of the fossil *Archaeopteryx* is that it

    a. is a link between the dinosaurs and the birds.
    b. was ancestral to the first mammals.
    c. was the largest of the flying reptiles.
    d. was the large predator that contributed to the extinction of amphibians.

17. The Cenozoic Era was a time of

    a. expansion of the amphibians into many niches.
    b. extinction of the dinosaurs during the glacial age.
    c. widespread seed ferns and naked seed plants.
    d. expansion of the modern varieties of mammals.

18. The species of man to which all modern humans belong appears in the fossil record starting about

    a. 400,000 years ago.
    b. 100,000 years ago.
    c. 35,000 years ago.
    d. 18,000 years ago.

19. The group that was directly ancestral to all living humans is called

    a. Australopithecus.
    b. Homo erectus.
    c. Cro-Magnon.
    d. Neanderthal.

20. Mass extinctions have been regarded as an agent of evolution. This is based on the

    a. idea that catastrophes cause faster mutations.
    b. fact that many ecologic niches will be vacated and new forms will evolve to occupy them.
    c. fact that major predators will perish allowing the less able forms to survive and repopulate Earth.
    d. concept of survival of the fittest, those species that lived through the mass extinctions.

MODULE IV

THE ROCK CYCLE

12

MINERALS: THE MATERIALS OF EARTH

GOAL

This lesson will help you understand the origin, classification, and importance of minerals.

LEARNING OBJECTIVES

After reading the textbook assignments, completing the exercises in this study guide, and viewing the lesson's video portion, you will be able to:

1. Distinguish among rocks, minerals, and chemical elements.

2. Explain what constitutes an atom and how atoms of various elements differ.

3. State how atoms bond together.

4. Name the eight most abundant elements in Earth's crust and relate these elements to the composition of most common minerals.

5. Recognize the significance of a silica tetrahedron and the principle of crystallinity in the structure of minerals.

6. Identify the conditions a substance must satisfy in order to be considered a mineral in the geologic sense of the term.

7. Describe the properties and structures of the most common rock-forming minerals.

8. Indicate the physical properties that are used to identify minerals in hand specimens.

9. Understand that minerals are the building blocks of rocks.

INTRODUCING THE LESSON

A mineralogy field trip is an exciting introduction to the world of minerals, but it doesn't always require spending the day in the mountains with pick and hammer. It could start in the kitchen with a cup of coffee and the morning paper. The coffee plants and practically all the food we eat have been treated with mineral fertilizers. Your cup, if it is made of the finest china, contains feldspar, the most abundant mineral in the crust of Earth. Your metal coffee pot may be an alloy derived from ores of aluminum, iron, or chromium, mined from some distant land. The icing on your morning sweet roll probably has a high content of gypsum. Even the paper in the newspaper is filled with kaolin clay and calcite.

The inquiring mineralogist will soon discover minerals in every aspect of daily life, including food, shelter, clothing, energy, transportation, recreation, and especially personal adornment. The Great Mogul Diamond, found in India in 1650, and the Cullinan, from South Africa, may conjure up visions of adventure, romance and wealth, but can diamonds, sapphires, and gold be included in our list of minerals?

Minerals are defined as naturally-occurring inorganic solids, having a definite chemical composition and characteristic physical properties. (So will those gemstones fit into this definition? . . . Absolutely!) In this lesson, you will learn how minerals are classified and identified and the circumstances under which they form. The properties of over 2,000 minerals have been described, but only a dozen or so "rock-forming" minerals make up over 90 percent of the materials of Earth's crust. Only the most abundant minerals and their properties will be discussed in this lesson.

Rocks are aggregates or mixtures of minerals. It is the minerals, however, that actually record the details of how and when a rock formed. Radioactive minerals, for example, are used to determine the age of the rocks in which they occur, acting as clocks that started millions to billions of years ago. Minerals in the most ancient Earth rocks offer a glimpse of our world over 3 billion years before the assembling of Pangaea. As we have learned in previous lessons, it is the magnetic minerals that tell of spreading oceans and gyrating continents and offer the best evidence for the theory of plate tectonics.

Minerals have also been prized aesthetically since the very beginnings of human history for their wondrous color and clarity. Diamonds, rubies, sapphires, emeralds, and the other colorful gems are justly called the "flowers of the inorganic world."

As all geologic processes leave their imprint on the rocks and minerals of the crust, a knowledge of Earth materials is essential to the understanding of all geologic phenomena. Thus, for many reasons, from every aspect of our well-being to a deeper appreciation of our planet home, minerals are well worth the study.

LESSON ASSIGNMENT

Completing the following eight steps will help you master the lesson objectives and achieve the goal for this lesson:

Step 1: Read the TEXT ASSIGNMENT, Chapter 9, "Atoms, Elements, and Minerals."

Step 2: Study the KEY TERMS AND CONCEPTS as noted in the study guide.

Step 3: Watch the VIDEO using the VIEWING GUIDE in the study guide.

Step 4: Read the study guide's PUTTING IT ALL TOGETHER section, which will help you summarize and integrate all of the information in this lesson.

Step 5: Complete any assigned lab exercises.

Step 6: Complete any assigned ACTIVITIES found in the study guide.

Step 7: Review the material for the lesson and complete the SELF-TEST found in the study guide.

Step 8: Go back to the LEARNING OBJECTIVES and make sure you have learned each one.

1. Read Chapter 9, pages 193-213.

2. Study Table 9.1, page 196, showing the crustal abundance of elements.

3. Carefully read Box 9.1, page 197, on bonding. Note Figure 9.5, page 196, showing the atomic structure of sodium and chlorine. Determine why these elements bond together to form halite (salt).

4. Learn Mohs' hardness scale, Table 9.3, page 205. Hardness is a basic property used in the identification of minerals.

5. Review Figure 10.10, page 221, to understand the relation between the rock-forming minerals and the common igneous rocks.

TEXT ASSIGNMENT

Read "Terms to Remember," page 213 in your text, admittedly a long list. Try to become familiar with a few at a time by referring to them in the chapter, the glossary, in Appendix A on page 499, and in the Index.

For easier understanding, try to re-organize the list by topics. One topic could be "chemistry" and would include atoms, bonding, ions, nucleus, internal structure and so on. Another list would include all the "rock-forming minerals." "Identification of minerals" using the physical properties could make up still another list. Here are additional definitions of some important key terms that will aid your understanding of the material in this lesson:

Compound: A substance containing two or more elements, chemically combined. Most minerals are compounds. Water (H_2O) is also a compound of two elements: hydrogen and oxygen. Air is not a compound, however, but a mixture of compounds and elements, including nitrogen and oxygen, and lesser amounts of argon, water vapor, carbon dioxide, smog, smoke, ash, dust, and whatever else is going around.

KEY TERMS AND CONCEPTS

Oxygen: The most abundant element in the crust. It is the odorless, colorless, invisible gas that makes up 21 percent of our atmosphere but is a hefty 46.6 percent by weight and about 93.8 percent by volume of the rocks of the crust!

Rock-forming minerals: About a dozen minerals and their related species that make up the greatest part of the rocks on the surface and probably within Earth's mantle as well. The most important rock-forming minerals include the silicates and the iron oxide, magnetite.

Silica: A term used to indicate a combination of silicon plus oxygen. Quartz, a common mineral, is composed of pure silica, SiO_2. Glass also contains SiO_2 derived from quartz sand plus other chemical elements. Silica is an interesting example of a chemical compound in which a colorless, odorless invisible gas (oxygen) combines with a gray opaque solid (silicon). The resulting product is a hard, clear transparent solid that is one of the most abundant minerals on earth (quartz) and one of the most useful products in daily life (glass).

Silicon: The second most abundant element in Earth's crust. Silicon is an opaque crystalline solid with a steel-gray shiny surface, but in spite of its abundance, it is a "hidden element." It is never found alone but is always combined with other elements, and the resulting compounds never resemble pure silicon.

VIEWING GUIDE

Once you have read the text material for the lesson and studied the terms, you are ready to view the lesson's video portion. Review the questions that appear in this section to help you watch for important points and to prepare you for what you will see in the video.

Before viewing, answer the following questions:

1. How does an element differ from a mineral?

2. How does a mineral differ from a rock?

3. What are the physical properties used in identifying minerals?

After viewing, answer the following questions:

1. Aside from their value and usefulness as part of everyday objects, what else can minerals tell us, according to Dr. Lawford Anderson?

2. Describe one common type of bonding occurring between elements in a crystal.

3. What are some of the physical properties used to identify minerals? What determines the properties of each mineral?

4. What are some techniques used in a petrology lab to examine the minerals contained in rocks?

5. Metallic minerals, such as gold, silver, and copper, form in the same way geologically. Describe that process.

6. How does common salt – called halite – form? And what minerals result from biological activities of such different organisms as oysters, corals, and sponges? Can you think of any other organisms that use this same mineral for shells, spines, or other hard parts?

7. Is silicon a rare mineral like gold or diamond? How are computers related to the source material of glass (SiO_2)?

PUTTING IT ALL TOGETHER

You have read the text assignment and seen the video portion of the lesson. This section, made up of a SUMMARY, a CASE STUDY, QUESTIONS, suggestions for ACTIVITIES, and a SELF-TEST, will help you pull the information together.

Summary

Exploring the world of minerals can be a fascinating adventure. Mineral collecting is a popular pursuit, but while the collector is usually seeking rare or unusual minerals, the geologist understands that minerals are literally the building blocks of our planet. To understand the nature of minerals and the chemical and physical properties that enable scientists to identify them, we turn to the very smallest particles of matter that make up atoms, elements, minerals, and the substance of life itself. How these unseen particles come together to form the orderly and often exquisitely beautiful crystals that make up rock is a remarkable story – the theme of this lesson on minerals.

Instant Chemistry

All materials are made of atoms that are composed of sub-atomic particles known as protons, neutrons, and electrons. Elements are basic substances, each made of its own kind of atom. The varying numbers of protons, neutrons and electrons in their atoms give the elements their essential properties and determine how they will combine with other elements.

Atoms bond together to form compounds because most atoms are unstable alone and seek stability by bonding. The outermost electrons surrounding an atom govern how the atom behaves and how it will bond with other kinds of atoms. Carefully read Box 9.1, page 196, on bonding. Note Figure 9.5, page 196, showing the atomic structure of sodium and chlorine. Determine why these elements bond together to form halite (salt).

Elements do not occur everywhere in equal abundance in Earth's crust. Oxygen is by far the most abundant, with silicon next and aluminum in third place. Five other elements complete the list that accounts for more than 98 percent of the weight of the crust: iron, calcium, sodium, potassium and magnesium. Carbon, the essential element of organic compounds found in all living things, is 17th by weight, a seemingly strange anomaly in view of the teeming life inhabiting the planet.

Mineralogy

Minerals are solids and occur naturally as either native elements or, more frequently, as chemical compounds of two or more elements. Minerals are by definition inorganic, meaning they have never been part of a living organism.

Minerals have a precise internal atomic structure and chemical composition or ratio of elements to each other. They also have specific physical and chemical properties such as crystal form, cleavage, hardness, color, and luster that are useful in mineral identification.

Only a few minerals like gold, silver, copper, diamond and graphite exist as pure elements. Most are compounds of several elements. Examples include halides (chlorine or other halogen elements plus a metal) such as the salt mineral halite; sulfides (sulfur plus a metal) such as pyrite, galena, sphalerite, and chalcopyrite; oxides (oxygen plus a metal) such as hematite and magnetite; sulfates (sulfur, oxygen and a metal) such as gypsum; carbonates (carbon, oxygen and a metal) such as calcite, aragonite and dolomite; and silicates, in which silicon and oxygen form quartz. However, other silicate minerals may contain iron, magnesium, aluminum, sodium, calcium, and potassium.

The internal building blocks of all the rock-forming silicate minerals are the silica tetrahedrons, consisting of four oxygen atoms and one small silicon atom. The tetrahedrons may be arranged in single or double chains, sheets, or frameworks, with other elements filling in some of the spaces.

Occurrence of Minerals

Just as elements vary in abundance, so do minerals, and the rock-forming minerals have the widest distribution. Feldspars (remember your coffee cup?), the most common minerals in the crust, occur in both intrusive and extrusive igneous rocks. The igneous rock known as granite sometimes contains as much as 70 percent feldspar.

Quartz can make up almost one-third the mineral content of granite, the most widespread continental igneous rock. Quartz is hard, tough, stable, and outlasts its competitors, eventually becoming the dominant mineral in sands and sandstone, a sedimentary rock, and eventually in quartzite, a metamorphic rock.

Augite – one of the pyroxene group of ferromagnesian silicate minerals (rich in iron and magnesium) – makes up about one-half of the mineral content of basalt, the rock material of the oceanic crust.

Minerals of considerable economic value are termed "ores" and are relatively rare. Gold, silver, and copper form from heated waters that originate from the crystallization of granitic magmas. Bauxite, an ore of aluminum, is a weathering product of rocks bearing aluminum oxide.

Minerals known as *evaporites*, including halite or rock salt, form as chemical precipitates from the evaporation of lakes or inland seas, frequently in an environment of tectonic crustal extension.

Minerals form in many environments, but aside from the rock-forming minerals, most are rare or occur in small percentages of the weight or volume of the crust. In some instances, this very scarcity increases their value. But specimens of a given mineral, no matter where they are found or how they originate, share the same distinctive properties. For example, all quartz crystals show the same six-sided pattern and lack of cleavage, while all micas show one perfect cleavage, and so on. Become familiar with Table 9.2, page 202, "Minerals of the Earth's Crust," especially with the group called the rock-forming minerals. In addition, review Figure 10.10, page 221, to understand the relation between the rock-forming minerals and the common igneous rocks.

Gem materials must have three principal qualities – splendor or beauty, durability, and rarity. Splendor in a gem depends upon transparency, brilliance, luster, and color. Durability is determined by hardness and toughness. The value of perfect gemstones is, of course, increased by their rarity. Diamonds exhibit all of these properties and are indeed splendid gems.

Gem stones are formed primarily by crystallization from water-rich solutions, crystallization from magmas, and metamorphism. Diamonds are composed of pure carbon, and occur principally within narrow, vertical pipelike bodies of igneous rock called *kimberlite*. Studies have shown that carbon requires extreme pressure such as that encountered at depths from 130 to 200 kilometers (78 to 120 miles) to become diamond. These depths, below the base of the crust, occur within the mantle of Earth. There is evidence that the rock in the pipes was blown upward through the crust in a powerful, jet-like blast, bringing the diamonds to the surface with the other rock materials. The diamonds are found in minute concentrations of less than one part per million. Kimberlite pipes occur principally in South Africa but are also found in Brazil, British Guiana, Australia, and Venezuela.

Diamonds have also been found in the United States in Arkansas, North Carolina, Virginia, southern Oregon, and in glacial tills in Indiana, presumably from hidden Canadian sources. Diamonds were found in California during the 1849 Gold Rush and were associated with gold-bearing placer deposits. Diamond deposits in the United States are considered insignificant, although a diamond found in 1934 in Amador County, California, weighed 2.65 carats. (One international metric carat equals 200 milligrams.) Most diamonds are not of gem quality, having darker colors or "flaws." They have many industrial uses, however, because of their extreme hardness. One interesting use is in drill bits for boring through hard rock in search of petroleum.

At least one variety of gem stone occurs in each of the United States. More than 60 gem minerals, nearly all semiprecious, have been produced

CASE STUDY: Looking at Diamonds, the Perfect Gem

domestically. They all have properties of beauty, durability, and rarity, and serve to enhance our lives in many ways.

Questions

1. **What is the relation among atoms, elements, and compounds? Where do minerals fit in this plan?**

   An atom is the smallest part of an element that still has the properties of that element. Elements are simple basic substances, made of one kind of atom that cannot be broken down into any other substance. Two or more elements chemically combined or bonded together form compounds. Minerals most frequently are compounds held together by ionic or covalent bonds. A few minerals, such as the native elements, consist of just a single element.

2. **What is the definition of rock? What two minerals are most abundant in granite? Are any rocks composed of just one mineral?**

   A rock, by definition, is a naturally formed aggregate of minerals, usually two or more, but some rocks do consist mostly of one mineral. The two most abundant minerals in granite are feldspar and quartz. Limestone, a sedimentary rock, is composed mostly of the mineral calcite (calcium carbonate).

3. **Is water a mineral, according to the accepted definition? What about ice? Could it be a mineral? Why isn't crystalline maple sugar considered a mineral? Why is crystalline table salt considered a mineral?**

   Is water a mineral? By definition, no, as it is not a solid. But ice would be a mineral. Crystalline maple sugar is not a mineral, even though it is solid and naturally occurring because it is organic in origin. Table salt, however, has all the defined properties of a mineral.

4. **What is the difference among silicon, silica, and silicate?**

   Silicon is an element; silica is a compound of silicon and oxygen. Silicates are substances that contain silica and usually one or more other elements.

5. **Name and describe at least five tests for physical properties and tell how they are used to identify minerals.**

   Luster. The quality or appearance of light reflected from a mineral. Minerals with a metallic luster look like a metal. Gold has a high metallic luster. Non-metallic lusters are described variously as glassy, pearly, silky, earthy or dull, and adamantine, meaning "like a diamond."

   Hardness. The resistance to being scratched. This is one of the most useful properties in the field for identifying minerals. Incidentally, if Mohs' scale indicated absolute hardness rather than relative hardness, diamond, instead of being listed as "10," would be "40"!

Streak. The color of the mineral after it is powdered by rubbing it on a plate of unglazed porcelain. In some metallic minerals, the streak is distinctly different from the color of the surface.

Cleavage. The tendency of minerals to split along planes where the bonding between the atoms is weak. Mica is an excellent example of a mineral with perfect cleavage in one direction; it easily splits into thin sheets. Mica has many industrial uses because of this distinctive property. Diamond, strangely enough, has four planes of cleavage even though it is the hardest known mineral. Diamond cutters spend many months studying large valuable stones before attempting to cleave them to remove any flaws and to obtain the largest perfect stone possible. If cut incorrectly, the diamond could cleave into small, almost worthless fragments rather than the desired gem of great value.

Natural magnetism. Some minerals such as magnetite are attracted to a magnet. Lodestone, a variety of magnetite, is a natural magnet with the attracting power and polarity of a true magnet. Lodestone was used by the Vikings to navigate during their far-reaching voyages across the Atlantic Ocean.

Crystal form. A geometric shape consistent for each mineral. The crystalline structure of a mineral depends on the internal atomic arrangement and the sizes of the ions involved. Crystal forms can be a diagnostic feature of certain minerals.

6. **Describe the mineral quartz, and indicate how you could identify it in the field.**

One of the properties of quartz is color, which ranges from clear, to white, gray, pink, lavender, yellow, green, smoky, even black, depending upon the impurities within the mineral. It may be clear, or opaque. Other properties are the hardness of 7 and the lack of cleavage, which are consistent. The luster is usually glassy, and the crystals are always six-sided. In the field, it could be identified by the hardness and by the lack of cleavage.

7. **Are any minerals not compounds? Explain.**

The native elements are composed of just one element but satisfy all the requirements to be considered minerals. These include the metals gold, silver, copper and the non-metals diamond and graphite, and sulfur.

8. **How do diamonds and graphite differ; how are they similar?**

Diamonds and graphite are both formed of carbon and are therefore similar chemically. However, their properties are very different; diamond has a hardness of 10, while graphite has a hardness of 1. Diamond is clear and brilliant, while graphite is black, opaque and dull. Look at the tip of your pencil to see graphite.

9. **What properties do gemstones have in common? What two gems are forms of the mineral corundum?**

   Gemstones must have splendor or beauty, durability or hardness, and rarity. Most gemstones are harder than 6 or 7 on the Mohs' scale and are of brilliant or unusual color. The gems ruby and sapphire are chemically corundum, which has a hardness of 9.

10. **Under what conditions do minerals originate? For example, explain the formation of some of the rock-forming minerals. Where does rock salt form? Where do diamonds form?**

    Minerals can form in many diverse environments. The rock-forming minerals originally form in an igneous environment, such as during the crystallization of granite or basalt. Other minerals form during metamorphism, in environments of high temperature or pressure. Some, such as the clay minerals, form as a result of weathering of the rock-forming minerals on Earth's surface. Rock salt is an evaporite mineral that forms from the evaporation of an inland sea or a saline lake. Diamonds form deep within Earth, under high temperatures and pressures.

Activities

1. Become aware of minerals. Look around your house, in your car, in the office. What mineral is the source of the glass you are looking through when you look out a window?

2. Start reading labels on cleansers, abrasives, canned food, frozen food, pet food, and soft drinks. Many of the chemicals will be unfamiliar, and many, especially in food, will be organic. But minerals do occur. Don't forget salt.

3. Visit a museum that has a mineral exhibit. Be sure to see the varieties of quartz and the different feldspars. Find the other rock-forming minerals because these are the most important geologically.

4. Visit the gem collection in the museum. These cut and polished stones can be magnificent, and they add another dimension to our enjoyment of minerals. Some labels on the specimens show the chemical formula. Try to identify the silicate minerals.

5. Stores that sell "natural" objects usually have some fine mineral specimens such as geodes filled with amethyst crystals or other forms of quartz. Also, look up "rock hound" shops in your telephone directory, and watch for local gem and mineral shows. The array of colorful specimens is tempting and may make you want to start your own mineral collection.

Self-Test

1. Coal is technically not a mineral because it

   a. is not naturally occurring.
   b. is black and opaque.

c. is organic in origin.

d. forms in a sedimentary environment.

2. Field geologists usually carry a small plate of unglazed porcelain. This is used to test for

   a. hardness.
   b. streak.
   c. cleavage.
   d. radioactivity.

3. Field geologists usually have a worn-down thumbnail. They are looking for minerals that

   a. cleave easily.
   b. have a hardness greater than that of feldspar.
   c. are softer than 2.5.
   d. show a colored streak.

4. Field geologists usually carry a bottle of weak hydrochloric acid. The acid will help them identify

   a. quartz crystals.
   b. silicate minerals such as feldspars.
   c. native elements such as copper and silver.
   d. calcite and other carbonate minerals found in limestone.

5. The angles of adjacent faces of quartz crystals

   a. will always be the same in relation to each other.
   b. will vary depending upon the size of the crystal.
   c. will depend upon the environment of origin.
   d. are related to the rock in which the quartz occurs.

6. Certain elements occur in nature uncombined with other elements. Which of the following should NOT be included in this group?

   a. Gold
   b. Sulfur
   c. Silicon
   d. Silver

7. Mica is a common mineral that is easily identified by its

   a. brilliant color.
   b. perfect cleavage in one direction.
   c. extreme hardness.
   d. unique red streak.

8. The smallest part of a gold nugget that still retains the properties of gold is

   a. any gold dust.
   b. an electron.

    c. any molecule of gold silicate within the nugget.

    d. an atom of gold.

9. Water is NOT considered an element because it

    a. is a liquid.

    b. occurs in living organisms.

    c. can be broken down into hydrogen and oxygen.

    d. is made of one kind of atom.

10. An atom that has lost two electrons from its outer shell becomes

    a. a mineral.

    b. a positive ion.

    c. electrically neutral.

    d. a stable element.

11. The atoms in certain minerals tend to stick together because of

    a. magnetic attraction.

    b. gravity.

    c. surface tension.

    d. electrical attraction.

12. A substance in which the atoms are arranged in an orderly repeating fashion is the definition of a(n)

    a. crystal.

    b. mineral.

    c. liquid.

    d. element.

13. In any list of the most abundant elements in Earth's crust, which of the following should NOT be included?

    a. Aluminum

    b. Iron

    c. Calcium

    d. Carbon

14. The basic building block of most of the rock-forming minerals is the

    a. calcium carbonate compound.

    b. silica compound.

    c. silica tetrahedron.

    d. iron-magnesium complex.

15. Potassium is not a mineral because it

    a. occurs in bananas.

    b. does not occur naturally as a pure element.

    c. is a common element in orthoclase feldspar.

    d. forms a halide salt substitute.

13

VOLCANISM

This lesson will help you understand how volcanoes are formed and how they affect Earth.

After reading the textbook assignments, completing the exercises in this study guide and the lab, and viewing the lesson's video portion, you will be able to:

1. Recognize the importance of volcanic activity to the science of geology.

2. Describe how volcanism relates to the origin of the atmosphere and affects Earth's climate.

3. Contrast the beneficial and catastrophic effects of volcanism on humans.

4. Indicate the factors that control the explosive violence of volcanic eruptions and influence the shape and height of volcanoes.

5. Differentiate among mafic, felsic, and intermediate lavas; describe the color and mineral composition of the volcanic rocks that result from each and their relationship to their intrusive equivalents.

6. Compare the three major types of volcanoes in terms of their size, shape, and composition and give examples of each.

7. Recognize the characteristics of volcanic domes.

8. Describe the origin of plateau basalts and pillow basalts.

9. Indicate factors that affect the texture of extrusive rocks, and relate these factors to the formation of obsidian, phenocrystic lava, tuff, breccia, and vesicular rocks.

10. Recall and discuss the relationship between volcanism and plate tectonics as it applies to the origin and alignment of andesitic vol-

canoes along the circum-Pacific rim and the origin of submarine basaltic volcanism along the mid-oceanic ridges.

Volcanoes are well-named for the ancient Roman god Vulcan, the god of fire in its more fearsome aspects, who manufactured the thunderbolts of Jupiter. In the eruptive phase, there is no force of nature more awesome or threatening than an exploding volcano. Volcanic eruptions, on the other hand, have created some of the most beautiful mountain peaks on Earth, such as the oft-painted Fujiyama, Kilimanjaro, Mount Rainier, Mount St. Helens, and the volcanic mountain islands of Hawaii, Samoa, and Tahiti.

Part of our fascination with volcanoes is their potential for destructiveness. The famous Krakatoa explosion in 1883 and the ensuing tsunami (seismic sea wave) off the coast of Java and Sumatra killed nearly 37,000 people. In 1902, Mont Pelée on the Caribbean island of Martinique erupted clouds of red-hot ash that destroyed the town of St. Pierre, and killed all but two of the 30,000 people who lived there.

In spring of 1980, Mount St. Helens in the Cascade Range, after a century-long sleep, erupted ash and steam before finally exploding in a furious blast that almost instantly destroyed the summit and north flank of the mountain. Fortunately, through the work of geologists with the U.S. Geological Survey and their prediction of an imminent eruption, only 63 lives were lost. Without their warning, it is estimated that at least 10,000 people would have died. In 1982, two years after the eruption of Mount St. Helens, a jungle-covered small volcano in Mexico called *El Chichon*, following many years of inactivity, erupted with a series of violent explosions. Towns nearby were either buried by the heavy fall of ash or charred by searing, gas-charged ash flows similar to those that incinerated the town of St. Pierre in Martinique. The number of people who perished can only be estimated but was placed in the thousands.

Volcanoes are worth studying not only because of their life-threatening potential but because they are a pipeline to Earth's interior and our only direct source of information about the rocks, temperatures, and pressures in the upper 250 kilometers (155 miles) of Earth's interior. The flowing lava is an important proof that rocks can and do melt in the natural environment. Volcanoes also create new landforms, even new land, as we see in the growing island of Hawaii, where orchids and papayas bloom on the fresh lavas. Volcanoes have influenced the chemistry of Earth's atmosphere and oceans since the beginnings of Earth time, and there is no evidence at present to indicate their activity is dying out.

In this lesson, you will learn that volcanoes differ from one another in shape and size, and even in the chemistry of their lavas, which in part determines their eruptive behavior. The three basic forms of volcanoes are the *shield volcanoes*, *composite volcanoes*, and *cinder cones*. Other examples of volcanic land forms include *lava domes*, constructed of such stiff, pasty lava that it does not flow at all, and *plateau basalts* such as the Columbia Plateau, where the fluid lava-floods reach a thickness of 3,000 meters (9,900 feet) and spread over much of the states of Washington, Idaho, Oregon, and parts of northern Nevada.

Actually, most of the world's volcanoes erupt unseen on the deep ocean floor at the *diverging* plate boundaries where they are quietly pouring out lava, forming mid-oceanic ridges, *pillow basalts*, and creating new sea floor. The most explosive volcanoes – as you remember from Lessons 5 and 6 on plate tectonics – are associated with *converging* plate boundaries such as the circum-Pacific belt. Yet some of the greatest volcanic mountains of all have formed right in the center of Earth's largest plate, the Pacific Ocean, more than two thousand miles from the nearest plate boundary.

LESSON ASSIGNMENT

Completing the following eight steps will help you master the lesson objectives and achieve the goal for this lesson:

Step 1: Read the TEXT ASSIGNMENT, Chapter 11, "Volcanism and Extrusive Rocks."

Step 2: Study the KEY TERMS AND CONCEPTS, as noted in the study guide.

Step 3: Watch the VIDEO using the VIEWING GUIDE in the study guide.

Step 4: Read the PUTTING IT ALL TOGETHER section in the study guide, which will help you summarize and integrate all of the information in this lesson.

Step 5: Complete any assigned lab exercises.

Step 6: Complete any assigned ACTIVITIES found in the study guide.

Step 7: Review the material in this lesson and complete the SELF-TEST found in the study guide.

Step 8: Go back to the LEARNING OBJECTIVES and make sure you have learned each one.

TEXT ASSIGNMENT

1. Before reading all of Chapter 11:
   a. Read the "Introduction" to Chapter 11 on page 239. Look at the photograph facing page 239 showing Hawaiian *pahoehoe* with the still-molten lava flowing beneath the hardened crust.
   b. In the diagram of the rock cycle on page 239, be sure to note the position of magma and igneous rocks.
   c. Read Box 11.1, page 240, about Mount St. Helens, as an introduction to volcanism. The diagram and photographs are outstanding.

2. Now read all of Chapter 11, pages 239-265, carefully noting the photographs and diagrams.

3. Learn the "Terms to Remember," page 265 This is a long list but describes some of the most interesting structures and events in the field of geology.

4. Try the "Questions for Review," page 265, and the "Questions for Thought," page 265. The answers are in the text and the study guide although it may take a little effort to find them.

KEY TERMS AND CONCEPTS

All of the "Terms to Remember" on page 265 in your text, are important to understanding volcanism, from *aa* (ah-ah) through *pahoehoe* (pah-hoy-hoy) to *welded tuff*. *Andesite* and *basalt* are familiar terms from the lesson on plate tectonics. The minerals you learned in Lesson 12 will help you understand the composition and differences in the lavas and associated rocks.

VIEWING GUIDE

Once you have read the text material for the lesson and studied the terms, you are ready to view the lesson's video portion. Review the questions that appear in this section to help you watch for important points and to prepare you for what you will see in the video.

Before viewing, answer the following questions:

1. What are the two factors that determine whether volcanic eruptions are explosive or quiet?

2. What are the two factors that determine the viscosity of lava?

3. What are the three major types of volcanoes?

4. If you wanted to see a composite cone, where would you look in terms of plate tectonic locations?

After viewing, answer the following questions:

1. In the video, watch for volcanoes on other planets and their moons. On what planet is the largest volcano? How can we interpret the presence of this huge volcano and the possibility of plate tectonic motions (or lack of them) on this planet?

2. The video states that 80 percent of the Earth's surface is volcanic rock. If this statement is valid, where are the greatest number of Earth's active volcanoes locate? Describe the Pacific "Ring of Fire."

3. If you discovered pillow lavas on land, what would this tell you about the history of the area?

4. How do composite volcanoes differ from shield volcanoes in eruptive history, appearance, materials erupted, and location?

5. In the video, you saw a rock float. What was the rock, and what conditions of origin led to this peculiar property (for a rock, that is!)?

6. What determines the viscosity and explosive character of lava?

7. The eruption of Katmai, Alaska – shown in the video – has been called the 20th century's most powerful eruption. Describe the nature of the eruption, materials ejected, the origin of the Valley of Ten Thousand Smokes, the pyroclastic flows, and the lava dome.

8. What research projects are being conducted at Katmai? How will the research be of value to vocanologists all over the world and especially to people living in volcanic regions?

9. What are the benefits to humanity from volcanic activity?

**PUTTING IT
ALL TOGETHER**

You have read the text assignment and seen the video portion of the lesson. This section, made up of a SUMMARY, a CASE STUDY, QUESTIONS, suggestions for ACTIVITIES, and a SELF-TEST, will help you pull the information together.

Summary

An erupting volcano, especially at night, is an unforgettable sight (Figure 11.13 B, page 249). It reminds us that the planet Earth, even after 4.6 billion years, still has a restless hot interior. Within the crust, the molten rock called *magma* works its way toward the surface erupting as *lava*, mostly at plate boundaries, through fissures in a spreading sea floor, or through hot spots or volcanic cones. The molten rock is our direct link to the hot interior and tells of the minerals within the crust, the temperatures and pressures at depth, and the processes that form black basalt on the sea floor in distinct contrast to the light colored rhyolite and andesite characteristic of continental volcanism. Volcanoes have created not only some of the tallest mountains on Earth but are still adding new lands. Volcanic activity, through geothermal installations, is producing electric power in Italy, Mexico, New Zealand, Argentina, and California. In Iceland, volcanically heated water warms the homes during the long winters in these Sub-Arctic islands. (Refer to Box 11.2, page 255.)

But the other side to volcanism is the destruction of lives and human habitation. The time span between eruptions may be so long that some volcanoes have been considered *extinct* and homes, farms, olive groves, and vineyards have been built on their fertile flanks. History has many notable examples – Mont Pelée, Mount Vesuvius, and Mount St. Helens – of the fearful toll taken by sudden violent eruptions of supposedly quiescent volcanoes (Figures 11.2, 11.22, and 11.33, pages 244, 253, and 261 and Box 11.2, Figure 1 A, page 255).

Distribution of Volcanoes

There are about 570 active or recently active volcanoes in many parts of the world, but their distribution is not uniform. Many occur within well defined linear belts such as the circum-Pacific belt, which surrounds the Pacific plate. (Follow this belt in Figure 11.17, page 251.) This is also a zone of earthquakes, which is not surprising since both earthquakes and volcanoes are responses to forces within the moving lithosphere. In the case of the Pacific Ring of Fire, this zone coincides with the convergence of major plates. Another belt, also a zone of plate convergence, is the Mediterranean belt, which includes Vesuvius, the destroyer of Pompeii in 79 A.D. (Figure 11.2, page 244).

The sea-floor volcanoes, on the other hand, are found in every ocean basin at sites of plate divergence such as those on Iceland. Such volcanoes have produced prodigious amounts of basalt over geologic time. The volcanoes of East Africa and the Red Sea also occur in zones of divergence but where a continent is rifting apart (Figure 4.26, page 81). There are still other volcanoes that are not obviously connected to belts, zones, or plate boundaries. In these cases, the volcanic activity apparently is related to *mantle plumes*, identified on the surface as *hot spots* (Figure 4.41, page 93) where columns of molten magma rise beneath the sea floor. They form volcanic islands such as the Hawaiian-Emperor chain or penetrate continents to form volcanic regions such as Yellowstone National Park.

Anatomy of a Volcano

While there is great variety in the structure of volcanoes, they have certain features in common. Most mark localized centers of eruption, which may result in a small steep cinder cone, a soaring composite volcano, the gentle slopes of a shield volcano, or no cone at all but a vast plateau or a small dome. In each case, however, the structure is built of its own eruptive materials ejected from and sometimes piled up around a vent. The summit *crater* is a depression marking the top of the conduit through which the eruptive materials were channeled. A *caldera* is similar in shape to a crater but is much larger and has a different eruptive history and origin. Scenic Crater Lake in Oregon should have been named *Caldera Lake*. The large opening was formed when magma drained away beneath the crater following an eruption. The summit collapse formed the caldera. A visual history of Crater Lake is seen in Figure 11.3 and Figure 11.4, A through D, page 245.

Materials Ejected During an Eruptive Cycle

Gases. During the early period of condensation, compaction, and differentiation of Earth, volcanic activity probably was at its greatest, and vast quantities of water vapor, nitrogen, carbon dioxide, and other gases were expelled from the hot interior. This long period of *de-gassing* of the mantle followed by cooling of the surface of Earth permitted the water vapor to condense into liquid water, forming the seas and adding moisture to the atmosphere. Although the atmosphere has changed through geologic time, and volcanoes probably erupt less frequently than they did 3 or 4 billion years ago, their effects are at least partly responsible for

Earth becoming a planet suitable for life. Without volcanoes, Earth would have no atmosphere or surface water.

Gases are still an important factor in *extrusive* igneous activity. Much of the explosive power of an eruption depends on the amount of gas contained in the magma or lava. The most abundant gas (about 50 to 80 percent) is water vapor, some of which may be original water trapped in the magma and appearing for the first time on Earth's surface. Most of the water, however, is from ground water, or in some cases from encroaching sea water, both of which form steam on contact with the rising molten magma. Other gases may be derived from the country rocks (rocks that are older than and intruded by an igneous body) the magma contacts as it moves upward in the crust.

Carbon dioxide, sulfur dioxide, hydrogen sulfide, and hydrochloric acid contribute to the gas content but in varying amounts. Around some vents, there is a coating of sulfur that was emitted as a gas with the steam and then condensed on the surface, forming beautiful yellow crystals. In some areas, the unmistakable aroma of rotten eggs announces the presence of hydrogen sulfide.

Lavas and volcanic rocks. All lavas contain silica (silicon dioxide or SiO_2) as a principal ingredient, and the amount of silica can vary from about 45 percent to about 75 percent of the total weight of the rocks (Figure 11.6, page 246). This is a crucial factor because the amount of silica in the lava is important in classification of solidified volcanic rock, the explosive behavior of the volcano, and to some extent, the shape of the volcanic structure.

Mafic lavas. Lavas low in silica or silica-poor (less than 50 percent), and higher in magnesium, iron, and calcium are termed *mafic* lavas. They tend to flow easily, have a low viscosity (are very fluid), and permit the contained gases, which are relatively low in abundance, to escape easily. Mafic lavas are responsible for the relatively quiet eruptions characteristic of oceanic islands such as Hawaii ("The Mellow Volcano") as well as the sea-floor eruptions along the mid-oceanic ridges and the flood basalts of the Columbia Plateau (Figures 11.1, page 244 and 11.24, page 254).

Basalt. One of the most ancient names in geology, basalt apparently dates back to Egyptian or Ethiopian usage. It was mentioned by name by Pliny the Elder, a Roman naturalist and author who perished in 79 A.D. during an eruption of Vesuvius. By far the most abundant of all volcanic rocks, basalt is dark gray to black in color. Although basalt is about 50 percent silica by weight, there is no excess silica in the form of quartz. Instead, the rock is composed of other silicate minerals including black augite and green olivine (ferromagnesian minerals) plus gray plagioclase feldspar. The rock is fine grained, however, and the individual minerals are often too small to be seen by the naked eye.

Thin flows of fluid basalt lavas, piled one on top of another, create shield volcanoes, which have almost flat slopes that average between 2 and 10 degrees from the horizontal (Figures 11.8 and 11.9, pages 247-248). The highly fluid nature of the basalts allows them to spread great distances away from their vents rather than build steep slopes. The basalt lava solidifies into two distinctive patterns. The first is *pahoehoe*, a very runny gas-rich lava which has a ropy or billowy surface formed when still molten lava below drags the surface into folds and wrinkles. (See the remarkable photograph facing page 239 and Figure 11.10, page 248.) If the lava is gas-poor, viscous, and has partly solidified while still moving forward, it will form *aa* – the second pattern – a hot pasty mass with a steep front that resembles a pile of jagged rocky rubble (Figure 11.11, page 248).

Felsic lavas. Silica-rich lavas containing by weight over 65 percent SiO_2 are called *felsic* and have a higher content of sodium and potassium and less iron and magnesium than do mafic lavas. They are extremely viscous and flow sluggishly like taffy or tar. In continental magmas of felsic composition, the gas content may be high, as much as 15 percent of the total weight of magma. When the lava is too viscous to flow, it may form a plug within a vent and prevent gases from escaping. When the pressure builds sufficiently, the volcano may erupt violently with sometimes disastrous results (Figures 11.20 and 11.21, page 252).

Rhyolite. Rhyolite (from the Greek, *lava torrent* and stone) is the most abundant felsic volcanic rock and is characteristic of continental volcanism. It contains quartz, potassium, plagioclase feldspar, and lesser amounts of the dark ferromagnesians, biotite and hornblende; therefore it is light colored, occurring as tan, light gray, and frequently pink to red, and may contain grains or crystals of quartz. Rhyolite frequently shows streaks called *flow banding*, which is a concentration of minerals in bands during a very viscous flow.

Intermediate lavas and andesite. The intermediate lavas have a chemical content between that of felsic and mafic. (Refer to Figure 11.6, page 246, noting that the boundaries are arbitrary.) *Andesite* rocks form from the intermediate lavas and are more abundant than rhyolite but less abundant than basalt. Neither quartz nor potassium feldspar is found in andesite. The andesite rocks are medium to dark gray, occasionally with a hint of lavender, and are fine grained. Most of the rock is comprised of plagioclase feldspar plus lesser amounts of pyroxene and hornblende. You may remember andesite as the volcanic rock characteristic of the convergent boundaries of the Pacific Rim, and hence the name: *Andesite*. (Study Figure 11.27, page 256, to understand the mineral content of rhyolite, andesite, and basalt, and see Figure 11.29, page 257, for a photograph of andesite.)

Other Eruptive Materials

In addition to the gases and lavas, enormous quantities of *pyroclastic* materials are ejected (from the Greek: *pyro* = fire and *clastic* = broken).

Violently escaping gases fracture the lava into droplets of various sizes, including *ash* that looks like black smoke but is actually tiny particles of dark, glassy, rock powder (Figure 11.13 A and B, page 249, and Box 11.1 Figures 4 and 5, pages 242-243). Volcanoes rarely send forth true fire or smoke, although some emitted gases may burn near the vents. Most of the pyroclastic particles fall back to Earth near the erupting vent and form steep-sided cinder cones; some fall on the volcanoes' flanks, and sometimes fine particles like ash are hurled high into the atmosphere, where they may circle Earth in the jet stream or with the prevailing winds. Following the eruption of Krakatoa, spectacular sunsets were seen over most of the world for two years because of the ash in the atmosphere.

Cinders are solidified lava fragments that range in size from about 4 to 32 millimeters (less than half an inch) and are full of small gas bubble holes. *Blocks* are large angular chunks of broken rock, too heavy to be carried far from the erupting crater, that fall on the summit or flanks. If blobs or clots of incandescent molten lava are ejected into the air, they become streamlined during flight and partially solidify before falling to the ground. These are called *bombs* and take various shapes; some look like strips of burned bacon and others form spindle- or lens-shaped objects (Figure 11.14, page 249). Some larger blobs develop a crust as they cool and look like a round loaf of French bread that is flat on the bottom.

Pumice is an ancient name from a Greek word meaning *worm-eaten* that was first mentioned about 325 B.C. It is a special variety of rhyolitic volcanic glass (obsidian) that resembles a solidified froth or foam. Pumice is formed when the gases in a magma come out of solution as the molten rock rises into regions of decreased pressure, much like the foam on beer. Because of the abundance of gas cavities (which also accounts for the old Greek name), pumice is extremely porous and of such low density that it floats (Figure 11.32, page 260). Enormous masses of light-colored pumice blanketed the Indian Ocean after the eruption of Krakatoa. With time, some of the pumice became water-logged and slowly sank into the sea, but rafts of pumice – some containing human skulls – floated onto the shores of East Africa!

Viscosity of Lava

You have learned that the viscosity of lava generally determines the violence of eruption and the type of ejected material, which in turn influences the structure of a volcano. Viscosity depends on the temperature of the lava and the silica content. If the lava is erupted at a high temperature, a characteristic of many mafic lavas, it will be fluid or less viscous. Gases are able to escape, and the lavas are emitted as flows rather than pyroclastics. If the lava is high in silica, it will be more viscous or sticky. The silicon-oxygen tetrahedrons in the silica tend to link together to form framework structures that cause the lava to stiffen. Basalt tends to be the most fluid of all eruptive lavas because of its high temperature (about 1,250° C) and low silica content (about 50 percent by weight).

In contrast, rhyolite is the most viscous because its eruptive temperature is lower (about 800° C) and its silica content (about 78 percent by weight) higher. Andesite is intermediate in character; it and rhyolite tend to have higher gas content than basalt, making them more prone to explosive eruptions. As the temperature drops, the magma starts to congeal. If the gases within the magma cannot escape, pressure can build within the volcanic cone to explosive levels. (Review Figures 9.7 through 9.11, pages 198-200, and Lesson 12 on minerals to understand how silica tetrahedrons link together in silicate minerals.)

The chemistry of igneous rocks is a continuum from felsic rhyolite to mafic basalt. Study Figure 11.27, page 256, to learn how the important minerals – quartz, potassium feldspar, plagioclase feldspar, and ferromagnesians – vary in proportion and how their abundance in the rock determines its classification as well as the properties of the erupting lava.

Textures

Rocks formed from solidified lavas, as you have learned, are classified in part by their mineral composition but also by their texture. The *texture* of extrusive rocks refers to the *grain size* of the mineral constituents. Most extrusive rocks are fine grained as compared to coarse grained *intrusive* rocks, which cool slowly from magma deep within the crust. Rapid cooling on the surface does not permit large crystal growth, and extrusive rocks tend to have a dense, stony appearance in which the individual minerals are too small to be visible to the naked eye. Viscosity also affects texture, since atoms in a very viscous lava cannot move around to form large mineral grains.

Obsidian is volcanic glass (Figure 11.28, page 257) much prized by early humans for ornaments, arrowheads, and other uses. Chemically, it has a very high silica content, similar to that of rhyolite, which makes the lava very viscous. During rapid cooling and solidification of the sticky lava, the atoms cannot organize themselves into minerals, and a natural glass forms rather than a crystalline rock. The black, and occasionally red, color is a result of impurities in the chilling lava.

Porphyries (por-fir-rees) are rocks that do not have a uniform texture but have large crystals, *phenocrysts* (from the Greek *phainein* meaning "to show" and *krystallos* meaning "crystal"), embedded in a matrix of finer grained material (Figure 11.29 A and B, page 257). Some black Hawaiian basalts have phenocrysts of brilliant green glassy olivine, making a striking color combination. If rate of cooling determines the size of the mineral grains in a rock, under what conditions can the same lava form two distinct sizes of crystals? (See Porphyritic textures, page 257 in your text for the clues.)

Volcanic *tuff* is a rock, frequently light colored, made of particles of consolidated pyroclastic fragments, primarily glassy ash. Cinders, bits of pumice or obsidian, blocks, bombs, and larger angular particles (Figure

11.30, page 260) accumulate in unsorted deposits and are known as *volcanic breccias*. Highly explosive volcanoes that pour forth great quantities of ash usually emit lavas of felsic to intermediate composition.

Most volcanic rocks are shot through with holes, caused by gases released in a lava as it solidifies. The holes are termed *vesicles*, and the texture is *vesicular*. Vesicular basalts are common (Figure 11.31, page 260) and are called *scoria*. Vesicular obsidian is called pumice, and the frothy texture is a result of released gases that churn the lava into a foam.

The Volcanoes

You have learned that cones, domes, mountains, shields, plateaus, and other landforms associated with volcanism are built of the various materials ejected by the volcano itself. *Shield volcanoes* are constructed of low viscosity basalt lava flows that intermittently erupt over a million or so years and build the gentle slopes, flow upon flow, which, in the case of island volcanoes, rise upward from the deep sea floor (Figure 11.1, page 244, and 11.9, page 249). The Pacific plate, moving over a hot spot (Lesson 6), has created the long chain of Hawaiian Islands, with current volcanic activity being limited to the southern portion of the Big Island of Hawaii (Figures 11.36, page 263, 4.45, and 4.46, page 96).

The eruptions, although spectacular and spewing out enormous volumes of lava – sometimes from lofty lava fountains (Figure 11.1, page 244) – are considered non-violent. (Kilauea has been called "the drive-in volcano" because it is so obliging to people who want to watch its fireworks.) As the name "shield volcano" suggests, the shape is in the form of a flattened dome or shield of ancient Norse warriors. The Hawaiian Islands are the largest shield volcanoes on Earth, and Mauna Loa, from sea floor to peak, is the tallest mountain on Earth, about 9,400 meters (31,000 feet) (Figure 11.8, page 247). Kilimanjaro in Africa is another fine example of a shield volcano.

Cinder cones are constructed entirely of pyroclastic basaltic fragments ejected from a central vent, generally over a short period of time. The slopes are steep, about 30 degrees from horizontal, and the height of the cones is rarely above 500 meters (4,900 feet). Pockets of gas within magma can cause local eruptions that will build a solitary cinder cone or several close together. Cinder cones are common in Hawaii – sometimes perched near the summit and more frequently on the flanks. Being of loose materials, these are not long-lived features and are rather easily eroded away (Figure 11.13, page 249).

An unusual cinder cone, Paricutin, began to erupt in Mexico in 1943 and continued to erupt for nine years, building a cone almost 369 meters (1,210 feet) high. In its early stages, the eruption of Paricutin emitted only pyroclastic blocks, bombs, cinders, and ash. Later, torrents of lava broke through the flanks and flowed outward, engulfing a nearby town. During the last days of Paricutin, huge fiery bombs and blocks weighing up to 100 tons were hurled out beyond the base of the cone.

Composite or *strato-volcanoes* are particularly interesting because they illustrate how the behavior of a volcano can alternate between explosive and quiet eruptions and thus build its cone of layers of both pyroclastics and lava flows (Figures 11.15 and 11.16, page 250). The slopes are intermediate in steepness between cinder cones and shield volcanoes, and the lava flows make the volcanoes less susceptible to erosion than cinders alone.

Composite volcanoes build slowly over long spans of time with sometimes thousands of years of inactivity between eruptive cycles. The lavas are predominantly andesite and less frequently rhyolite. If the temperature of the erupting lava is high, a flow of low viscosity andesite will be emitted; but if there is considerable gas in the lava, or if the temperature is cooler, the eruption will be of andesitic pyroclastic material. Composite volcanoes are individualists and vary in composition of lava, in eruptive history, and in the nature of their eruptive cycles.

Most of the great scenic volcanoes are composite volcanoes (Figures 11.16, 11.19, 11.20, and 11.37, pages 250, 252, and 263). They are generally distributed around the Pacific basin. Mount St. Helens (Box 11.1, Figure 5, page 243), the other volcanoes in the Cascades (Figure 11.5, page 246), and the great peaks of Mexico and Central and South America are part of this belt. In the western Pacific basin, the comnposite volcanoes of New Zealand, Indonesia, and particularly Mount Mayon in the Philippines and Fujiyama in Japan are noted for their beautiful, symmetrical cones.

Volcanic Domes

Volcanic domes are steep-sided, dome- or spine-shaped masses of volcanic rock formed from highly viscous lava that solidified in or above a volcanic vent (Figure 11.21, page 252). Domes are more common than many people think. Lassen Peak in northern California was last active in 1914-1917, and a protrusion of blocky lava was pushed about 770 meters (2,526 feet) above the crater rim. This is one of the largest volcanic domes known. A volcanic dome grew within the caldera of Mount St. Helens after the eruption of May 1980 (Figure 11.20, page 252). Mono Craters in east-central California are a series of cinder cones whose craters are filled with volcanic domes. Some of the volcanoes of the Valley of Ten Thousand Smokes in Alaska also have domes within their craters.

An astounding spine-shaped mass arose within the volcanic neck of Mont Pelée on the Island of Martinique in the Caribbean after the devastating eruption of 1902. The sticky lava solidified in the neck and prevented the gases from escaping; the gases trapped below pushed the plug straight up like a piston, at a rate of about 20 meters (66 meters) per day! Part of the sides of the plug broke off, leaving a spine that reached over 300 meters (1,000 feet) high before collapsing into a pile of rubble (Figure 11.22, page 253). The complete destruction of the town of St. Pierre on Martinique was caused by an explosion that blew out clouds of incandescent gas, ash, and dust that descended as a 700°

C glowing avalanche (*nuee ardente*) travelling at about 100 kilometers (60 miles) per hour. In one minute, and with hardly a sound, it engulfed the city and all life perished. The particles were so hot that upon settling, they became welded together forming a rock called *welded tuff*.

What is the Source of the Lava?

As you know from your study of Lesson 6 on plate dynamics, theories for volcanism and magmatic intrusions are closely related to plate movements and our understanding of Earth's interior.

Basaltic magmas are produced by dry partial melting of the asthenosphere, a zone in the upper mantle in which the temperatures are sufficiently high to be above the melting point of mantle rocks (Figure 11.34, page 262). The magma formed from melting of the mantle minerals (largely olivine and pyroxene) will have the composition of basalt. Being molten and less dense, the basalt magma will drift upward into areas of less pressure; these will be under the rifting sea floor (a divergent plate boundary), where the basalt will fill the rift, create new sea floor, form the mid-oceanic ridges, and pile up as pillow basalts (Figure 11.26 and 11.35, pages 254 and 262). In the rift zones, the hot asthenosphere rises to within a few kilometers of the sea floor, and the erupted basalt will form new oceanic crust. The Columbia flood basalts are explained as eruptions of the fluid basalt from what was a spreading center beneath a rifting continent (Figure 11.24, page 254). Evidently, after a few million years, the process stopped. Basalts are the rocks of the seas, although they do occur on the continents in more limited exposures.

In contrast, andesites and rhyolites are never erupted in oceanic island volcanoes, but they occur abundantly in island arcs, active continental margins and within continents. Andesites are characteristic of convergent plate boundaries, where subduction of the sea floor occurs (Figures 11.37 and 11.38, pages 263-264). The small oceanic plate subducting beneath Washington and Oregon is responsible for the Cascade volcanoes. While the origin of andesite is not yet fully understood, it is believed that subducting sea floor, basaltic in composition, carries down with it a veneer of silica-rich marine sediments that are saturated with water. In laboratory experiments, partial melting of wet basalt yields, under suitably high pressure, a magma of andesitic composition. The fact that andesite occurs landward from subduction zones is supportive of this theory of its origin, because this is where wet basalt is found at suitable depths and the temperatures are high enough for partial melting to occur. Andesite is not known to occur seaward of the deep sea trenches.

Predicting Eruptions

A great part of the world's population lives in or near volcanic belts. It is important to understand how volcanism is activated and what the precursors to an eruption might be. For an interesting account of how one small city in Iceland – Heimaey – dealt with the threat of a neighboring volcano, read Box 11.2, "An Icelandic Community Battles a Volcano – and Wins," page 255.

The study of volcanoes is a relatively new science. It has come a long way, thanks to the concepts of plate tectonics, but complete understanding and accurate prediction of eruptions is still ahead of us.

CASE STUDY:
The Long History
of Vesuvius

Vesuvius, one of the world's most famous volcanoes, is the only active volcano on the European mainland today. Its fame results from the well-publicized eruption of 79 A.D. and the destruction of the cities of Pompeii and Herculaneum. Vesuvius provided a time capsule of Roman art, architecture, and artifacts for its 19th- and 20th-century excavators.

Vesuvius had been active in pre-historic times but had been dormant for so many years that the Romans were unaware of its real character. In fact, the present mountain is superimposed in large part on the wreckage of an older and lower crater, to which the name Monte Somma is given. In about 63 A.D., the volcano showed some signs of life when a succession of earthquakes caused some of the damage still to be seen around Pompeii. That was a prelude to the historic eruption of August 24, 79 A.D.

A fascinating description of the destructive event is included in letters from Pliny the Younger, describing the death of his uncle, the famous Pliny the Elder – a leading philosopher of the day – as well as an admiral of the Roman Navy. Pliny the Elder was the world's first volcanologist, and as a staunch Roman met his death on the erupting mountain. From the description, Pliny the Elder was aboard one of his ships when the first clouds of gas and ash arose from the mountain. When ashes began to fall about his ships, Pliny ordered a small boat to take him ashore to investigate what was happening. Cinders, pumice, and black fragments of rock fell about them. By the next morning, the sky was still darkened with the erupting material. They tried to return to the ships, but the sea was too unsettled. Apparently hot ash and sulphurous vapors overcame the philosopher, and he never reached the safety of his ships.

During the destruction of Pompeii and Herculaneum, about 2,000 of the 20,000 inhabitants perished. Most were suffocated by falling ash, hot volcanic mud, or incandescent volcanic gases. The temperature of the ash was so high it charred away the bodies as it covered them with the fine particles. Centuries later, when plaster of Paris was poured into the cavities once occupied by the bodies, their shapes, as well as those of dogs and cats, loaves of bread, and all sorts of objects in similar cavities, stood revealed. In fact, in the molds of the bodies found in Pompeii, the ash preserved the bodies so well that you can actually see the expressions on the people's faces as they took their last breaths!

Visiting Pompeii is certainly a memorable experience. Imagine walking down a 1st century city with many of the buildings looking as they did in 79 A.D. Because of the type of ash that fell, much of Pompeiian life was preserved for centuries. For instance, we have a good idea of the way the rich citizens of Pompeii furnished their homes because much of the stone furniture and beautiful artifacts – plates, goblets, statues – were covered and saved in the volcanic ash. Their homes were decorated with elaborate wall paintings and the original colors still remain. In some of the buildings, you can walk in and get the feeling that you are back in 79 A.D.

Pompeii had all of the amenities of 1st century Roman life at its best – paved streets, sewer systems, and even running water. We have learned a great deal about life in the Roman Empire by studying Pompeii. It's like walking into a 1st century ghost town – and we have a geologic event to thank for that experience.

Vesuvius has continued its activity sporadically from 79 A.D. to the present. An especially violent eruption in 1631 is estimated to have killed 18,000 people, and came after such a long period of dormancy that the volcano was overgrown by vegetation. A major eruption occurred in 1944 in the midst of the Italian campaign of World War II. During that eruption, there were extensive lava flows from the crater, one of which rolled through the town of San Sebastiano, burying it under a front of aa lava. After the lava was apparently exhausted, the eruption of 1944 ended with a few days of strong explosions of dark ash.

The two cities of Pompeii and Herculaneum remained undisturbed, buried under the volcanic pyroclastic debris, until one of the outer walls was discovered in 1748, ushering in the period of modern archaeology. Records of the 79 A.D. eruption do not include any descriptions of lava flows. The first lava is said to have appeared in 1036 A.D., and since that time, lava has been part of most of the eruptions.

Thus the present structure of Vesuvius is similar to that of most composite volcanoes with solidified lava flows alternating with pyroclastic material. This is not unexpected, as Vesuvius is within a converging plate boundary, the location of most composite volcanoes.

1. What are some of the beneficial effects of volcanism?

There are several beneficial effects of volcanism. For example, weathering converts volcanic ash very rapidly to exceptionally fertile soils. (Some volcanic ashes found in New Zealand, however, have produced impoverished soils lacking essential trace elements for plant growth.) In some areas, crops can be grown within a year after an eruption of volcanic ash. Thus the soil can be a strong attraction for agricultural people, although the specter of another violent eruption is always present. In other places, volcanic steam is tapped by deep drill holes and is used to drive electric generators; this is known as *geothermal energy*. New lands are added, especially in Hawaii, by lava flowing down to the sea.

2. Why are volcanoes important to the science of geology?

Volcanoes are the pipeline to Earth's lower crust and mantle. From the analysis of minerals and gases in the emitted lavas, geologists can deduce the nature of rocks in the interior and the temperatures that exist below Earth's surface. The gases present in erupting volcanoes offer clues to the composition of the early atmosphere and enable scientists to note the changes since the period of de-gassing. The chemical composition of volcanic rocks helps geologists understand the nature of Earth below the weathered surface. The variations in volcanic activity are evidence of plate motions and explain the different geologic effects seen at plate boundaries.

3. Why are some volcanoes explosive and others relatively quiet?

The behavior of a volcano depends in large part on the nature of the magma being emitted. Lavas that are viscous (having an internal property that offers resistance to flow) tend to produce the more explosive eruptions. The gases contained within the lava cannot be easily expelled through the viscous mass and can build up to enormous pressures that may eventually blow out in sudden massive eruptions.

Viscosity is related to temperature – the higher the temperature, the less viscous or more fluid the lava and the quieter the eruption. Another factor is the silica content of the lava. In lavas containing a high proportion of silica (SiO_2), the silicon-oxygen tetrahedrons form irregular groupings of chains, sheets, and networks. The greater the number of tetrahedrons in the magma, the more viscous and the more resistant it will be to flow. Violent eruptions are associated with silica-rich lavas such as andesite and rhyolite, while lavas that are silica-poor and have low viscosity, such as basalt, release their enclosed gases easily and have quiet eruptions.

4. Why do some volcanoes form steep-sided cones and others form flat plateaus or have very gently sloping slides?

The difference in the shape of the volcano is related to the nature of the material erupted. Steep cones are built by cinders, loose pyroclastic fragments that will lie at a steep angle around a central vent. Pyroclastic materials tend to form from more explosive magmas or from gas pockets in a rising magma. The gentle slopes of the shield volcanoes or the basalt plateaus form from basaltic or silica-poor lavas that are so low in viscosity that they will flow great distances rather than pile up around the vent or fissure.

5. What do the terms mafic and felsic mean?

Mafic and felsic are terms used to describe the chemical or mineral composition of lavas. Mafic refers to silica-poor lava, in which the silica is about 50 percent by weight and the remaining elements are predominantly magnesium, iron, and calcium. *Mafic* is from magnesium and iron. The most abundant mafic extrusive rock is basalt, which is colored dark by its large proportion of iron magnesium minerals.

Felsic refers to silica-rich lava, containing over 65 percent silica and very small amounts of calcium, magnesium and iron. The remaining constituents are minerals containing potassium, sodium, and aluminum. *Felsic* is from the mineral feldspar and silicon. Rhyolite is the most abundant felsic rock, and it is light colored, even pink or red, from the abundance of potassium feldspar in its composition.

6. Do volcanic bombs cause explosions?

Not at all! The term *volcanic bomb* refers to the blobs of lava that are ejected into the air during an eruption. As they fly through the air, they become streamlined and spindle- or lens-shaped before they solidify. Sometimes a large clot of molten lava will be emitted that will preserve its generally globular shape as it chills. Some blobs hit the ground before solidifying and form round splatters.

7. What are the three major shapes taken by volcanoes? Which are the largest and tallest, and how do they form?

The smallest of the volcanoes are the *cinder cones*, formed entirely of pyroclastic materials. Cinder cones can occur singly or in clusters and rarely reach over 500 meters (about 1,650 feet). Cinder cones have formed near many volcanically active areas.

Shield volcanoes are broad, gently sloping cones, constructed almost entirely of fluid, low viscosity basaltic lavas. The Hawaiian Islands are the largest shield volcanoes, and Mauna Loa, from its base on the sea floor to its summit, is taller than Mount Everest, making it the tallest mountain on Earth. Shield volcanoes are usually oceanic island volcanoes, formed over a mantle plume or *hot spot* and characterized by quiet eruptions.

Composite volcanoes have cones of intermediate steepness between cinder cones and shield volcanoes. Since they are active over very long periods of time, they construct very large cones. Aconcagua, in the Andes, over 6,960 meters tall (22,970 feet), is a composite volcano and is the highest peak in the western hemisphere. As the name implies, composite volcanoes are built of lava flows alternating with layers of pyroclastic materials. The lavas are intermediate in composition, and the rock is predominantly andesite. Composite volcanoes are aligned with the zones of plate convergence that ring the Pacific basin and also follow the length of the Mediterranean. Of all the volcanoes throughout history, composite volcanoes have been the most explosive, the most dangerous, and most destructive of human life and habitation.

8. Are pillow lavas soft? Why are they called pillows?

Pillow lavas are hard rocks, usually basalt, that form on the sea floor from submarine eruptions. They are called *pillows* because they have a cylindrical shape like pillows. When molten lavas are erupted from vents on the sea floor, the flow contracts in the cold water into a generally cylindrical shape, and a solid layer congeals around the flow. Inside, the lava is still molten, and squeezes out like toothpaste through the chilled skin, also to congeal in the cold water and extend the length of the pillow. Another molten blob may also squeeze out, until the whole cylinder has solidified. During the next eruption, the new lava pillows settle on those below and so build up the sea floor. Pillow lavas are interesting because they only form under water. Thus, when pillows are discovered on a mountain top or in the middle of a continent, we have evidence of submarine eruptions in the geologic past, followed at some later time by extensive uplift (Figures 11.26, 3.16, pages 254 and 55, and 3.25 A, B, and C, page 60).

9. **What is the origin of porphyritic textures, and what are phenocrysts?**

Most volcanic rocks are evenly fine grained because of rapid cooling, but some have large crystals enclosed in a fine-grained *matrix* or *ground mass*. Since the size of the crystals in an igneous rock is determined to a large extent by the rate of cooling, porphyritic rocks are something of a puzzle. How can there be two sizes in one rock? Two stages of cooling seems to be the answer. Slow cooling takes place deep underground in the still fluid magma. The minerals that crystallize at higher temperatures will start to grow first and will form the *phenocrysts*, which are the large crystals. If the entire mass is later erupted near or on the surface, the liquid portion of the magma, now lava, will cool rapidly and form the fine-grained matrix around the previously solidified phenocrysts.

10. **How hot are lavas?**

Taking the temperature of an erupting volcano is a risky business. Using a *pyrometer*, an optical measuring device, scientists have determined temperatures of mafic magmas ranging from 1,040° C to 1,250° C. Temperatures as low as 800° C have been obtained from felsic lavas that were just barely flowing.

Activities

1. There are many very good books about volcanoes, Mount St. Helens in particular, that are worth reading. You can find material on Katmai (seen in the video portion of this lesson) Paricutin, Vesuvius, Mont Pelée, Krakatoa, Surtsey near Iceland, and of course Hawaii. Ask your local librarian or visit the geology library of your nearest college or university.

2. The best thing would be to take a field trip to a volcano. With your new understanding of volcanoes, Hawaii would be a perfect place to visit. Kilauea might still be in its present eruptive cycle and will provide you with some thrilling sights.

3. Read "Astrogeology" Box 11.1, "Extraterrestrial Volcanic Activity," pages 258-259. Some of the pictures recently transmitted from

Venus show many small domes, evidence of widespread volcanic activity in the past. Where, according to your text and the evidence sent back from space craft, is active volcanism occurring in the solar system (besides Earth, of course)?

1. The widespread volcanic activity on Earth indicates the interior of the planet is still very hot. The heat is most likely caused by

   a. impact of space debris.
   b. compaction due to gravity.
   c. radioactive decay of certain elements such as uranium.
   d. settling of denser elements toward the center.

2. Most of the world's present volcanic activity is located

   a. on transform plate boundaries.
   b. on the sea floor at divergent plate boundaries.
   c. in volcanic island arcs.
   d. in subduction zones at convergent plate boundaries.

3. Some of Earth's tallest and most consistently active volcanoes occur far from any plate boundary. These are

   a. shield volcanoes.
   b. cinder cones.
   c. composite volcanoes.
   d. lava domes.

4. All but one of the following statements describe the most destructive and explosive volcanoes.

   a. They are located in the circum-Pacific belt.
   b. They are landward of subduction zones.
   c. They have long lives and build very large structures.
   d. They emit enormous amounts of basalt.

5. From a chemical analysis of gas emitted during volcanic eruptions, it has been discovered that the most abundant gas is

   a. hydrogen sulfide.
   b. nitrogen.
   c. water vapor.
   d. carbon dioxide.

6. If you examined a volcanic rock and from chemical tests learned that its content of silica was fully 50 percent of the total, which of the following would NOT be true?

   a. It formed from a silica-rich lava.
   b. It was probably part of a highly explosive eruption.
   c. It was characteristic of an eruption from a continental volcano.
   d. The lava had a relatively low viscosity.

7. Taking a world-wide view, the volcanic rock that we associate with most of Earth's crust is

    a. basalt.
    b. andesite.
    c. rhyolite.
    d. breccia.

8. The relation between silica-oxygen tetrahedrons and the properties of a lava are that the tetrahedrons

    a. increase the temperature at which a lava is emitted.
    b. increase the viscosity or stickiness of the lava.
    c. allow gases in the lava to escape more easily.
    d. allow a greater percentage of water to be held in solution.

9. Lavas of low viscosity usually have all of the following characteristics except that they

    a. are very hot.
    b. form volcanoes with gentle slopes.
    c. produce *nuees ardentes* and ash falls.
    d. have a high content of iron magnesian minerals.

10. Laboratory studies have shown that melting of wet basalt under pressure can produce

    a. pillow lavas.
    b. pumice.
    c. rhyolite.
    d. andesite.

11. The only environment in which melting of basalt can occur is

    a. at divergent plate boundaries on the mid-oceanic ridges.
    b. at convergent plate boundaries.
    c. above mantle plumes where basalt rises from shallow asthenosphere.
    d. on the sea floor where pillow lavas are forming.

12. The origin of phenocrysts in volcanic rocks is best explained by

    a. two periods of cooling.
    b. rapid release of dissolved gas in emitted lavas.
    c. silicon-oxygen tetrahedrons forming a ground mass.
    d. rapid ascent of magmas through the volcanic neck.

13. Mount St. Helens and Vesuvius emitted great clouds of ash during eruption. Analysis of the ash showed it to be

    a. ash from vegetation burning on the slopes.
    b. drops of liquid lava.
    c. tiny chilled fragments of volcanic glass and rock.
    d. smoke composed of carbon particles included in the lava.

14. Volcanic domes, although a relatively small feature, are not uncommon. They are most likely to be found

    a. on the flanks of shield volcanoes.
    b. within the craters of cinder cones or composite volcanoes.
    c. on the sea floor, within the rift valleys at spreading centers.
    d. near plateau basalts.

15. The volcanic rock derived from melting of mantle material is

    a. andesite porphyry.
    b. vesicular rhyolite.
    c. plagioclase feldspar.
    d. basalt.

14

INTRUSIVE IGNEOUS ROCKS

This lesson will help you appreciate the importance of igneous activity, the rock-forming processes of magmas that do not reach Earth's surface but solidify underground.

After reading the textbook assignments, completing the exercises in this study guide, and viewing the lesson's video portion, you will be able to:

1. Distinguish among the major classifications of intrusive structures (dikes, sills, plutons, batholiths, and stocks.)

2. Explain how the texture of an igneous rock gives clues about its origin.

3. Describe how igneous rocks are classified, including mineralogical and chemical distinctions among granite, diorite, gabbro, and ultra-mafic rocks.

4. Name the sources of heat that may contribute to a rock's melting.

5. Discuss the effects that dry pressure, water pressure, and mixtures of different minerals have on the temperatures at which rocks melt or partially melt.

6. Explain the significance of Bowen's reaction series.

7. Describe how variations in the composition of igneous rocks may be accounted for by differentiation, assimilation, and partial melting.

8. Explain the origin of gabbroic (basaltic), dioritic (andesitic), and granitic (rhyolitic) magmas.

9. Relate plate tectonic theory to the distribution of various types of igneous rocks and to various kinds of intrusive and volcanic activity.

10. Describe how intrusive activity and volcanism are related.

INTRODUCING THE LESSON

While volcanoes belch out great clouds of black ash and cinders or emit streams of flowing lava, beneath the roaring vents and feeding the eruptions are great chambers of churning molten magma – unseen, mysterious, incredibly hot, and enormously large. Here, deep in the crust or in the upper mantle, the molten magma waits or works its way up into the cooler rocks of the shallow crust, sometimes to erupt on the surface but most often to chill and slowly solidify several kilometers underground.

In this lesson, you will learn of the similarities and differences between *intrusive* and *extrusive* igneous rocks. Intrusive rocks are the builders of continental crust and the mountain belts, while extrusive rocks are the volcanics that created the oceanic crust – the subject of the previous lesson. If intrusive rocks form deep underground, where can we see them to study their properties?

As you know from the lessons on plate tectonics, plates not only glide horizontally on the partially molten asthenosphere, but during plate convergence (episodes of mountain building), great slabs of continental crust are uplifted to incredible heights. The raised portions of continental crust often contain intrusive rocks, which after long periods of erosion are *unroofed* and finally exposed to view. The much photographed Half Dome in Yosemite Park in the Sierra Nevada of eastern California, for example, was once part of a deep molten mass that rose to cooler levels, slowly chilled to a granite about 100 million years ago, and was later uplifted during a plate collision. Erosion stripped away the overlying *country rock*, and ice age glaciers carved Yosemite Valley into one of the most scenic places on Earth. In fact, the whole Sierra, an enormous granite mountain range, forms an immense complex of intrusive rocks that once fed volcanoes long eroded away.

In your study of igneous rocks, you will discover some curious anomalies. Basalt, as you learned, is the most widespread extrusive igneous rock, but intrusive *gabbro*, having the same mineral composition as basalt, is relatively uncommon. Likewise, rhyolite is one of the least common extrusive rocks, but its subsurface counterpart, *granite*, is the most abundant intrusive rock and makes up much of the continental crust. Any theory on the origin of igneous rocks must account for this distribution.

In this lesson, you will learn about the origin of intrusive rocks, how they are classified, and the various structures they form. These are the primary and most abundant rocks of the crust and upper mantle and probably of the inner terrestrial planets of the solar system as well. According to ancient Greek and Roman myths, the hot unseen interior of Earth is Pluto's realm, and it offers us tantalizing glimpses of processes and events that at best we can only infer.

Completing the following eight steps will help you master the lesson objectives and achieve the goal for this lesson:

Step 1: Read the TEXT ASSIGNMENT, Chapter 10, "Intrusive Activity and the Origin of Igneous Rocks."

Step 2: Study the KEY TERMS AND CONCEPTS as noted in the study guide.

Step 3: Watch the VIDEO using the VIEWING GUIDE in the study guide.

Step 4: Read the study guide's PUTTING IT ALL TOGETHER section, which will help you summarize and integrate all of the information in this lesson.

Step 5: Complete any assigned lab exercises.

Step 6: Complete any assigned ACTIVITIES found in the study guide.

Step 7: Review the material in this lesson and complete the SELF-TEST found in the study guide.

Step 8: Go back to the LEARNING OBJECTIVES and make sure you have learned each one.

LESSON ASSIGNMENT

1. Before reading the whole chapter, read the introduction to Chapter 10 on page 215, noting the position of *magma* in the rock cycle.

2. Study the photograph of Granite Peak, Sequoia National Park, California, facing page 215. This is part of the massive granite intrusion that forms the Sierra Nevada.

3. Now read Chapter 10, pages 215-237. Study Figure 10.10, page 221, and compare it with Figure 11.27, page 256, noting the similarities and differences.

4. Turn to Table 9.2, page 202, "Minerals of Earth's Crust," and carefully review the rock-forming minerals. This table will help you understand the origin of igneous rocks.

5. Learn the sequence of crystallization shown in Figure 10.15, Bowen's reaction series, page 228. This is one of the most important concepts in *Earth Revealed*.

TEXT ASSIGNMENT

The key terms listed under "Terms to Remember," page 236 in your text, are important and will supplement your knowledge of igneous rocks learned from the chapter on volcanism. Use the glossary and the summary in the text to help you understand the definitions.

KEY TERMS AND CONCEPTS

VIEWING GUIDE

Once you have read the text material for the lesson and studied the terms, you are ready to view the lesson's video portion. Review the questions that appear in this section to help you watch for important points and to prepare you for what you will see in the video.

Before viewing, answer the following questions:

1. If granite and rhyolite are chemically similar, how do you tell them apart?

2. Scientists have been unable to form granite in the laboratory. Can you explain this?

3. What does Bowen's reaction series illustrate?

After viewing, answer the following questions:

1. What is the evidence that granite was once a molten magma that intruded into the country rock?

2. Granite and rhyolite are both igneous rocks and are chemically similar, but have different textures. Describe the textures of each and tell what these textural differences indicate about their origin.

3. All of the following are igneous intrusions, but they have different characteristics. From what you saw in the video, how would you describe a:

   a. dike?
   b. sill?
   c. laccolith?
   d. batholith?

4. From a careful examination of an intrusive igneous rock, how would you tell which minerals were the first to crystallize? According to Bowen's reaction series, which minerals will be the last to crystallize in a cooling magma?

5. According to plate tectonic theory, where would you most likely encounter diorites and andesites?

PUTTING IT ALL TOGETHER

You have read the text assignment and seen the video portion of the lesson. This section, made up of a SUMMARY, a CASE STUDY, the QUESTIONS, suggestions for ACTIVITIES, and a SELF-TEST, will help you pull the information together.

Summary

In the preceding lesson, you learned that extrusive igneous activity, or volcanism, created landforms of various sizes, shapes, and eruptive history – from domes, cinder cones, shield volcanoes, and composite volcanoes to lava plateaus and the oceanic crust. Intrusive igneous activity also forms structures that differ in size and shape in relation to the surrounding rocks. Some intrusions that solidify very slowly at great depth will be coarsely crystalline, with crystals at least 1 millimeter in size (Figure 10.9, page 221). Other magmas chill more quickly in small shallow intrusive bodies and will be fine-grained, similar to their extrusive counterparts.

Classification of intrusive rocks depends on mineral content and texture, but intrusives are usually easier to identify than extrusive rocks because the mineral grains are large enough to be seen with the naked eye (Figure 10.10, page 221).

The source of the immense heat required to melt rock and form magma is discussed in this lesson as well as several factors that control the temperatures at which various minerals melt. The origin of different magmas is still subject to debate, but plate tectonics and concepts of partial melting and mineral differentiation provide some viable answers.

Intrusive Structures

Intrusive structures are all formed of liquid magma that intruded into older rock and solidified. Magma is a natural, mobile, hot melt composed of rock-forming minerals (mainly silicates) and volatiles (mainly steam held in solution by pressure). Intrusives are classified by their size, their shape, depth of solidification, and relation to country rock.

Shallow intrusions solidify at depths of less than 2 kilometers (1.6 miles or 8,450 feet) and are generally small. One example would be a *volcanic neck*, in which the magma chilled within the vent or throat of a volcano, usually during the last stages of an eruption. The texture of the rock is fine-grained because it cooled quickly. Ship Rock, New Mexico (Figure 10.3, page 217), is a well-known example. Devil's Tower, Wyoming – used in the movie *Close Encounters of the Third Kind* – is the preserved neck of a former volcano in which the magma cooled and formed vertical joints in the rock.

Dikes are also shallow intrusions of magma, emplaced into fractures that cut across the layers of the country rock and are thus termed *discordant* (Figures 10.4 and 10.5, pages 218-219). Dikes in three dimensions are shaped like table tops (tabular) and, in some areas where they are more resistant to erosion than the country rock, project above the land like stone fences. Dikes vary greatly in size but are seldom more than 30 meters (98 feet) wide, although some are hundreds of kilometers long. The longest in the world is the so-called Great Dike of Zimbabwe in southeast Africa. More than 480 kilometers (298 miles) long, it has an average width of only about 8 kilometers (5 miles). The walls are essentially vertical, and the rock in the dike is *gabbro*, the coarse-grained equivalent of basalt.

Sills are also tabular but are termed *concordant* since the molten magma was intruded between and parallel to the layers of country rock (Figure 10.6, page 219). These shallow intrusions, which are small and cool rapidly, are fine-grained and generally are composed of basalt or andesite. One of the most famous sills in the United States is the Palisades sill that follows the New Jersey shore of the Hudson River. This is an unusually large sill, more than 300 meters (984 feet) thick, and is formed of a gabbro that cooled slowly and formed columnar joints, similar to those at Devil's Postpile in California (Figure 11.25 A and B, page 254).

Deep intrusions, also called *plutons* (named for *Pluto*, god of the underworld), crystallize several kilometers deep in the crust and are characterized by coarse-grained textures. Most plutons are irregular in shape and are discordant with the country rock. The largest plutons are called *batholiths* (from the Greek word *bathos* meaning "deep" and *lithos* meaning "stone"), and by definition must have a surface exposure of over 100 square kilometers (36 square miles) (Figures 10.7 and 10.8, page 220).

Most batholiths are composed of granite and solidified at depth from a felsic magma. Granite batholiths occur in large elongate bodies along the western margin of North America, and include the Peninsular Ranges batholith, which runs almost the full length of Baja California in Mexico, the Sierra Nevada batholith, the Idaho batholith, and the Coast Range batholith, which extends from British Columbia into Alaska. The Cretaceous period, about 100 million years ago, was a time of increased granitic magma production, and these great batholiths were emplaced then above an ancient subduction zone.

Dikes and sills often occur near batholiths, since the intruding mass will fracture the rock above and the molten material will invade the openings. Water-rich fluids from the solidifying magma often deposit ores of great economic value, such as the famous gold-bearing *Mother Lode* district of the Sierra foothills – the site of the gold rush of 1849. (Note in Figure 10.10, page 221, that granite has the same mineral composition as rhyolite but is coarse-grained, a characteristic of plutonic rock and slow cooling.)

Classification of Intrusive Igneous Rocks

Learning the classification of intrusive igneous rocks should be much simpler now that you have learned the mineral composition of the three basic volcanic rocks: rhyolite, andesite, and basalt. In Figure 10.10, page 221, the coarse-grained equivalents have been added: *granite* (rhyolite), *diorite* (andesite), and *gabbro* (basalt). As we noted before, granite is the most abundant intrusive rock and is predominant in continental crust, whereas basalt is the most abundant extrusive rock and constitutes most of the sea floor. Further description is presented in Table 10.1, page 222, including the color most commonly seen in rock.

In the right hand column of Table 10.1, *ultramafic* rocks have been added, and from Figure 10.10, you will note that they are composed entirely of ferromagnesian minerals such as olivine and pyroxene. These are dark rocks, sometimes green or black, and are very heavy or dense. From laboratory experiments it was determined that high temperatures (almost 2,000° C) are required to melt ultramafic rocks, which comprise most of Earth's mantle.

One other group of intrusive igneous rocks that should be mentioned is *kimberlite*, found in pipe-like conduits. Kimberlite, composed predominantly of ferromagnesian minerals, contains diamonds as phenocrysts and is the only source of these precious gems. Diamonds require very high pressures and temperatures to form, and scientists believe kimberlite magmas must have come from the mantle at depths near 250 kilometers (155 miles). Most of the pipes are only about 90 meters (295 feet) across at the surface. A kimberlite pipe resembles an extremely long, very thin drinking straw.

There are many other intrusive rocks representing the mineral variations in the continuum between granite and ultramafic rocks. One batholith may be made of many separate intrusions occurring at different times, each slightly different in mineral composition, hence having a different name. The huge Andean batholith, stretching along the western coast of South America, consists of over 800 individual plutons!

Heat to Melt Rock

Rock becomes molten when the temperature is higher than the melting points of all the minerals in the rock. The question is: where does all the heat come from that forms magma in the crust and mantle, and why isn't all mantle rock molten, rather than just in limited areas or pockets?

The Geothermal Gradient. Geologists believe that early Earth was much hotter than it is now because of heating from impacts of meteors, from gravitational compression, from crystallization, and most importantly from the heat generated by radioactive decay of certain elements such as uranium. Earth is still very hot inside; some small part of the heat may be residual, left over from the days of Earth's origin. Most of the heat is believed to be from radioactive decay of uranium, thorium, and potassium, although the amount of radioactive elements in rocks is now very small.

The geothermal gradient is the rate at which the temperature increases downward from the surface; it averages about 25° C for each kilometer. The rate is not the same everywhere but is hotter at shallower depths in volcanic regions than in plate interiors (Figure 10.12, page 226). The temperature also rises faster beneath the oceans than beneath the continents because the oceanic crust is thin and transmits heat more readily from the mantle. Radioactive elements are more abundant, however, in felsic continental crustal rocks than in mafic oceanic rocks. Continental rocks seem to act as a thermal blanket to the hot mantle rocks below, preventing transmission of heat.

Local *hot spots* are caused by mantle plumes, narrow upwellings of hot material from the upper mantle (Figures 11.33 and 4.41, pages 261 and 93). The origin of the Hawaiian Island chain has been previously discussed in terms of the Pacific plate moving over a hot spot.

Factors that control melting points of minerals. In trying to explain why only some of the mantle rock is molten, scientists note that temperature alone does not determine the melting point of every mineral. *Pressure* is an important factor, because an increase in pressure will raise the melting point of most minerals. Although the temperature within Earth increases with depth, so does pressure caused by the increasing load of rock above. As the pressure rises, the temperature at which the minerals will melt also rises. (This is the principle of the pressure cooker: water boils at 100° C at atmospheric pressure, but in the pressure cooker it can reach 200° C before boiling.)

The effect of pressure on temperature of melting is straightforward provided the mineral is dry. *Water vapor* under pressure, on the other hand, acts to lower the melting point of most minerals. (Refer to Figure 10.13, page 227, which shows the effects of water pressure on the temperature of melting of sodium plagioclase.) In other words, wet minerals will melt at a lower temperature than dry minerals of the same composition. As the pressure rises, the effect of water in the melt also rises. This effect is exactly opposite from the effect of pressure on the melting of dry minerals.

In some cases, *mixed minerals* will melt at a temperature lower than either of the combined minerals alone. (Study Figure 10.14, page 227. At what percentages of quartz and potassium feldspar will the mix have the lowest temperature?)

The Origin of Magmas of Different Composition

Although only three major families of magma exist (mafic, intermediate, and felsic), there are literally hundreds of different kinds of igneous rocks. The reason a single magma can crystallize into many different kinds of igneous rock is because magma is a complex silicate liquid, unlike water, which can only crystallize into ice. As the temperature falls and the magma starts to solidify, different minerals will start to crystallize at different temperatures. At any one time, the magma will consist of solid or *frozen* crystals and still unfrozen liquid, something like a *slush*.

The first crystals to form will have a composition different from that of the remaining liquid. As the temperature drops, the composition of the remaining liquid changes as various elements are removed from the melt by the solidifying minerals. If the early formed crystals are separated from the still liquid magma, they will form a rock of different composition from that formed by crystallization of the remaining magma.

This process is called *differentiation* and explains the formation of different igneous rocks from one magma. For example, it is possible for a gabbroic magma to have early crystallizing olivine, pyroxene, and plagioclase removed by gravity settling to form diorite. Likewise, diorite can further differentiate to form granite.

Bowen's Reaction Series

The importance of magmatic differentiation was first recognized by Norman L. Bowen, a Canadian scientist working at the Geophysical Laboratory in Washington, D.C., in the early 1900s. He investigated the melting and crystallization properties of minerals and in particular the order in which minerals crystallize in cooling magmas of different compositions. He found a plausible way for both felsic and mafic rocks to form from a single basaltic parent magma, a very significant discovery that explained to a large extent the vast array of igneous rocks.

The discontinuous branch. From his experiments, Bowen discovered that mineral crystallization in a basaltic magma follows two series, the *discontinuous branch* and the *continuous branch*. The minerals in the discontinuous branch are all ferromagnesians, dark minerals rich in iron and magnesium that are abundant in mafic magma. As the molten magma starts to cool, the first mineral to crystallize at a high temperature is *olivine* (Figure 10.15, page 228). Continued cooling causes the olivine to react with the silica in the melt to form a more silica-rich mineral, pyroxene. The simple crystal structure of olivine changes to a single chain of silica-oxygen tetrahedrons characteristic of pyroxene.

As the temperature lowers, pyroxene continues to form until it reaches the point at which amphibole forms. Amphibole is a mineral that contains more silica than pyroxene and has a double chain structure. If there is still liquid in the melt and the temperature is continuing to drop, the last mineral to form in this sequence is biotite, which is black mica with a sheet structure. Each step has removed iron and magnesium, and theoretically the residual magma will contain less of both elements. (Review Figure 9.9, page 199, showing common silicate structures especially of the ferromagnesian minerals in the discontinuous series.)

The continuous branch. Bowen knew from his research that plagioclase feldspar in basalts and gabbros are usually calcium-rich, while those in diorites and granites are usually sodium-rich. He discovered that as crystallization proceeded in a cooling magma over the same temperature range as in the discontinuous series, the calcium-rich feldspar reacted with the melt and continuously changed toward a composition higher in sodium. He termed this series of reactions the *continuous branch*, which contained only one mineral – plagioclase feldspar. As the temperature slowly drops, more and more plagioclase forms until most of the calcium and some of the sodium in the magma are used.

The residual magma, after crystallization is completed for the two series, is very rich in silica and contains potassium and aluminum, which combine to form potassium feldspar and muscovite mica. The pure silica that is left over will crystallize into quartz.

Rarely in nature will the full series of reactions take place. Only if the early forming crystals can be removed from the magma will the reactions continue. *Crystal settling* is the method proposed to remove the crystals. In crystal settling the heavier, denser crystals such as the olivine and calcium plagioclase settle through the still liquid magma to the bottom of the magma chamber to form a gabbro. The left-over magma, depleted in mafic constituents but rich in silica, aluminum, and potassium, will solidify into a granite (Figure 10.16 A, B, and C, page 229).

Differentiation does take place in nature and has been observed in the thick Palisades sill. The problem is that there is no way to account for the origin of granite from a mafic magma alone because less than 10 percent of a crystallizing basalt will form a granite. Yet granite is the most abundant intrusive rock in the continental crust, whereas gabbro, the intrusive equivalent of basalt, is relatively rare.

Assimilation, Mixing, and Partial Melting

In order to explain the differences in composition of igneous rocks, other processes besides differentiation have been suggested. *Assimilation* refers to the contamination of magma by other rocks such as those comprising the continental crust. Portions of the crust will melt, and the high-silica rock will be assimilated and become part of the magma. This will increase the silica content to the level of intermediate composition, which may account for the andesites and diorites that form in the subduction zones of the Pacific Rim (Figure 10.17, page 230). Mixing of magma occurs when two molten bodies meet, such as a mafic and a felsic magma, and merge within the crust; the resulting combination will differ from either of the two original melts (Figure 10.18, page 230).

The concept of *partial melting* holds that a granitic magma could be formed by partial melting of a more mafic rock. If there is sufficient heat to cause rock to start to melt, the first minerals to become liquid at relatively low temperatures would be the last ones to crystallize in Bowen's reaction series. Starting at the bottom of Figure 10.15, page 228, you can see that orthoclase feldspar and quartz – the major constituents of granite – would be the first to melt, forming a felsic magma. Continued heating at low temperatures would cause more orthoclase and quartz to melt, plus smaller amounts of muscovite mica, biotite mica, hornblende (amphibole), and sodium plagioclase feldspar. Check Figure 10.10, page 221, for the mineral composition of granite.

Partial melting of the lower part of the continental crust is a viable source of felsic magma for granite batholiths. The magma is liquid and less dense than the surrounding rock. Although very viscous, it moves slowly upward into the cooler parts of the crust, where it solidifies into a granite. Rarely will a granitic magma reach the surface; this accounts for the lack of abundant rhyolite extrusive features. When it does reach the surface, it will erupt violently and blanket the surface with ash to form rhyolite tuff. Some rhyolitic lavas congeal into plug domes. Granite is generally emplaced within mountain belts.

Basalt may be a product of partial melting within the mantle at very high temperatures. Since the mafic magmas that form basalt are very fluid, the basalt will rise rapidly to the surface, erupt at mid-oceanic rift zones or as mantle plumes, and rarely pause long enough to solidify within the crust. This would account for the rare occurrence of gabbros – the intrusive equivalent of basalt – in batholiths and other large intrusions.

Plate Tectonics and the Origin of Igneous Rocks

The theory of plate tectonics offers some explanations for the variety of igneous rocks and for their rather distinctive distribution. The actual processes such as partial melting or crystal settling cannot be observed in the field as easily as cooling of lavas on the surface. However, some of the newer theories of magma formation are well accepted and are reviewed below.

Basalt, as you learned, is derived from partial melting of ultramafic rock of the mantle. These rocks form at the highest temperatures, in the range of 1,100 to 1,300° C and are the most fluid form of igneous activity within Earth. Basalts consist of the minerals at the top of Bowen's reaction series and are the principal rocks of the sea floor. Basalts, however, also occur on continents, and there is some evidence that the lower continental crust is also basaltic (Figure 10.19, page 232).

The intermediate rocks – andesite and diorite – will crystallize at lower temperatures. If you examine Figure 10.10, page 221, you will note that their mineral composition ranges from about two-thirds plagioclase and one-third ferromagnesian minerals to about half of each. Andesites are usually lighter in color than basalt, a light gray rather than black.

Intermediate rocks form at subduction zones, where the sinking wet basalt and wet marine sediments partially melt as the temperature increases downward. The presence of water here lowers the melting point, causing the minerals higher in silica to melt and separate from the more mafic minerals.

Another way intermediate magmas form is through the mixing of the molten basalt with the granitic continental crust. Thus, andesites and diorites occur in island arcs, active continental margins, and in older mountain building belts within continents that were once the site of ancient subduction zones (Figure 10.20, page 234).

We are still left with the problem of the source of granite, the most abundant igneous rock in the continental crust. As you learned, partial melting of a basalt would produce only about 10 percent felsic magma. Melting of the marine sediments subducting with the basalt sea floor, while creating rocks higher in silica, is too small in volume to account for the massive granite intrusions (Figure 10.20, page 234). Partial melting of the andesitic crust would generate a silica-rich magma that would rise into the continental crust and have the composition of the minerals in the lower temperature portion of Bowen's reaction series.

Another explanation concerns mafic magmas rising and pooling at the base of the continental crust, heating the lower crust in what is termed *magmatic underplating*. Partial melting of the overlying crust would result in a granitic magma. As the molten material rose, there would be further melting of the quartz and feldspar, and it would expand in volume, creating more granite (Figure 10.21 A and B, page 235, and Figure 4.34, page 85). At present, granite plutons are still puzzling, but given our knowledge of the various processes of partial melting, assimilation, and differentiation, combined with what is understood about plate dynamics, geologists are approaching some very plausible theories.

CASE STUDY:
Yellowstone
Park Has It All

Yellowstone Park, covering about 8,950 square kilometers (3,215 square miles) in northwestern Wyoming, has Old Faithful geyser, hot springs, mud pots, fumaroles (steam vents), huge calderas, active faults, and earthquakes. Yellowstone is underlain by a shallow chamber that contains molten silica-rich magma, both of rhyolite and basalt, resulting from a hot spot anchored in the mantle beneath the Park. This is a unique natural laboratory for studying the interactions between intrusive and extrusive igneous activity resulting from tectonic plate interactions, specifically the motions of the North American, Pacific, and Juan de Fuca plates. Explaining all these features is challenging; most volcanic activity, as you have learned, takes place at plate boundaries, but Yellowstone is 2,000 kilometers (1,200 miles) from the western boundary of the North American plate. The igneous activity is believed to be related to a hot spot, similar to that under Hawaii, except Yellowstone is surrounded by a continent, while the Hawaiian Islands are surrounded by an ocean.

The stage was set for the volcanic and tectonic evolution of Yellowstone in the late Mesozoic era, a time of intense deformation of the crust in the western United States, culminating in the formation of the middle and southern Rockies. Volcanic activity, consisting mostly of andesite flows, was common until about 40 million years ago, then ceased. In the past two million years, however, magma has filled immense chambers under Yellowstone Plateau, an area of about 6,500 square kilometers (2,362 square miles) with an average elevation of 2,000 meters (6,600 feet).

The now partially crystallized magma is the source of heat for the numerous hydrothermal features in the Park: geysers, hot springs, mud pots, and so on. During the last volcanic cycle, thousands of cubic kilometers of rhyolitic magma erupted to the surface. The rate of magma production is comparable to that of Hawaii, Iceland, and the mid-oceanic ridges. In Yellowstone, periods of voluminous eruption lasting for hours to months are separated by quiet intervals lasting for as long as hundreds of thousands of years.

The rhyolite eruptions were not principally of lava (as you would expect) but were mainly flows of volcanic ash and hot gas as a consequence of violent eruptive explosions. The hot ash flows were on a scale unknown elsewhere in geological history. The massive eruptions drained the sub-surface magma chambers, causing the chamber roofs to collapse to form huge calderas, tens of kilometers across. Fragments of the glassy volcanic material were jetted high into the atmosphere and have been found as far away as Saskatchewan, Texas, and California. During the last 200,000 years, fractures have penetrated the old calderas and the underlying magma bodies, and basalt has erupted through the fractured caldera floors – all the result of tectonic movements.

Tectonic extension taking place in the western United States has resulted in both rhyolitic and basaltic volcanism similar to that observed in Yellowstone. During the last 15 million years, basalt and rhyolite have erupted in a systematic sequence propagating northeastward from near the borders of Idaho, Nevada, and Oregon. If you could go back there in time, you would see a "Yellowstone," a topographically high area of rhyolitic volcanism with hot springs and earthquakes. The next visit, years later, would find the "Yellowstone" farther northeast than it had been. The zone of rhyolitic eruptions has moved northeastward at the rate of about two to four centimeters per year followed by a period of basaltic volcanism. What mechanism accounts for these two different magmas erupting in the same place, and why is there a systematic change in location?

The basalt forms from partial melting of Earth's upper mantle, presumably in a hot spot or in an area of mantle uplift where the pressure is reduced and minerals with low melting points begin to liquify. The flow of basalt toward the crust transports heat upward and allows silica-rich materials of the crust to melt. The silica-enriched magma rises and accumulates as a large hot viscous mass in the upper crust, accounting for the rhyolitic eruptions. The progression northeastward of the volcanism indicates movement of the plate over a hot spot, similar to that occurring in the Hawaiian Islands but in the opposite direction. A recent alternative is that the hot spot itself is moving, a phenomenon as yet not well documented.

The high heat flow, studies of earthquake waves, and other data indicate the present existence of magma in the crustal chamber under the Yellowstone Plateau. What will the future be? Most likely more of the same: earthquakes, explosive hot ash and gas emissions preceding basaltic emissions, and new eruptive locations to the northeast. For geologists, time will bring a better understanding of this remarkable Park that "has everything."

1. What is the evidence that intrusive rocks were once molten?

Questions

There are several lines of evidence that indicate intrusive igneous rocks solidified from molten magma. Most of the minerals found in intrusive rocks can form only at high temperatures and pressures. The presence of dikes, sills, and veins intruded into country rock show that the magma must have been liquid when it forced its way in. The margins between the country rock and the intrusion appear "baked" (contact metamorphism) by the heat of the magma. Within the intrusion itself may be *xenoliths* or inclusions of nearby country rock (Box 2.2, page 31) that fell into the magma and were not completely melted (Figure 10.1, page 216). And finally, within the intrusive body itself there will be *chill zones* at the contact with the country rock, where the crystals will be finer because of the more rapid loss of heat to the cooler surrounding rock.

2. What is the difference between a dike and sill? What is a laccolith?

Both dikes and sills are shallow intrusions, often above or near a deeper batholith. Dikes are discordant intrusions, where magma has been injected into a fracture that cuts across sedimentary rocks or older igneous rocks. Sills are concordant, meaning that the intruded magma squeezed in between and is parallel to sedimentary layers (Figures 10.4, 10.5, and 10.6, pages 218-219). A laccolith is a type of sill in which the magma accumulates in one area like a blister and raises the sedimentary rocks above into a dome.

3. What equipment does a field geologist need to quickly differentiate between intrusive and extrusive igneous rocks?

Since the texture of igneous rocks is a reliable clue to their origin, all the geologist needs is a good magnifier or hand lens. Extrusive rocks cool quickly on the surface of Earth; this results in the individual minerals being too small to be seen with the naked eye. The short time available for the crystals to grow, combined with the increasing viscosity of the cooling lava, prevents the atoms from moving about and assembling into a crystal lattice larger than microscopic.

Intrusive rocks – especially in a deep batholith where cooling may take thousands of years – are characteristically coarse-grained, and the minerals are large enough to be distinguished with a simple magnifier. The mineral grains are irregular and interlocked in igneous rocks. During the final stages of crystallization after most of the minerals have grown together, the last to solidify, such as quartz, fill all the spaces. This creates an interlocking network between the grains that holds the igneous rocks together and makes them solid and durable (Figure 10.9 A and B, page 221).

4. What does granite really look like?

Granite has a "salt and pepper" appearance, with the light colored minerals consisting of quartz, potassium feldspar, and muscovite mica and the dark minerals consisting of elongate crystals of hornblende (amphibole) and biotite mica, which are black flakes (Figure 10.11 A, page 222). Granites can vary, however, depending on the amount of quartz and hornblende and the color of the feldspar. Some orthoclase (potassium feldspar) varies through the warm colors of pink, peach, and rose to red and brownish red. When you are in a bank or public building, look for granite in counter tops, walls, and near the elevators. Granite is a colorful and lasting building stone.

Granite in a mountain range is light colored, with the same general salt and pepper texture. Notably lacking in layering or stratification, it is massive and often fractured or exfoliated into curved slabs. The photograph of Granite Peaks, facing page 215, will give you a good sense of a granite mountain.

5. How are granite and rhyolite related?

First, how do granite and rhyolite differ? Granite never occurs as a lava flow but shows evidence, such as the large size of the crystals, of long slow cooling deep underground. Rhyolite is never seen in intruding rocks that formed at depth but, judging from the fine-grained texture, must have cooled quickly on the surface or in shallow intrusions. Chemical analysis of granite and rhyolite shows they are of similar composition and that both rocks formed from the same parent magma.

A similar relationship is seen in basalt, which is usually erupted as a lava, and gabbro, an intrusive rock with a coarse-grained crystalline structure but the same composition as basalt. What other pair of igneous rocks has the same relationship?

6. What are pegmatites?

Pegmatites are extremely coarse-grained intrusive igneous rocks that often occur in dikes and contain large crystals of quartz, feldspar, muscovite mica, and occasionally rare elements. The crystals in a pegmatite frequently grow to several meters across, a result of both very slow cooling and the low viscosity of the fluid from which they formed (Box 10.1 Figure 1, page 223). During the last stages of cooling, usually of a felsic magma, the last materials to solidify will be quartz, feldspar, mica, and certain other elements that could not fit into the crystal structure of the minerals in the cooling magma. The key to the coarse texture, however, is the water in the magma that will stay fluid, forming a residual solution of very low viscosity. The atoms are able to move freely and over great distances to become part of the crystals, which can grow to be very large before finally solidifying.

7. What is the effect of pressure on the melting point of minerals?

Pressure increases with depth in Earth's crust and has the effect of raising the melting point of most minerals. In other words, the temperature at which a feldspar will melt in the laboratory will be much lower than the melting point under pressure deep in the crust or mantle. Thus, in spite of the increase in temperature that comes with depth, the pressure acts to keep most of the crust and mantle in the solid state. This is true for dry magmas but changes drastically if water is present under pressure, since water will tend to lower the melting point of minerals.

8. How do the textures of some igneous rocks support the laboratory evidence of Bowen's reaction series?

A close examination of intrusive rocks reveals that all minerals do not solidify at the same temperature or at the same time. The early forming minerals, such as calcium plagioclase and olivine, crystallize at high temperatures while the rest of the magma is still fluid. The floating crystals develop unconstrained in the liquid and form beautiful complete crystalline forms. As the temperature drops, the next group of minerals crystallizes in the remaining spaces but shows irregular shapes, and the crystals are rarely perfect. The last minerals to solidify, such as the potassium feldspar and quartz, crystallize from small pockets of left-over magma and act as space fillers, unable to develop their own distinctive crystal form. Thus, the shapes of the crystals in the solidified rock support the order of crystallization in Bowen's two reaction sequences.

9. How do granite magmas make room for themselves in the solid rock of the crust as they rise into shallower depths?

Magmas have great upward buoyant strength, and they make room for themselves in several different ways. At depth, the surrounding solid rocks are hot and under pressure and will become plastic, slowly flowing around the rising body. At shallow levels in the crust, the rocks are less hot and more rigid and brittle. The overlying rock may be upwarped or it may shatter, and some of the broken rock will become assimilated by the hot magma. The rising magma can also dislodge large blocks of the overlying country rock by a process known as *stoping*. Dislodged blocks, usually more dense than the magma, will sink and most likely be assimilated at depth. Not all fragments of rock will melt, however, and those that are enclosed by the solidified granite are known as *xenoliths*, (from the Greek words *xenos* meaning "stranger" and *lithos* meaning "stone").

10. If molten magmas are less dense then the country rock and are very buoyant, why don't they all reach the surface?

Some magmas, of course, do reach the surface and erupt as volcanoes. Most magmas, however, solidify at depth for various reasons. Resistant rock layers in the crust may prevent the magma from reaching the surface. If the magma enters rock layers of low density, it will lose its buoyancy and cease to rise. Most probably, the magma is crystallizing in the cooler upper layers of the crust and has lost its early mobility; it becomes too viscous to flow.

Magmas containing appreciable amounts of water will crystallize early and form large intrusive bodies in the middle crust of continents. Most of the wet magmas are granite, accounting for the abundance of granitic intrusions and the scarcity of rhyolite eruptions on the surface. Basalt magmas are very hot and very fluid and will rise more rapidly through the crust to erupt as basalt volcanoes, accounting for the abundance of basalt on Earth's surface and the relatively small amounts of gabbro found in intrusive bodies.

11. How does plate tectonic theory account for the distribution of igneous rocks? What does not fit the theory?

The origin of igneous rocks is an ongoing process that is still forming oceanic and continental crust and producing active volcanoes around the world. The source materials, either directly or indirectly, are mantle rock. The different magmas result from differentiation, partial melting, crystal setting, assimilation, and magma mixing.

The story starts with partial melting in the mantle asthenosphere, where the temperature is high enough to cause melting (Figure 10.19, page 232). Basalt is formed by dry partial melting of mantle rocks and is erupted in the mid-oceanic rift valleys as pillow lavas, flows, and volcanoes, filling the rift, building the ridges, and forming new sea floor. The unmelted solid mafic minerals, such as olivine and pyroxene, are left behind in the mantle that was partially depleted of calcium, aluminum, and silicon oxides by the removal of the basalt. The residual mantle rock thus formed is a variety of ultramafic rock. (See Table 10.1, page 222.)

The basaltic oceanic crust will become partially molten in a subduction zone where the lithosphere plunges back into the asthenosphere. As the oceanic plate heats up, the wet basalt with its capping of sea-floor sediments will become partially molten, releasing minerals higher in silica and forming an andesite magma of intermediate composition (Figure 10.20, page 234). The composition of andesite and diorite is approximately that of the average rock of the crust formed in ancient or present continental margins.

Partial melting of the andesitic crust generates a silica-rich magma characteristic of rhyolite or granite. As the magma rises through the continental crust, it will become further enriched in those felsic minerals at the low temperature end of Bowen's reaction series (Figure 10.21 A, page 235).

The abundance of granite in the continental crust has several explanations but still is not completely understood. Mafic or intermediate magmas intruding the continental crust will cause heating and melting of quartz, potassium feldspar, and other minerals abundant in felsic magmas. Underplating by magmas from the asthenosphere will also cause heating of the lower continental crust and generate felsic magmas (Figure 10.21 B, page 235). The molten magma, being of a low density, will work its way up into a shallower level of crust to solidify into a pluton. The tremendous volume of granitic intrusions is hard to explain but undoubtedly is related to partial melting and differentiation of andesitic rocks in the lower crust.

Activities

1. Igneous rocks, as you learned, are excellent building materials. As you travel through business sections of your city, take a closer look at the stone facades and the floors or walls of the buildings. Most frequently, granite, gabbro (dark and crystalline), and sometimes diorite are the rock of choice. Metamorphic rock such as marble and sedimentary rocks such as travertine also are frequently used, but the display of strong color and the crystalline texture will help you identify the igneous rocks.

2. If you live in a mountainous area, ask your local geological survey office or college geology department where you can see granites or other igneous rocks.

3. If there is a planetarium where you live, you might be able to see an exhibit of moon rocks. (See "Astrogeology" Box 10.2, pages 231-232, "Extraterrestrial Minerals and Igneous Rocks.") Now that you understand igneous rock, you will find the exhibit interesting and provocative and very much related to the origin of Earth and its moon.

Self-Test

1. In the field, geologists can generally separate intrusive igneous rocks from volcanic rocks by the

   a. color of the rocks.
   b. texture.
   c. mineral content.
   d. hardness.

2. Igneous rocks composed primarily of quartz and feldspar are most likely to be encountered in

   a. divergent plate boundaries and oceanic ridges.
   b. subduction zones.
   c. continental mountain ranges.
   d. hot spots and mantle plumes.

3. According to the classification system used in your text, which of the following minerals are least abundant in a diorite?

   a. Potassium feldspar and quartz
   b. Sodium and calcium plagioclase feldspar
   c. Olivine and pyroxene
   d. Amphibole and biotite

4. If a piece of coarse-grained granite is melted in a lab and allowed to cool rapidly, the resulting rock will be

   a. coarse-grained.
   b. pegmatitic.
   c. glassy.
   d. mostly ultramafic in composition.

5. If a piece of basalt is melted in a lab and slowly cooled, which of the following would be the first minerals to crystalize

    a. Quartz and muscovite
    b. Olivine and calcium plagioclase
    c. Amphibole and hornblende
    d. Biotite mica and potassium feldspar

6. Which of the following is NOT true about batholiths?

    a. Batholiths are generally concordant with the overlying country rock.
    b. Batholiths are most often granitic in composition.
    c. Batholiths occur in elongate belts paralleling continental margins.
    d. Batholiths are the largest intrusive bodies.

7. Laboratory geologists know that, of the following, one condition is most likely to lower the melting point of a rock. It is

    a. high pressure.
    b. the presence of a single mineral, such as quartz, in the rock.
    c. lack of water in the magma.
    d. abundant water under pressure.

8. The presence of inclusions or xenoliths "floating" in a rock body is considered evidence of

    a. solidification at great depths.
    b. intrusion of liquid molten magma into solid rock.
    c. rapid solidification in near-surface locations.
    d. explosive eruptions on the surface.

9. Granites have not yet been created in laboratories because

    a. high enough temperatures cannot be achieved.
    b. the pressures are greater in nature than can be duplicated in a lab.
    c. the minerals found in granites are not available in most labs.
    d. the rate of cooling needed to form granites requires too long a time span.

10. The boundary dividing felsic magmas from other magmas is the presence or absence of

    a. silicate minerals.
    b. plagioclase feldspar.
    c. quartz.
    d. ferromagnesians.

11. A study of the location of very ancient andesite-diorite belts can indicate

    a. sites of former oceanic-continental plate convergence.
    b. former mid-oceanic ridges and rises.
    c. extinct chains of shield volcanoes.
    d. collision of two continental plates.

12. Field work has shown there are relatively few batholiths of gabbroic mineral composition. This circumstance is best explained by which of the following?

    a. There are very few minerals in the mantle that contribute to the composition of a gabbro.
    b. Basaltic magmas are too viscous to rise into the upper crust.
    c. Gabbros, being denser than crustal material, are not buoyant.
    d. Basaltic magmas are very fluid and travel through fissures to erupt on the surface.

13. The heat for melting rocks comes from all of the following except

    a. geothermal gradient.
    b. solar radiation.
    c. friction from rocks sliding past rocks.
    d. radioactive decay of certain elements.

14. The importance of Bowen's reaction series is that it shows how

    a. basalt can be differentiated from gabbro under high pressure.
    b. granite develops a coarse-grained texture.
    c. partial melting and differentiation can form an andesite from a basaltic magma.
    d. granites, by partial melting, can form directly from mantle rock.

15. An igneous intrusion that resembles the filling in a layer cake is called

    a. a stock.
    b. a sill.
    c. a dike.
    d. batholith.

15

WEATHERING AND SOILS

This lesson will help you understand how minerals and rocks change when they are subjected to the physical and chemical conditions that exist at Earth's surface.

GOAL

After reading the textbook assignments, completing the exercises in this study guide, and viewing the lesson's video portion, you will be able to:

LEARNING OBJECTIVES

1. Recognize the relationship among weathering, erosion, and transportation.

2. Compare and contrast mechanical and chemical weathering.

3. Describe three processes that account for most of the mechanical weathering of rock.

4. Explain why many rocks and minerals are out of equilibrium at Earth's surface and how this relates to chemical weathering.

5. Describe how atmospheric gases and water decompose rock.

6. Compare the chemical weathering of the minerals feldspar and quartz and indicate why there is a difference.

7. Indicate how soil forms and why it is the basis for life on land.

8. Describe the characteristics of each major soil type, and explain how the parent rock, time, and slope affect the soil profile.

9. Differentiate between residual and transported soils.

10. Explain how climate affects the formation of pedalfers and pedocals.

11. Describe the origin of bauxite and laterite.

INTRODUCING THE LESSON

Weathering, the disintegration of rocks on the surface of Earth by natural means, is generally a quiet, unobtrusive process. The chemical and mechanical effects of weathering, however, have been factors in the creation of a planet unique in the solar system: a planet suitable for life. Since the first drop of water fell on the barren surface of the young rocky planet, rocks and minerals have been fragmented and altered, slowly forming soils and making life-giving nutrients available on the land and in the sea.

In this lesson you will learn how mechanical weathering breaks up the rocks of the surface into smaller particles but does not change their chemical composition. For example, the much-photographed Half Dome in Yosemite is a spectacular example of a form of mechanical weathering called *exfoliation*.

You will also learn that chemical weathering decomposes rock by exposure primarily to water, oxygen, and carbon dioxide. Most of the original minerals that formed deep in Earth's crust are changed by chemical weathering into new compounds that are stable in the new environment – the atmosphere and surface conditions of temperature and pressure.

Since the products of weathering of the rock-forming silicate minerals are mostly clays or muds and quartz sands, they are of particular interest to geologists. These are the source materials of sedimentary rocks, as well as the soils that support all life on land. Fertile soils are Earth's greatest natural resource but have been neglected, abused, and destroyed since the rise of agriculture early in human history. Scientists are working to better understand weathering and to prevent depletion and degradation of the world's soils.

LESSON ASSIGNMENT

Completing the following eight steps will help you master the lesson objectives and achieve the goal for this lesson:

Step 1: Read the TEXT ASSIGNMENT, Chapter 12, "Weathering and Soils."

Step 2: Study the KEY TERMS AND CONCEPTS as noted in the study guide.

Step 3: Watch the VIDEO using the VIEWING GUIDE in the study guide.

Step 4: Read the study guide's PUTTING IT ALL TOGETHER section, which will help you summarize and integrate all of the information in this lesson.

Step 5: Complete any assigned lab exercises.

Step 6: Complete any assigned ACTIVITIES found in the study guide.

Step 7: Review the material in this lesson and complete the SELF-TEST found in the study guide.

Step 8: Go back to the LEARNING OBJECTIVES and make sure you have learned each one.

1. Read Chapter 12, pages 267-283, being sure to include the chapter's "Introduction" and "Summary" in your reading.

2. Review Chapter 9, pages 193-213.

3. Review the description of Bowen's reaction series in Chapter 10, on pages 227-228 and Figure 10.15, page 228, to compare the order of crystallization of minerals with the order of decomposition during weathering.

3. Take special note of Table 12.1, page 276, "Chemical Equations Important to Weathering," and Table 12.2, page 278, "Weathering Products of Common Minerals."

The key terms listed under "Terms to Remember," page 283 in your text, are important and will supplement your knowledge of weathering. Use the glossary and the summary in the text to help you understand the definitions. Here are some other definitions to help you with this lesson:

Carbonic acid: The most important naturally occurring acid, especially effective in the weathering of the mineral calcite (limestone) and in the alteration of feldspars to clay minerals. (See Figure 12.14, page 274.) Carbonic acid results from the chemical combination of water and carbon dioxide in the air or in the soil. This is a weak acid, but it is widespread, abundant, and effective over very long periods of time.

Humus: Dark-colored, well-decomposed organic matter found in the upper horizons of fertile soils. It supplies nutrients to growing plants. Humus usually contains particles of decayed leaves, roots, twigs, animal remains, and abundant microorganisms.

Leaching: The removal of fine particles and soluble materials by rain water percolating downward through the upper soil horizon. Leaching may deplete the upper horizon of certain minerals which will be deposited in the lower or B horizon of the soil profile.

Once you have read the text material for the lesson and studied the terms, you are ready to view the lesson's video portion. Review the questions that appear in this section to help you watch for important points and to prepare you for what you will see in the video.

Before viewing, answer the following questions:

1. What are the three most effective processes of mechanical weathering?

2. Define chemical weathering, and state the most abundant weathering products of the principal rock-forming minerals.

After viewing, answer the following questions:

1. How does mechanical weathering affect rock? How did the granite domes in Yosemite, seen in the video, respond to mechanical weathering?

2. What are the principal agents of chemical weathering and what are some of the responses of rock to this form of weathering?

3. According to Dr. Lawford Anderson, all minerals in a rock such as granite do not weather at the same rate. How do you account for this?

4. Explain how human activities have affected rates of weathering. Give examples from the video.

5. What are the greatest beneficial effects of weathering?

6. What is meant by soil horizons, and what are the general properties of A, B, and C horizons?

7. What are the basic factors that determine what kind of soil will develop?

8. How did human activities lead to the dust bowl disaster of the American Midwest in the 1930s? What do you think should be done in terms of soil conservation and improvement in the Midwest?

9. Who does Rick Aguayo work for? Why did he visit farmer Wayne Soppeland's farm? What were some of the suggestions made by Aguayo to protect the farm?

PUTTING IT ALL TOGETHER

You have read the text assignment and seen the video portion of the lesson. This section, made up of a SUMMARY, two CASE STUDIES, the QUESTIONS, suggestions for ACTIVITIES, and a SELF-TEST, will help you pull the information together.

Summary

Weathering is a natural process on the surface of Earth that breaks rocks into smaller particles and changes them chemically into new compounds. The process is most effective on rocks that formed at depth at high temperatures and pressures in an environment lacking water, carbon dioxide, or free oxygen. During *chemical weathering*, the original minerals will react with water, acids, and atmospheric gases to form new minerals that are stable on Earth's surface. The products of weathering will include both solid materials, such as clay minerals and quartz sands, and substances in solution, such as calcium carbonate and other salts. Rusting of steel tools left in the rain is a well-known example of chemical weathering. (You might want to review the sections in Chapter 9 in the text and Lesson 12 in the study guide that refer to the rock-forming minerals. Also review Chapter 12 in the text and Lesson 13 in the study guide if you need a refresher on the minerals that make up igneous rocks.)

Mechanical (or *physical*) *weathering* breaks the solid rock into small fragments without changing the rock's chemical composition. By breaking the rock and exposing greater surface area to water and atmospheric gases, mechanical weathering speeds up the rate of chemical decomposition. Frost action or frost wedging, the growth of plant roots, abrasion, relief of pressure, and exfoliation are agents of mechanical weathering. Table 12.2 on page 278 shows the weathering products of common minerals.

The rate of weathering is related to the order of crystallization of minerals in a magma as shown in Bowen's reaction series. (Review Bowen's reaction series Figure 4.15, page 228, for an explanation.) The first minerals to crystallize at high temperatures will be the least stable on Earth's surface and the first to weather. Quartz, the last to crystallize at the lower temperatures, is the most resistant and is a widespread residual weathering product.

Soil Formation

Soils are composed of the unconsolidated products of long periods of mechanical and chemical weathering. A mature, fertile soil will usually consist of clay minerals, quartz sand, some iron oxide, calcium compounds, many microscopic living organisms, and decomposed parts of plants and animals.

Mature soils will exhibit three distinctive layers or horizons that differ from each other in chemical and physical properties, as follows:

1. At the surface is the A horizon, a dark-colored layer rich in humus and characterized by the downward movement of water. It is known as the *zone of leaching*.

2. Some of the more soluble substances or finer particles of the A horizon will be carried down and deposited in the B horizon, known as the *zone of accumulation*. This layer tends to be enriched with both clay and iron oxides, and in some soils, particularly those in arid and semi-arid climates, there will be a build-up of calcite.

3. The C horizon consists of only slightly-weathered parent rock and is transitional between unweathered bed rock and the developing soil above.

Soils and the character of the soil horizons vary greatly from place to place and depend on the parent rock, the climate, the moisture available, the vegetation, and the slope of the land. In the same climate, no matter whether the original rock is igneous, metamorphic or sedimentary, the soils will eventually become quite similar as all the less stable minerals weather away.

CASE STUDY I:
Limestone

Limestone is composed primarily of the mineral calcite (calcium carbonate), which is soluble in the presence of carbonic acid. We will see in Lesson 21 on ground water that carbonic acid in ground water percolating through limestones will dissolve out caves and caverns and form sinkholes that can swallow up whole buildings.

Many of the great cathedrals of Europe were built of limestone, which was not only plentiful but sufficiently soft to shape into intricate stone carvings and adornments. Today – because of modern urban activities – emissions of carbon dioxide and oxides of nitrogen and sulfur dioxide from the burning of coal, oil and other fossil fuels surround these buildings. In the moist climate of central Europe, rain reacts with the carbon dioxide in the air to form carbonic acid. As was mentioned above, the limestone is particularly susceptible to solution by carbonic acid. The oxides of nitrogen and sulfur also react with water to form acidic rain that is damaging to the structures. As a result, many of the facings and fine stonework on the buildings are literally dissolving or crumbling away.

On the other hand, the Pyramids were constructed of limestone about 3,000 years before the great European cathedrals but appear fresh and strong and show little evidence of weathering. The difference, of course, lies in the climate and the arid conditions of Egypt. But conditions are changing in this ancient country. Since the construction of the Aswan Dam, additional irrigation along the Nile has been increasing the moisture in the air. This may prove to be disastrous to magnificent antiquities that have endured for thousands of years.

CASE STUDY II:
Laterite Soils

Laterites – the deep red soils of the Tropics – are characteristic of old landscapes in regions where the temperatures are high and the rainfall is abundant and seasonal as well. In these soils, we see the end product of extreme weathering. All the soluble elements such as calcium, sodium, and potassium have been leached out. Even such resistant substances as silica have been removed. All that is left are the highly insoluble iron oxides or aluminum oxides and some clays. If the parent rock had a high content of aluminum, then bauxite – the principal ore of aluminum – will form. Bauxite is mined today in Arkansas, Jamaica, and certain South American and Caribbean lands. Iron oxide is mined in Cuba and the Philippines.

Tropical soils such as laterite support lush jungle vegetation long adapted to poor soils, but are not fertile for crop plants. The humus layer is thin, and the organic materials decompose quickly in the warm moist environment. If the trees and humus are cleared, laterite cannot be used for agriculture for more than a few years. After that, its meager supply of nutrients is exhausted; it then becomes barren and has to be abandoned. Great sections of the Amazon rain forest have already been cleared for agriculture. This results in the depletion of vegetation which absorbs carbon dioxide, creating a major environmental disaster that some scientists believe may be of more consequence than the ozone hole in the atmosphere. In addition, destruction of the rain forest is endangering many species of birds, plants, and animals – leading to possible extinction of many of them.

When laterite is quarried and exposed to the sun and rain, it hardens and is often used for building material. In fact, the name "laterite" is from the Latin word *latere*, or brick. As a construction material, laterite has served to build enduring monuments such as the ancient city of Angkor Wat in Cambodia. The building blocks of laterite are highly resistant to weathering because they are composed of residual iron oxide; all the other materials that can weather have already been removed.

Weathering is a natural process, but the rates can be drastically altered by human intervention and usually not for the better, as we have seen in these two case studies. But we have to remember that weathering also produces the life-sustaining soils of Earth.

Questions

1. **What is the difference between weathering and erosion? How does weathering increase the rate of erosion?**

   Weathering is the natural disintegration and decomposition of rocks on the surface of Earth. Erosion is the removal of the weathered products. If there is no erosion in a specific region, the products of weathering will accumulate and form residual soils. Weathering increases the rate of erosion by producing small fragments that can be easily carried away by running water, wind, or glaciers.

2. **What is the major difference between mechanical and chemical weathering? How does mechanical weathering speed up the effects of chemical weathering?**

   During mechanical weathering, the rock may be fragmented, but no chemical changes in the minerals will occur. During chemical weathering, new minerals will form that are stable under Earth's surface conditions. Mechanical weathering, by breaking the rock into small pieces, exposes more surface to the effects of water and atmospheric gases.

3. **What are the properties of water that make it effective in the mechanical weathering of rock?**

   Water has the interesting property of expanding as it freezes. If water enters cracks in bed rock, it can freeze during the night and expand, wedging open the cracks. This action repeated many times can cause blocks to loosen or can fracture the rock (see Figures 12.7 and 12.8, pages 271-272).

4. **Explain the steps that change a granite block to a rounded form.**

The rounded form of granite blocks results from "spheroidal weathering" and occurs when chemical weathering attacks the corners and edges faster than the smooth sides of a block. The block will eventually form a rounded shape. See Figure 12.2 and 12.3, pages 269-270.

5. **Why does the Grand Canyon exhibit a series of cliffs and slopes?**

The distinctive profile of the Grand Canyon is an example of "differential weathering." The steep cliffs are mostly limestone and sandstone, which are resistant to weathering in an arid climate. The slopes are usually shale, a softer and less resistant formation (see Figures 12.5 and 12.6, pages 270-271).

6. **What weathering process is so beautifully illustrated in the photograph of the exfoliation dome, Figure 12.12, page 273? Why does granite in particular weather in this manner?**

Half Dome in Yosemite illustrates exfoliation, in which great curved slabs of rock spall off (break loose) from the underlying rock. This condition is believed to result from the relief of pressure due to tectonic uplift, erosion, removal of the overlying rock, and the subsequent expansion of the granite.

7. **Why are clay minerals and quartz sand some of the most widespread surficial deposits on earth? Trace their origins back to the parent rock.**

Clay minerals are the most widespread surficial deposits because of the abundance of the rock-forming minerals and the products produced by their chemical weathering. Under the influence of water and carbon dioxide, the feldspars, ferromagnesians, and micas all decompose, and the main solid product of the reactions is the clay minerals, hydrous aluminum silicates. Quartz grains break out of granite during weathering of the rock and rarely decompose because quartz is extremely stable in the surface environment. The grains will accumulate without further alteration.

8. **What is the origin of the terms pedalfer and pedocal? In what areas of the United States do pedalfer and pedocal soils occur? What are the major factors contributing to their different soil profiles?**

Pedalfers are soils formed in humid regions; they have a high content of aluminum (Al) and iron oxides (Fe). Pedocals are soils of arid or dry climates, and they contain calcium salts (Ca). In the

United States, soils in the higher rainfall regions east of the Mississippi River are pedalfers. The soils of the western states are usually pedocals because of low rainfall. The difference in their soil horizons is related to the amount of rain that leaches soluble minerals from the A horizon in the pedalfers and the amount of evaporation that draws up water and deposits calcium salts in the B horizon of the pedocals.

9. **What are some of the sources of carbonic acid? Why is it more effective as a weathering agent in cities than in rural areas?**

   Carbonic acid results from the chemical combination of carbon dioxide and water. Sources of carbon dioxide include a small amount in the atmosphere (about 0.03 percent); fumes from the burning of coal, petroleum and wood, combustion of gasoline in automobiles, factory and manufacturing gases, and natural decay of organic materials in the upper layers of soil. Obviously, in cities there are many more sources of carbon dioxide than in rural areas so, as a result, there is more opportunity for carbonic acid to form.

10. **Discuss some of the uses of and problems with laterite soils. If you were to inherit a jungle property in the Tropics, what would be some of your options in using the land to best advantage?**

    Laterite soils, in spite of their lack of nutrients, support a lush plant growth that is the home of many species of plants and animals that are rare or endangered and may include some that have not yet been described. Many of these plants are slow-growing and well-adapted to the nutrient-deficient soils in which they grow. It is believed that some of the plants may provide important medicines and other substances of great benefit to humans, but it will take time to study them. The problem with using laterites for crop plants is that they have already been leached of nutrients, and the crop plants will remove even more. They do not contain adequate clays or sand and when dried and sun-baked become hard and worthless for agriculture. The best use for the tropical soils at present is to leave them alone or to form the soil into bricks for inexpensive housing.

Activities

1. Notice in the chemical equations in Table 12.1, page 276, that bicarbonate HCO_3 is a product in each of these reactions. What common household item is used to absorb odors in refrigerators, to clean drains, act as an efficient toothpaste, and settle an upset stomach? Hint: there is picture of an "arm and hammer" on the yellow box. Read the contents on the label. Drop a spoonful of this wonder chemical into a glass of water and stir. Taste the solution. Is this material a clay mineral? What property will answer the question?

2. Examine the diagram in Figure 12.20, page 280. Do you live in a pedalfer or pedocal soil region? What factor determines the difference? Pick up a handful of soil in your back yard and examine it with a magnifying lens. Can you identify quartz grains, or perhaps humus? Why will it be more difficult to identify clay minerals?

   There are actually many kinds of soil, the descriptions of which are far beyond the scope of this course. Soils may change almost from one backyard to the next as a result of grading and transporting, watering, fertilizing, use of soil amendments, and the influence of microclimates.

3. Watch for weathering on old buildings, tombstones, and of course on granite and other rock exposures. You can watch weathering take place as the paint on your car or house slowly dulls and exposed sections of metal turn to rust.

Self-Test

1. Mechanical weathering results in

   a. formation of new stable minerals.
   b. removal of fragments of rock.
   c. disintegration of rock without creating new minerals.
   d. decomposition of silicates.

2. The property of water that is important in mechanical weathering is its ability to

   a. dissolve many substances.
   b. facilitate chemical changes.
   c. expand as it is heated.
   d. expand as it freezes.

3. The unusual surface of Half Dome in Yosemite is caused by

   a. decomposition of the orthoclase feldspar.
   b. relief of pressure from removal of overlying rocks.
   c. plant roots.
   d. solution by carbonic acid.

4. Mechanical weathering increases the rate of chemical weathering by

   a. eroding the soft calcite grains out of the granite.
   b. abrading away the upper layers of rock exposures.
   c. breaking rocks into smaller fragments and exposing more surface area.
   d. leaching out certain minerals that hold the rock together.

5. The physical removal of rock by glaciers, wind, or streams is called

   a. exfoliation.
   b. abrasion.
   c. weathering.
   d. erosion.

6. The abundance of clay minerals on the surface of Earth is most closely related to the

   a. abundance of oxygen in the atmosphere.
   b. abundance of the silicate rock-forming minerals.
   c. widespread effects of mechanical weathering.
   d. widespread limestone formations.

7. Of all the gases in the atmosphere, which of the following is the least important agent of chemical weathering?

   a. Oxygen
   b. Nitrogen
   c. Carbon dioxide
   d. Water vapor

8. Which of the following does NOT weather into a new mineral?

   a. Quartz
   b. Potassium feldspar
   c. Mica
   d. Pyroxene

9. The minerals in granite that solidify at the highest temperatures are

   a. immune to chemical weathering.
   b. the most stable on Earth's surface.
   c. more durable than the other minerals in granite.
   d. the most susceptible to decomposition.

10. Which of the following is NOT a product of weathering?

    a. Bauxite
    b. Clay minerals
    c. Feldspar
    d. Bicarbonates in solution

11. Buildings in modern urban areas are more subject to weathering than buildings in rural areas because of

    a. more extreme temperature changes.
    b. fewer plants to take up acid rain.
    c. more sources of carbon dioxide in the air.
    d. greater use of metals that rust.

12. If a soil consisting of quartz sands rests primarily upon limestone bed rock, you might guess it is a

    a. residual soil.
    b. transported soil.
    c. laterite soil.
    d. hardpan.

13. The zone of leaching in soils is characteristic of

    a. slightly weathered bed rock.
    b. pedocals.
    c. A horizons in mature soils.
    d. C horizons in youthful soils.

14. The distinction between pedalfers and pedocals is based on the

    a. parent bed rock.
    b. agent of transportation.
    c. angle of slope of the land.
    d. amount of rainfall.

15. The common mineral that leaves behind no solid product after weathering is

    a. mica.
    b. quartz.
    c. calcite.
    d. a ferromagnesian mineral.

16

MASS WASTING

The purpose of this lesson is to help you understand the causes, types, and effects of mass wasting.

After reading the textbook assignments, completing the exercises in this study guide and the lab, and viewing the lesson's video portion, you will be able to:

1. List controlling factors or causes of mass wasting and describe how each affects slope instability.

2. Discuss the various types of mass wasting in terms of type and rate of movement.

3. Recognize features of the landscape that indicate slope instability or previous episodes of mass wasting.

4. Compare the various techniques for controlling mass wasting, especially in populated areas, explaining why some work and some fail.

5. Explain how humans can act to reduce or accelerate mass wasting.

The mass downhill movement of rock and soil is not as dramatic as Kilauea lighting up the sky or earthquakes shaking whole cities off their foundations. But in terms of human experience, some of the most devastating events in history resulted from great landslides where whole towns were almost instantly destroyed and thousands of inhabitants were killed. These episodes are part of an endless cycle in which tectonic forces raise the land and erosion wears it away.

This process of moving rock and soil downhill is referred to as *mass wasting* because during transport the fragments tend to stay together and move as a unit in bulk. The moving material varies widely and includes everything from great slabs of solid rock to loose unconsolidated rock debris or soil.

Human intervention both causes mass wasting and mitigates its effects. As urban development and road building are expanded into new and less stable sites, conditions are created which may seriously affect the well-being of people and property.

In this lesson, you will learn that the ultimate driving force for all types of mass wasting – from rapidly moving landslides to an almost imperceptible but continuous motion called creep – is *gravity*. But you will also see that the angle of slope and thickness of the mass of debris play important roles in downhill movement. Finally, you will learn about the role humans play in both causing and controlling mass wasting.

LESSON ASSIGNMENT

Completing the following eight steps will help you master the lesson objectives and achieve the goal for this lesson:

Step 1: Read the TEXT ASSIGNMENT, Chapter 13, "Mass Wasting."

Step 2: Study the KEY TERMS AND CONCEPTS as identified in the study guide.

Step 3: Watch the VIDEO using the VIEWING GUIDE in the study guide.

Step 4: Read the study guide's PUTTING IT ALL TOGETHER section.

Step 5: Complete any assigned lab exercises.

Step 6: Complete any assigned ACTIVITIES listed in the study guide.

Step 7: Review the material for the lesson and complete the SELF-TEST found in the study guide.

Step 8: Go back to the LESSON OBJECTIVES and make sure you have learned each one.

TEXT ASSIGNMENT

1. Read Chapter 13, pages 285-301.

2. Make sure you read the chapter's "Introduction" and "Summary." And note the spectacular photograph facing page 285 in your text which shows a landslide in Hong Kong.

3. Review "Avoiding Geologic Hazards" in Chapter 1, page 11, noting Figure 1.7 showing Nevado del Ruiz before and after the disastrous mudflow. Compare with the disaster at Yungay in Peru.

4. Carefully study the figures and tables in the Chapter.

5. In Chapter 16, study Figure 16.40 on page 371 showing the extension and widening of a gully by undercutting and slumping of the

banks. This is one of the most important effects of mass wasting in nature.

6. Review the eruption of Mount St. Helens in Chapter 11, Box 11.1, pages 240-243.

The key terms listed under "Terms to Remember," page 301 in the text, will aid your understanding of the lesson's material. Make sure you look up the definitions of these terms in the glossary, the chapter, or the study guide: *bedding plane, bed rock, colluvium, lahar,* and *seacliff*

KEY TERMS AND CONCEPTS

Once you have read the text material for the lesson and studied the terms, you are ready to view the lesson's video portion. Review the questions that appear in this section to help you watch for important points and to prepare you for what you will see in the video.

VIEWING GUIDE

Before viewing, answer the following three questions:

1. Recall from your text the contributing factors in mass wasting.

2. Review the conditions that caused the great slide at Yungay. (See Box 13.1, pages 286-287 in your text.) What triggered this landslide?

3. What are the three types of mass wasting? Which is the slowest, and which is the fastest?

After viewing, answer the following questions:

1. Describe *creep*. Tell where it is most likely to occur and why it causes economic damage.

2. What natural conditions caused the slump at Point Fermin?

3. List the natural factors mentioned by the narrator that are causing the instability in the Big Rock area.

4. What were some of the indications of previous landsliding that the geologists and engineers were looking for in the prospective building site?

5. Summary: What did you see in the video that showed you

   a. creep?
   b. slump?
   c. landsliding?
   d. mud and debris flow?
   e. rock slides and rock avalanches?
   f. mass wasting caused by human activities?

PUTTING IT ALL TOGETHER

Now that you have read the text assignment and seen the video portion, here are the MAIN CONCEPTS, a SUMMARY, a CASE STUDY, the QUESTIONS, suggestions for ACTIVITIES, and a SELF-TEST to help you integrate your learning of the material in this lesson.

Main Concepts

Mass wasting. A natural phenomenon that occurs wherever tectonic movements have raised the land surface.

Gravity. A force that never gives up, gravity acts to bring the previously raised materials to a lower, stable level. Mass wasting is an important part of the erosional cycle and actually works with streams in valley erosion. The streams undercut their banks and the edges "cave in" or slump; the debris is carried off by the stream to be deposited downstream in a delta or an ocean.

Landslide. A general term that refers to the descent of rock and rock debris or soil – either slowly or very rapidly. No two slides are exactly alike. Mass wasting is a continuum of processes between creep and rock fall, between movement of dry particles and of mud that is almost like stream flow, and between the movement of solid rock alone and of unconsolidated debris alone.

Summary

This lesson focused on mass wasting – its definition, appearance, and consequences. Specifically, you learned about the classification of mass wasting and the factors that contribute to it. The two lists below summarize much of the material you have learned. By this time in the lesson, you should feel comfortable discussing each point in the following lists in more detail.

The classification of mass wasting – as you have learned – is based on three things:

1. The *rate of movement*, which can be hundreds of miles per hour or an imperceptible but continuous creep;

2. The *type of material involved*. This varies from bed rock seen in rock slides or rock falls (such as those you saw at the base of the cliff at Point Fermin) to loose, unconsolidated debris or soils (such as that you saw underlying the hummocky topography of Portuguese Bend); and

3. The *type of movement*, which may be a:

   a. *flow* as in earthflow or mudflow that you saw on the video at Wrightwood,
   b. *slip* in which a coherent mass of material moves downslope on a well defined surface, as noted in the slump areas of Point Fermin and Portuguese Bend, or
   c. *rockfall* in which slabs of bed rock are in free fall from a cliff.

In order to understand why there are so many varieties of landslides, you should know the basic factors that contribute to mass wasting:

1. *Water* – as you have seen at Big Rock – is a critical factor. Water makes saturated debris heavy, and forces the grains of soil and rock particles apart so they readily glide one over the other. Damp debris may not move much at all or it may contain so much infiltrated water that it becomes a fast moving mudflow.

2. *Steep slopes* – as seen along the Pacific Coast Highway – are always susceptible to mass movement.

3. *Thick unconsolidated weathered debris* or soil mantle tends to be unstable.

4. *Dip slopes* – in which the bedding planes of the rocky layers are parallel to the hillside slopes – tend to slide with the addition of moisture. Part of the problem at Point Fermin was created by the sedimentary shales that dip toward the sea and slump on slippery layers of clay.

5. The importance of *climatic factors* is seen in solifluction moving over permafrost layers in the Arctic. You have learned that even in arid regions, heavy rains following long dry periods can move great masses of water-saturated debris as a mudflow. Mudflows can wrench homes from their foundations and move objects as large as bridges that might be in their path.

6. Earthquakes can dislodge rocks and soil – especially in steep road cuts in mountains and on steep mountain slopes such as those in Yungay.

7. *Vegetation* can anchor soils and loose particles in humid regions, especially if the roots are deep. Forest fires, however, can burn off natural vegetation and make an area that was previously stable subject to mass wasting. Fire prevention in hillside urban locations sometimes requires removal of surrounding grasses and shrubs; this loss of vegetation can lead to increased instability.

CASE STUDY: Big Rock Re-visited

A re-examination of the causes of landsliding, particularly in the areas we have studied, will help you get a feeling for the processes and problems involved in mass wasting. Using Big Rock as a case study, we see that this area is interesting not only geologically, but also because of its human habitation and costly property values on this part of the Malibu coast.

The causes of the mass movements here include the steep angle of the slopes above the Pacific Coast Highway. These bluffs were originally natural seacliffs that were being undercut by waves and retreating landward behind a narrow beach. Construction of the highway required the bulldozing and cutting back of the bluffs to make room for the road. The slopes are now very steep and retaining walls and wire screens have been erected in places to keep the loose rocks from falling on the road and disrupting the flow of traffic. Another factor is the presence of the Malibu Coast fault which runs parallel to the coast in this location and is near the base of the cliffs. The bed rock adjacent to the fault is severely fractured, distorted, and displaced. It also contributes to the slope instability. The broken rock can also store more ground water than solid bed rock – an important consideration at this site.

At Big Rock, the problem became more acute as construction of homes on the mesa above the bluffs added weight to the already weakened rock. Sprinklers, irrigation, and private sewage disposal systems such as cesspools added water which – as you have seen – also adds weight in addition to "lubricating" the rock particles.

A rise in ground water level noted as early as 1961 coincided with residential development and an increase in septic-tank effluent. The rains of 1979 and 1980 probably started a process that produced the well-defined massive movement in 1983. Heavy rain was most likely the trigger for this slide. The material that fell on the highway consisted of an assortment of large solid rock and boulders combined with finer rock debris, soils, and muds. In the residential section above, 250 homes collapsed, cracked, or slipped off their foundations. The movement was rapid and unexpected, and would probably be classified someplace between a slide and a slump.

Most geologists feel that the best method for control in Big Rock is to de-water the ground. There are vertical wells drilled into the upper slump area to depths of 150-800 feet which pump out the water. Horizontal drains have also been installed that reach as far as 1,500 feet into the hillside. The question now is whether present methods of control can stabilize the slide area, especially during seasons of abundant rainfall. Sewers have not been installed. And legal battles involving millions of dollars are expected to go on for years.

Questions

1. **Various classification systems for mass wasting are used by geologists, engineers, and others. What three factors are used as the basis of the system of classification used by the textbook authors?**

   Classification of mass wasting is based on rate of movement, type of material and type of movement.

2. **What are the three main types of movement used in this classification system? Which is the slowest and which is the fastest?**

   The three types of movement are basically flow, slip, and fall. A type of flow called creep – which may move at less than one centimeter per year – is considered the slowest while rockfall or debris fall is the fastest. Rock avalanches and debris avalanches are also rapid.

3. **Describe the various effects of water on unstable slopes. Why does water seem to make dry sand more cohesive?**

The presence of water – especially in unconsolidated particles – forces the grains apart and allows them to move freely over one another. Water also adds weight to a deposit and increases the rate of movement. But a small amount of water will form a thin film around the grains which will then adhere together by surface tension.

4. **Where do mass movements most commonly occur in nature? If there were no highways or residences in coastal zones, would any form of mass wasting occur?**

Mass wasting is important in stream erosion, in steep glacial valleys, and along mountainous coasts where waves undercut the cliffs causing retreat of the land. This process takes place even in the absence of homes and other human structures in the coastal zone.

5. **What were the four or five factors that led to the destructive landslide at Yungay? Could this landslide have been prevented by proper human intervention? Explain.**

The factors that contributed to the slide at Yungay were its location at the base of a still-growing mountain range with steep slopes, the presence of loose rock and debris on the slopes, the high relief between the summits and the base of the slopes where the town was located, the frequent earthquakes, and the peaks covered with ice and snow. The town was built on the site of a previous landslide. In this case, there are no methods that could control or divert the slide. The only thing to do is not to build in this site again.

6. **How have human activities contributed to certain landslides? If there were no people living at Point Fermin or Portuguese Bend, would there still be landslides there? Explain.**

Human intervention – as at Big Rock – consisted of cutting back the already steep bluffs and increasing the angle of slope; another factor is the amount of water added through sprinklers, irrigation, and sewage disposal systems. The weight of buildings on the upper mesa also contributes to the instability. Point Fermin and Portuguese Bend are natural slides because of the erosion by waves at the base of the slides, the shale layers that dip toward the sea, and the presence of a slippery clay layer within the shales. These are ancient slides that moved long before the presence of humans.

7. **What would you look for as evidence of previous episodes of mass movement in an area? Describe "hummocky topography." Why is it significant?**

Evidence of previous slides might be tilted telephone poles or fence posts where the surface has undergone slow creep. There might be large piles of talus at the base of cliffs, or a land cover of unsorted boulders and smaller particles. Great lobes of vegetated debris would be an indication of earthflow; steep scarps at the head of a flow would indicate slump. In the Arctic, whole forests that have slid downslope or a wrinkled land surface are indications of solifluction. A channel of mud and debris would tell of mudflows. "Hummocky topography" is the name given to an irregular land surface consisting of shallow undrained depressions and low hills. It is an indicator of past landslides. (See the text photograph of Yungay after the slide on page 287.)

8. **Under certain circumstances climate may increase slope instability. For example, what is solifluction and where does it occur? At what time of year do you think solifluction becomes most active?**

Solifluction is the flow of water-saturated debris over an impermeable layer which can be solid bed rock or, most frequently, permafrost. Solifluction is characteristic of high latitudes and, sometimes, high altitudes. Movement of the land surface is most active in summer when the layers above the frozen permafrost melt, become water-saturated, and move even on relatively gentle slopes.

9. **What type of mass wasting is most likely to occur in arid regions? What are the contributing factors and what can be some of the effects?**

Mudflows are most likely to occur in arid regions where the cover of vegetation is thin and where rocky debris accumulates on the slopes. Sudden thunderstorms can saturate the loose debris, and mudflows can develop quickly. Forest fires and the attendant destruction of vegetation can increase the possibility of mudflows.

10. **Describe methods of control of mass movement of soils and rockfalls. What methods of control are being used at Big Rock?**

Mass movement of soil may be controlled by construction of retaining walls, by putting in drains through the walls, by reducing the angle of slope through construction of terraces, and by reseeding with grasses or other plants whose roots anchor the slope. Builders frequently will sink pilings through unstable debris down to firm bed rock. Rockfalls are sometimes prevented by removal of sliding rock layers or by inserting bolts through the unstable rocks. The best solution is not to build or cut roads through sites susceptible to rockslides. At Big Rock, where water is increasing the possibility of slides, vertical wells have been drilled into the water table to pump out the water and horizontal drains have also been inserted into the cliff faces.

1. Episodes of mass wasting are rare within urban areas or on geographically flat lands. But if you live in a rural district or on a hillside, you may see many evidences of mass movement. Take a walk after a rain and watch for small earthflows or slumps. Walk along a river and note how the river widens its channel by undercutting its banks.

2. Mountainous areas are sometimes a laboratory for observing mass wasting and the efforts made to control the effects of mass wasting. Watch for oversteepened slopes resulting from bulldozing or blasting – with piles of talus at their bases. You will frequently see terracing of slopes to reduce the steep angles, retaining walls, and homes built on pilings. Also watch for planting of grasses or other vegetation to stabilize the slopes.

3. Suppose you are planning to purchase a lot and build your dream house. What would you look for in a building site? What if the developer admits there has been sliding in the past but says it can be controlled easily? What would be your response, based on what you have learned in this lesson?

1. If, on a rather gently sloping site, there were tilted fence posts and telephone poles and broken stone walls, you might infer that the

   a. area was a lahar.
   b. site was on a rockfall.
   c. surface was the site of a prehistoric slump.
   d. surface was undergoing creep.

2. Any unconsolidated material on Earth's surface – such as soil and rock fragments – is described as

   a. bed rock.
   b. debris.
   c. scarp.
   d. solifluction.

3. The force that drives mass wasting is

   a. freeze and thaw.
   b. angle of slope.
   c. gravity.
   d. surface runoff.

4. Of the following, the type of mass wasting that involves the *least* amount of soil water would be

   a. creep.
   b. mudflow.
   c. slump.
   d. solifluction.

5. A small amount of water in soil may

   a. increase the rate of downslope movement.
   b. add to the slope instability by lubricating the particles.
   c. force apart the grains allowing the mixture to flow easily.
   d. allow the grains to adhere together by surface tension.

6. Talus slopes are usually the result of

   a. earthflow.
   b. slump.
   c. rockfalls or rock slides.
   d. mudflows on young glaciated volcanoes.

7. The term "hummocky topography" refers to a(n)

   a. irregular land surface of undrained valleys and low hills.
   b. fan-shaped deposit at the base of a desert mountain.
   c. moving mass of soil above a frozen layer.
   d. series of oversteepened cliffs.

8. The term "mass wasting" implies a

   a. fast moving flow of mud.
   b. movement of rock particles in bulk or as a unit.
   c. slow wasting away of a hillside.
   d. mass of broken rock at the base of a cliff.

9. Coastal landslides are common because

   a. coasts are always along faults.
   b. coastal rocks are usually soft slippery clays.
   c. waves undercut the seacliffs.
   d. rock layers always dip landward.

10. Of the following, the most stable condition in terms of mass wasting is

    a. an area of high relief.
    b. where the debris is thickest over bed rock.
    c. where there is seasonal freezing and thawing.
    d. where bedding planes are at right angles to hillside slopes.

11. The form of mass wasting known as slump can be recognized

    a. as a flow of mud in a defined channel.
    b. by slow movement on a gentle slope.
    c. by a scarp at the head of a slide.
    d. by a pile of boulders at the base of a cliff.

12. The method of control considered the most effective at Big Rock is

    a. removal of ground water by wells and drains.
    b. removal of all the homes that are adding weight to the slope.
    c. terracing of the bluff above the highway.
    d. planting of grasses on the steep slope.

13. Lahars on the slopes of extinct volcanoes indicate

    a. quiet Hawaiian-type eruptions.
    b. eruptions under the sea.
    c. explosive eruptions with flows of mud and volcanic debris.
    d. extensive lava flows with small amounts of ash and cinders.

14. Coastal slides such as those at Portuguese Bend and Point Fermin are most damaging after

    a. roads through the area are removed.
    b. storm waves undercut the seacliffs at the shore.
    c. the clay layers are exposed to the drying effects of air.
    d. the slopes are weakened by drilling of water wells.

15. The rise in the ground water table at Big Rock has been correlated with

    a. water wells lifting the water to the surface.
    b. the last big rainstorm of 1983.
    c. installation of individual sewage disposal systems.
    d. the installation of a sewer system in 1961.

17

SEDIMENTARY ROCKS: THE KEY TO PAST ENVIRONMENTS

This lesson will help you appreciate the importance of sedimentary rocks in understanding Earth's history.

After reading the textbook assignments, completing the exercises in this study guide and the lab, and viewing the lesson's video portion, you will be able to:

1. Explain the influence of weathering on the composition of sedimentary rocks.

2. Describe how clastic sediment particles are classified by grain size.

3. Indicate why rounding, sorting, and deposition occur.

4. Explain how loose sediment becomes sedimentary rock.

5. Differentiate among clastic, chemical, and organic sedimentary rocks.

6. Recognize the origins and characteristics of conglomerate, sandstone, shale, limestone, chert, and evaporites.

7. Given the following list of depositional environments, predict the types of sedimentary rock that might be formed: glacial, river, delta, lake, beach, dune, reef, shallow marine, and deep marine.

8. Give examples of sedimentary rocks that are considered valuable and the ways in which they are used.

9. Describe the sedimentary structures found within sedimentary rocks, and explain how those structures are formed.

10. Describe how sedimentary rocks are used to explain Earth's history.

11. Relate the origin of petroleum, natural gas, coal, and other resources to sedimentary rocks.

INTRODUCING THE LESSON

Nature has been compared to a relentless housekeeper, ceaselessly sweeping weathered rock debris off the unweathered bed rock below, carrying off the sweepings, and depositing them as *sediment* in a lake, river bed, or eventually the sea. Sediment is the dust on the window sill, the sand on the beach, pebbles in a river, clam shells on the sea floor – survivors of the rigors of weathering and battering by glaciers, streams, or wind, on their way downhill to a resting place. The "sedimental journey" doesn't end in the basin of deposition, however; for it is there that the soft, unconsolidated particles begin to transform, by natural processes, into hard, durable, sedimentary rocks.

In this lesson, you will learn to read the language of sedimentary rocks, which form a kind of geologic "Rosetta Stone." From an examination of their particles, you will be able to understand the source materials of sediments and the weathering that produced the shales, sandstones, and limestones that make up over 99 percent of all sedimentary rocks. Each mode of transportation – the streams, glaciers, and wind – leaves its distinctive imprint in the sediments. The environment in which the materials are finally deposited – be it sea floor, beach, continental shelf, lagoon, lake, river bed, glacial moraine, or a dry lake bed on the floor of a desert – also has an impact on the sediments. The fossils in the rocks tell us the time of deposition of the sediments. Movements of Earth's restless crust, which ultimately control all phases of the sedimentary process through uplift and erosion, allow us to glimpse deeply buried sedimentary rock, as in the Grand Canyon, and to discern the life and history of Earth in the past.

You will explore the colorful Grand Canyon in this lesson, a truly awesome display of sedimentary rock exposed in a mile-deep gorge. When viewing this tremendous accumulation, you may find it difficult to realize that sedimentary rocks make up only about 5 percent of Earth's crust. But they cover about 75 percent of the surface of the land because they are relatively thin and form a widespread blanket over igneous and metamorphic rocks beneath the surface. Most of the sedimentary rocks in the Grand Canyon and elsewhere on the continents, including some of the oldest rocks ever found, were deposited in a marine environment. They provide evidence of the vast and yet changing extent of ancient oceans resulting from plate movements.

As you move through this lesson, you will become aware of how the geologist interprets the past history of an area from the kinds of rocks exposed and from the structures, such as fossils, ripple marks, ancient rain drops, mud cracks, and cross-bedding preserved within them. Learning to identify the various kinds of sedimentary rocks and to realize their significance will greatly enhance your enjoyment of the spectacular scenery provided by these layered deposits.

Completing the following eight steps will help you master the lesson objectives and achieve the goal for this lesson:

Step 1: Read the TEXT ASSIGNMENT, Chapter 14, "Sediments and Sedimentary Rocks."

Step 2: Study the KEY TERMS AND CONCEPTS as noted in the study guide.

Step 3: Watch the VIDEO using the VIEWING GUIDE in the study guide.

Step 4: Read the study guide's PUTTING IT ALL TOGETHER section, which will help you summarize and integrate all of the information in this lesson.

Step 5: Complete any assigned lab exercises. Even if you have not been assigned to do the lab for this lesson, if you have the AGI lab manual used in this course, look at diagrams and photographs in the manual for Exercise Three. They are excellent and will help you learn about sedimentary rocks.

Step 6: Complete any assigned ACTIVITIES found in the study guide.

Step 7: Review the material in this lesson and complete the SELF-TEST found in the study guide.

Step 8: Go back to the LEARNING OBJECTIVES and make sure you have learned each one.

1. Read Chapter 14, pages 303-326. Note the position of *sediment*, *lithification*, and *sedimentary rock* in relation to igneous and metamorphic rocks in the rock cycle diagram, page 303.

2. The diagrams and photographs are important for acquiring a visual understanding of sedimentary rocks. The photograph taken in the Grand Canyon (facing the introduction to the chapter) really is worth "a thousand words."

3. Make sure you re-read the "Summary" in Chapter 14.

4. The "Questions for Review" and "Questions for Thought" are not difficult and are worth trying as a review of the lesson.

Read "Terms to Remember," page 325 in your text, and use the glossary for any words you don't understand. Many of the terms will be familiar from previous lessons. It is important to know these words before viewing the video. One other term to note for this lesson:

Rosetta Stone: A piece of black basalt found in 1799 near the mouth of the Nile River, bearing a bilingual inscription (in ancient Egyptian hieroglyphics and in Greek). The Stone, which is on display in the British Museum in London, is famous for providing the key to deciphering hieroglyphics. The term *Rosetta Stone* is commonly used for any device that permits deciphering or interpreting material of difficult or unknown content. In the case of sedimentary rocks, the rocks are a *Rosetta Stone* to understanding the past.

VIEWING GUIDE

Once you have read the text material for the lesson and studied the terms, you are ready to view the lesson's video portion. Review the questions that appear in this section to help you watch for important points and to prepare you for what you will see in the video.

Before viewing, answer the following questions:

1. What is a *sediment*? List the depositional environments in which sediments are likely to accumulate.

2. How do loose sediments become sedimentary rock?

3. What are the two most abundant clastic sedimentary rocks?

4. What sedimentary rock is composed of calcite?

5. What sedimentary rock is composed of silica?

6. What sedimentary rock is composed of plant material?

7. What sedimentary rock is the most frequently used seasoning for food?

After viewing, answer the following questions:

1. How are the environments of deposition recorded in the characteristics of sedimentary rocks?

2. According to Dr. Cathy Busby-Spera, what do the sedimentary structures indicate about the history of the ridge basin?

3. How does the geologist use the principle of uniformity to interpret the history of sedimentary layers?

4. Review how sediments and sedimentary rocks are formed.

5. What are some of the economic uses of sedimentary rocks?

6. Briefly describe the story of the Grand Canyon, starting with the oldest sediments. How many distinct periods of uplift and erosion are recorded in the rocks of the Grand Canyon? What is the evidence for these periods?

You have read the text assignment and seen the video portion of the lesson. This section, made up of a SUMMARY, a CASE STUDY, the QUESTIONS, suggestions for ACTIVITIES, and a SELF-TEST, will help you pull the information together.

Summary

Sedimentary rocks are *secondary rocks*, derived from previously existing igneous, metamorphic, and even older sedimentary rocks. Most sedimentary rocks are made of the products of weathering of rocks and minerals and therefore consist either of bits and pieces of other rocks or of *clays* (that will become *shales*), quartz sands (*sandstone*), or calcium carbonate (*limestone*). (Review the end products of weathering described in Lesson 15.)

Sedimentary rocks are the key to geologic time and place. The rocks contain fossils (locally of startling abundance) that are used to determine relative age of the enclosing layers and to provide a means of correlating widely separated rock units (see Lessons 8 and 10). The definitive evidence for the existence of Gondwanaland and its subsequent fragmentation exists in similar fossil assemblages found in sedimentary rocks on different continents (see Lesson 5).

The rock particles are important clues in the search for the rock type and location of the *source* area. The source might have been a mountain range that has been completely worn away by long erosion; the sedimentary rocks are the only evidence left of that topographic feature. The sediments might represent deposition from great continental glaciers of past ice ages and tell of the frigid climate and the extent of the long-gone ice sheets. Tropical climates and vast swampy lands populated by lofty trees are marked by beds of coal; the trees are extinct and the swamps have been buried under perhaps 200 million years of deposition. Parts of an ancient sea floor subducting under a continent are preserved in the surviving sedimentary rocks that were thrust over the continental margin and are a reminder of that tectonic event. Thus, from these layered rocks, geologists are able to reconstruct the geologic past and the ancient landscapes. And finally, the long history of evolving life on Earth is read in the sequence of species preserved in the thin cover of sedimentary rocks (see Lesson 11).

Valuable Sedimentary Rocks

Apart from their scientific value, many sedimentary rocks yield great economic resources. These include crude oil, natural gas, and water, which are trapped in the pore spaces of sedimentary rocks and whose value is incalculable. Other resources include coal, iron ore, and the sand and gravel widely used in construction, and limestone. The great cathedrals and palaces of the European Middle Ages were constructed of limestone – and in some cases sandstone. Certain shales serve as a basic raw material for pottery, brick, tile, and china. When mixed with limestone, shale is used to make Portland cement. Halite or rock salt has so many uses besides seasoning that people have sought, traded, and fought over it for much of human history. (Refer to Box 14.1, page 315.)

What are Sediments?

Sediments are loose particles that originate from weathering (both chemical and mechanical) and erosion of pre-existing exposed rocks. Because clays and quartz sands are the insoluble end products of chemical weathering of feldspars, ferromagnesians, and other widespread rock-forming minerals, much of the volume of sedimentary rock is shale or sandstone.

Chemical precipitation from solution is another source of sediments – usually calcium carbonate or *calcite* (the mineral content of limestone) or salt – which forms when seawater evaporates and leaves the minerals behind. Marine organic sediments form from the shells of both plants and animals, usually microscopic and planktonic, which settle to the sea floor in great abundance and form certain kinds of limestone. (Refer to the section on the origin of *chalk* in Lesson 11.) Whatever the source, the sediments are unconsolidated, loose, separate grains, such as sand on the beach or mud on the sea floor or a lake bottom.

How are Sediments Classified?

Clastic sediments (from the Greek word *klastos*, meaning "broken") are composed of broken fragments or mineral grains of pre-existing rock and are classified by the size of the individual particles. (See Table 14.1, page 304, showing size, name of sediment, and type of clastic sedimentary rock.)

The smallest are the *clay* particles. To get a feeling for clays, put your fingers in chocolate pudding (mentally, of course) and rub your fingers together. It should feel smooth, no grit or grains, just slippery. Clay particles feel about the same. They are small, about $1/256$ millimeters, and settle to the sea floor some distance from land and on lake bottoms. Fine clays or muds form *shale*, the most abundant sedimentary rock (Figure 14.14, page 310). (Why do you think it is no surprise that shale is the most abundant sedimentary rock? Lesson 15 will give you the answer.)

Now, pick up some beach sand and rub it between your fingers (again, mentally, unless you have a beach nearby). Beach sand grains, from $1/16$ to 2 millimeters in diameter, are usually composed of quartz, a very durable product of weathering of granite and lesser amounts of other minerals. The sedimentary rock formed of sand grains is, of course, *sandstone* (Figure 14.10, page 308).

The next larger group of particles is *gravel*, but the range of sizes for this group is wide – from 2 millimeters (small pebble size) to 256 millimeters in diameter (about 10 inches or a little larger than the size of your head). The sedimentary rock called *conglomerate* is composed of gravel particles of different sizes which are rounded in shape. If the particles are angular, the rock is called *breccia* (Figures 14.3, 14.8, and 14.9, pages 305 and 308.)

Transportation: Carrying the Particles to the Site of Deposition

Sediment may be transported in several ways. The shortest transport may be weathered rock that slides downhill in one of the various processes of mass wasting. The resulting deposit will consist of angular particles of all sizes. Streams are the most important transportation agent, bringing most of the sand to the beaches of the world and the fine sediments to the sea floor. Glaciers, at high latitudes and high altitudes, can move weathered rock debris which is transported frozen in the ice. Continental glaciers of the Ice Age left rocky unsorted glacial deposits called *ground moraines* over vast areas of North America and northern Europe. Wind, particularly in arid regions, can move enormous quantities of dust and sand (Figure 19.14, page 442). Each agent of transport leaves its own mark on its deposited load.

Sorting by size. Some transporting agents *sort* or separate the rock particles according to size. Streams sort particles by leaving the larger, heavier particles such as boulders upstream near a mountain face (Figures 14.2 and 14.3, page 305), while the finer sands, silts, clays, or mud are carried to the mouth of the river (Figures 14.4 and 14.5, page 306). Glaciers, however, do not sort sediment, and glacial debris is a mixture of all sizes, from fine *rock flour*, clay, and sand to gravel and even huge boulders (Figures 18.24 and 18.25, page 418). The meltwater draining the front of the glacier, however, will rework the glacial debris, which will then be sorted according to river processes. Wind sorts particles by its limited carrying power; fine dust and volcanic ash can be carried high in the air great distances, across continents and oceans, while the larger sand-size fragments, moving near ground level, are left behind to accumulate in sand dunes or in the great *sand seas* of the Sahara and other deserts (Figures 19.14 and 19.28, pages 442-448).

Rounding of particles. As the rock fragments are carried downstream in a turbulent river, they will be abraded as a result of rubbing against each other and against the stream bed. The sharp corners will become rounded. Rounding also takes place on beaches, where the waves tumble the stones against one another (Figure 14.1, page 305). Some rounding may also appears in glacial debris, but the larger fragments may also show grooving and scratching, a feature useful in identifying solid ice as the agent of transport.

The Environments of Deposition

Deposition occurs when the transporting agent (streams, wind, waves or glaciers) loses its energy and can no longer transport its load. Deposition also refers to the accumulation of organic or chemical sediment, such as fossil shells on a sea floor or chemical precipitation of calcium carbonate. Salt, gypsum, and silica (silicon dioxide), may be deposited following a change of temperature, pressure, or chemistry of sea water or a desert lake. (Refer to Table 14.2, page 313, and study Figure 14.38, page 323.)

Continental environments. Continental environments of deposition include the channels and flood plains of streams. As the rate of flow decreases downstream, the ability to transport sediment also decreases. The largest fragments – boulders and cobbles – will drop out first. Sands and gravels will be rolled or bounced along to be deposited as elongate lenses within the channel, while the finer sands, clays, and muds, easily buoyed up by the water, continue downstream to settle out on a delta, in a quiet lagoon or estuary, or on a beach (Figures 14.4, 14.5, 14.36, 14.39, 14.40, pages 306, 321, and 323; and 16.31, pages 366-369).

When the transporting agent is a glacier, the rocky load, which is frozen into or carried on top of the ice, will be dropped as a result of glacial melting and left at the lower end to form a looping hill called a *glacial moraine*. Refer to Figures 18.24 and 18.25, page 418, which show glacial debris. Other continental environments include alluvial fans, lakes, desert sand dunes, and widespread deposits of dust and fine silt that cover much of the central United States and China (Figures 19.14, 19.21, and 19.22, pages 442 and 445).

Marine environments. Beaches, deltas, lagoons, and estuaries are actually transitional environments, influenced by marine waves and currents as well as entering streams. The true marine environment of deposition includes the shallow waters of the continental shelf, the continental slope, and the final resting place of the finest muds and clays, the deep sea floor.

Clues to ancient environments. A glacial moraine deposit may be simple to tell from a well-sorted beach sand. But the moraine material may not be as easy to distinguish from an ancient landslide. Deep sea and continental shelf deposits are easier to tell apart. (How would you tell them one from the other?) Each environment has its own physical, chemical, and biological characteristics. One task of the geologist is to determine from the sedimentary rocks the original environment of deposition, which may be very different from the landscape where the rocks occur today. For example, marine sandstone and limestone deposited on the floor of some Precambrian sea over 600 million years ago form the mountains that rim the arid desert of present day Death Valley. Ancient Ice Age deposits occur in subtropical India, coal beds containing tropical plants are found in Antarctica, and tropical coral reef deposits occur in Alaska! These occurrences have been used as compelling arguments for the validity of the theory of plate tectonics.

From Loose Sediment to Sedimentary Rock

Once the sediment is deposited and buried by other sediments, major changes start to take place that will harden the soft sediment into rock. These changes are called *lithification*, a process that usually entails *compaction* and *cementation*, or *crystallization* from a solution.

When a thick pile of sediments accumulates in a basin, the particles near the bottom become *compacted* because of the weight of the overlying material. The mineral grains are pressed closer together, and the water in the pore space between the grains (which may occupy between 60 and 80 percent of the volume of clays and muds) will be squeezed out. This reduces the volume and re-orients clay minerals parallel to each other like cards in a deck (Figure 14.15, page 310). Thus *compaction* is the primary factor in the formation of shales, although some cementation may take place.

Lithification of sands occurs primarily by *cementation*. Sand grains are fairly well compacted when deposited, and further compaction is slight even when the grains are deeply buried. Natural cements are chemicals (also products of weathering) carried in solution by circulating ground water that percolates through the open spaces between particles. In time, the cement precipitates around the sediment grains, more or less fills the pore spaces, and binds the particles together (Figure 14.6 A, B, and C, page 307), effectively changing the loose sediments into solid sedimentary rock. (Remember: this process takes place over geologic time.)

The most common cements are *calcite*, which is calcium carbonate, and *silica*, which is silicon dioxide. Other cements are iron oxide and sometimes clay minerals.

Sedimentary Rocks: Clastic, Chemical, and Organic

Sedimentary rocks are classified by their original sediments and by their mode of origin. *Clastic* sedimentary rocks are formed from fragments or mineral grains of pre-existing rock. The size of the particles ranges from the finest clay particles that make up shales to boulders that occur in conglomerates. Clastic rocks are lithified by cementation, compaction, or a combination of both. When they are combined with limestone, they make up almost all sedimentary rocks.

Chemical sedimentary rocks are formed of minerals precipitated from solution. Thick layers of rock salt are formed when sea water evaporates from an ancient basin of deposition. Salt also forms within desert playa lakes (Figure 19.12, page 441). These deposits are called *evaporites* and are crystalline in texture. Limestone can be precipitated by the activities of coral reef organisms and by certain marine plants. (See Table 14.2, page 313.)

Organic sedimentary rocks are composed of the remains of organisms, such as the plant material in coal, or the actual shells of sea creatures that form shell limestones such as *coquina* (Figure 14.17, page 311).

Sedimentary Structures

Sedimentary structures are features within sedimentary rocks that form during or shortly after deposition but before lithification. The most notable feature of an outcrop of sedimentary rocks is the *bedding*, a series of layers that were originally deposited in a horizontal position. Remember the law of original horizontality? (See Figure 14.26, page 315, and page 302.) A *bedding plane* is a nearly flat surface separating two layers or *beds*. The bedding planes reflect any changes or breaks in the process of deposition, such as a change in grain size that might occur during flood time or a time of non-deposition as during a long drought; it might be an erosional surface resulting from a local uplift or a new cycle of deposition resulting from the lowering of a basin. Climatic change, tectonic movements, or changes in the source region would all affect the sediments being deposited. Great thick beds such as the *Coconino Sandstone*, described in Box 14.2, Figure 1, page 320, represent a long period of stability and uninterrupted deposition of sands.

Cross-bedding is a special case in which the bedding is not originally horizontal but is deposited at an angle to adjacent bedding planes (Figure 14.27, page 316). Cross-bedding – seen most frequently in sandstones – can develop in sand dunes deposited by wind, in sand bars and dunes deposited by rivers in their channels, in the foreset beds of deltas (Figure 16.32, page 367), and on the sea floor in sand ridges shaped by bottom currents (Figure 14.28, page 317). *Graded beds* are noted for the change in grain size within one layer, from coarse at the bottom to fine toward the top. Graded bedding is evidence of turbidity currents flowing down the continental shelf and slope and across the deep sea floor (Figure 14.13, 14.29, and especially 14.30, pages 309, 317, and 319).

Mud cracks indicate drying by air and form in sediments periodically above water on tidal flats, exposed stream beds, playa lakes, and similar environments that have wet and dry cycles. (See Figure 14.31, page 318, and especially page 432 for a spectacular photograph of mud cracks.)

Ripple marks on bedding planes are excellent indicators of water moving either as waves or currents or of wind blowing across sand dunes. Moving water or wind form asymmetric ripples in soft sediment perpendicular to the direction the wind or water is moving. In shallow water where wave action is moving water up toward the beach and back, the ripples in the sand will be symmetric (Figure 14.32 and 14.33, page 319).

The colors of sedimentary rock vary considerably. Chemical weathering may impart to the exposed surface of rocks a color that may be quite different from the color of a fresh surface. The great coloring agent in nature is iron oxide, derived from oxidation of iron-rich minerals. The deep red, yellow-brown, and orange tones seen in desert formations are the result of staining by iron oxide (page 302). Black color in sedimentary beds is generally caused by the presence of carbon of organic origin. This implies that organisms were deposited faster than they could be destroyed by bacteria and oxidation. Such situations are common in stagnant water.

Fossils – the remains of plants and animals preserved in the rock – are surprisingly common sedimentary structures, especially in shallow marine deposits. As you learned in Lesson 11, animals with hard parts that are rapidly buried in sediments are the best subjects for fossilization. Because many limestones are composed of shell and coral debris and are characteristically formed in biologically productive warm shallow seas, fossils in limestone are common, although organic remains do occur in sandstone and sometimes in shale (Figures 14.16 to 14.21, 14.34, and 14.35, pages 311, 312, and 319). (Why are fossils rare in breccias or conglomerates, or in river flood plain deposits?) The great value of fossils, as we mentioned before, is that they are the chief basis for age correlation of sedimentary strata separated over great distances.

Origins of Common Sedimentary Rocks

Clues to the ancient environments of deposition, as you have now learned, come from the composition and structures within the sedimentary layers.

Clastic rocks. *Conglomerates* and *breccias* seen in Figures 14.8 and 14.9, page 308, are coarse-grained, poorly sorted rocks that may form in several different environments. Some breccias may have been part of a landslide and were carried only a few miles before deposition. Other conglomerates form where mountainous areas are rising rapidly and shedding rock debris into fast moving streams. Alluvial fan deposits in arid regions contain sands and gravels from the occasional flash floods that will lithify into conglomerates (Figures 14.39, 16.33, and 16.34, pages 323 and 368). Glacial sediment contains unweathered particles of all sizes, many of which are scratched and grooved as they grind over one another under the weight of the ice. In general, in the case of clastic sediments, the larger the particles, the poorer the sorting, and the shorter the distance of travel to the basin of deposition.

Sandstone is a medium-grained rock formed by the cementation of sand grains. Sand deposits also occur in many environments of deposition, and sandstones are one of the most abundant of the sedimentary rocks. Beach sands and shallow water deposits, composed usually of quartz sand grains, are derived from the weathering of granite or pre-existing sandstone. Rivers also deposit sand in their channels and on their flood plains, and winds pile up sand into dunes. Even on the deep sea floor, turbidity currents spread sand deposits in areas where fine muds usually settle. The turbidity current deposit is called a *graywacke* and generally contains as much as 15 percent of fine-grained sand and clay between the larger sand grains. Thus, there is considerable variation in sandstones in mineral composition and degree of sorting and rounding, the reflection of the different source areas, the transporting medium, and the environments of deposition. (Refer to Figure 14.10 A, B, and C, page 308, and Figures 14.11, 14.12, and 14.13, page 309.)

Shale is a fine-grained rock consisting of fine clays, muds, and silts, and it is noted for its ability to split into very thin layers (Figure 14.14, page 310). The thin-bedded shales form partly by compaction – which squeezes out the water and orients the clay particles parallel to each other – and partly by cementation (Figure 14.14 and 14.15, page 310). Clays, like quartz sands, are the end products of weathering and thus are not subject to further chemical change at Earth's surface. The fine particles accumulate in quiet waters, such as on lake bottoms, in deltas (Figures 16.30 to 16.32, pages 366-367), on river flood plains following episodes of high water (Figure 14.40, page 323), on tidal flats, lagoons, and in quiet areas of the deep ocean floor. Clays accumulate very slowly, especially in deep water, and thick shale beds tell of hundreds of thousands of years of deposition.

Chemical and organic sedimentary rocks. Refer to Table 14.2, page 313 for information about these rocks.

Carbonate rocks. *Limestone* is composed mostly of the mineral *calcite*, which chemically is calcium carbonate ($CaCO_3$). Limestone, an organic sedimentary rock, generally forms in warm shallow seas from life activities of organisms. Some limestones are extremely fine-grained and dense from the accumulation of microscopic fragments of algae (Figures 14.19, 14.20, and 14.21, page 312). Others, called *coquina*, are formed of broken shells or bits of algae and corals cemented together (Figures 14.16, 14.17, and 14.18, page 311). Limestones are subject to recrystallization, the process by which calcite grains recrystallize especially under the weight of overlying sediment. During crystal formation, the fossils and original texture may be destroyed, making the geologic history of the rock difficult to decipher. (Remember: Calcite is an important mineral in lithification of sediments.)

Chert. Chert is a very hard, compact, fine-grained rock composed almost entirely of *silica*, which chemically is silicon dioxide (SiO_2). Chert occurs as layered deposits and as lumps called *nodules*, which are found in some limestone and chalk deposits. Chert nodules, when dark gray or black, are called *flint*; flint was prized by early humans for its hardness and toughness and was used in making arrow points, hand axes, and other stone tools. Chert may be partly a chemical deposit and partly the accumulation of microscopic marine organisms (radiolaria) that use silica in forming their tiny shells (Figure 14.23 A and B, page 314).

In short, each environment influences its sedimentary deposits, if any, and a geologist can reconstruct this original environment of deposition by studying today's exposed rock types and fossils. (Review Figure 14.38, page 323.) Thus a story of Earth is revealed, layer by layer, in the colorful and varied sedimentary rocks.

A visit to the Grand Canyon, whether teetering on the rim, looking down into the gorge, or hurtling through the rapids of the Colorado River in a small boat, inspires awe and can leave one speechless. The dimensions are indeed impressive: 1.6 kilometers (over 1 mile) deep in places, 6.5 to 29 kilometers (4 to 18 miles) wide, and 450 kilometers (280 miles) long. The first trip down the Canyon in a small boat was made in 1869 by Major John W. Powell, a government geologist, and his party of nine men in four small rowboats. That perilous journey took three months and cost the lives of two men (who presumably were attacked by Indians after they climbed out of the Canyon).

For the geologist, the Canyon offers an unparalleled glimpse of Earth history, extending back into Precambrian time. The bedding planes of the horizontal sedimentary rocks can be traced for great distances, which is remarkable in itself, since the uplift of this section of the Colorado Plateau equals over one mile yet shows very little folding or faulting.

To start our journey through time, we begin at the bottom of the Grand Canyon, where we will encounter the oldest exposed formation. This is the *Vishnu Schist*, a severely deformed rock unit that is folded and faulted, intruded by granite, and so altered by heat and pressure that the original nature of the rocks is almost impossible to determine.

The Vishnu has no fossils, and its exact age is uncertain but, based on radioactive dating methods, is assumed to be Precambrian, about 1,400 to 1,500 million years old. The Vishnu in the lower Gorge represents the roots or remnants of a great mountain range that was deeply eroded before the land subsided and younger sediments were deposited on top. The *Grand Canyon Series*, consisting of sandstones, limestones, and shales, is separated from the Vishnu by an unconformity. These younger sedimentary rocks are not metamorphosed but are folded and faulted and lack fossils. In this Series, we see another episode of mountain building, uplift, and, deep erosion – still in the Precambrian Eon. (Refer to Figure 8.14, page 181.)

An angular unconformity separates the tilted Grand Canyon Series from the overlying horizontal brown pebbly *Tapeats Sandstone*, which is early Cambrian (Lower Paleozoic) in age. This formation, which is not fossiliferous, represents an encroaching sea that flooded the land from the west, covering the ancient Precambrian rocks with beach-type sediments. Looking higher in the Canyon walls, we can see that the Tapeats gradually gives way to a formation called the *Bright Angel Shale*, originally a muddy deposit, that contains a few fossils, most of them trilobites, extinct arthropods distantly related to present day crayfish (Figure 8.16, page 182). The base of the Bright Angel is about 550 million years old. The land subsided and the sea continued to enter from the west; support for this comes from evidence that part of the Bright Angel Shale is older in the west than it is in the east (based on the differences in trilobite species). As the sea advanced, beach deposits (Tapeats) and then deeper-water muds (Bright Angel) accumulated.

CASE STUDY: "Earth Revealed" in the Grand Canyon

The Bright Angel Shale is about 140 meters (450 feet) thick and grades upward into the overlying *Muav Limestone*, which is about the same thickness and contains trilobites of the same general age as those in the Bright Angel Shale. The Muav is part of the same sequence of the transgressive sea, and occasional ripple marks indicate a deposition in a shallow sea but probably deeper than the environment of the Bright Angel Shale (Box 14.3, Figure 1, page 324).

From the Canyon Rim, these three formations (Vishnu Schist, Grand Canyon Series, and Bright Angel Shale) can be easily recognized. The Tapeats Sandstone forms a well-marked wall above the contorted Precambrian rocks. The thin bedded Bright Angel Shale, being a less durable formation, is partly hidden by rubble and slopes back to the base of the Muav Limestone, which forms a *step* or rock wall (Figure 12.6, page 271). There are caves in places in the Muav, which is relatively soluble since it is limestone.

Continuing our journey up the walls of the Canyon, the next formation we see is a thin limestone, the *Temple Butte Limestone*, that is either missing in part of the Canyon or is inconspicuous. The important thing about the Temple Butte is that it contains fossil skeletons of primitive fishes, animals that lived long after the trilobites of the underlying Muav. All of the layers deposited between Muav and Temple Butte time are missing, as are the remains of animals that lived in this long interval. This includes all of the late Cambrian, Ordovician, Silurian, and early to mid-Devonian (Table 8.4, page 189), a gap of over 100 million years! This either represents a long period of non-deposition in this region or, after deposition, a time of uplift and long erosion. There is another gap in geologic time between the Temple Butte (Devonian) and the overlying *Redwall Limestone*, which contains fossil marine animals of Mississippian age. The Redwall and Temple Butte are both cliff-forming and stand almost vertically.

The Redwall cliffs are imposing features that stand more than 500 feet high. The limestone on a fresh surface is a medium gray; the exposed surfaces, however, are stained red from iron oxide washing down during rainstorms from the overlying red Supai formation. The purity of the limestone indicates it was formed in a relatively wide, shallow, quiet sea, far from sand and clay deposition near the shore. Sea shells and a wide variety of other marine fossils, including corals, are found well preserved within the Redwall. Beautiful red jasper of gem quality (a form of chert) is also found within this formation. The Redwall is susceptible to solution by groundwater and has produced many *temples*, pillars, caves, solution caverns, arches, and springs within the Canyon.

Above the Redwall Limestone and separated from it by another unconformity is the *Supai Formation*, which is Pennsylvanian-Permian in age (the time of the assembling of Pangaea). This deep red formation consists of shaley siltstone and sandstone. The Supai lacks marine fossils but contains fossils of land plants similar to those of the famous coal beds in the southeastern United States and in the Ruhr Valley in Europe. The direction of cross-bedding indicates that the source of the sediments was in southern Utah (Figure 14.37, page 322). From the tracks of reptiles and amphibians and from mud cracks, geologists believe the Supai may have been part of a large delta with the sea to the west. It was a low lying, near-shore environment, with marine and non-marine deposition alternating as the shore was displaced during cycles of sea level change. The Supai is one of the great terrestrial deposits found in the Canyon.

The terrestrial depositional cycle continues with the *Coconino Sandstone*, a prominent, white-colored, homogeneous cliff-forming formation that caps most of the buttes in the Canyon. The sand is composed of fine-grained, almost pure quartz grains that are well rounded, pitted, or frosted and show excellent sorting by size. Cross-bedding is common, and many vertebrate animal tracks are present. There is some uncertainty about the origin of the Coconino; it may have been a wind-deposited dune field like the modern Sahara Desert or it may have originated in large river sandbars (Box 14.2, Figure 1, page 320).

A massive formation of limestone and sandstone forms the top of the cliffs at the upper rim of the Canyon; this is the *Kaibab Formation*, Permian in age, with a thickness of about 500 to 700 feet. In this cream-colored fossiliferous limestone can be seen the end of the lengthy period of terrestrial deposition and another transgression of the sea. As we examine the very top of the Kaibab, however, we see terrestrial red beds with layers of gypsum, signals that the environment of deposition was changing from completely marine to possibly lagoonal. The top of the Kaibab has been dated at about 225 million years.

Is this the end of the Grand Canyon story? Not at all. If we were to travel over the highlands above the canyon rim, we would encounter formations younger than the Kaibab, less than 200 million years old, which are red, brown, yellow, and gray sandstones, conglomerates and shales that contain petrified forests of tree trunks, and in some places, dinosaur remains. The beautiful cross-bedded Navajo sandstone of Zion National Park, which resembles the Coconino, is actually a much younger formation but was part of the great depositional history of the area. (Refer to Figures 8.12, 8.13, and 8.14, page 181.)

The rocks of the Grand Canyon have many stories to tell. In these colorful layers we have seen the advance and retreat of the seas over the continent, the appearance and disappearance of different kinds of organisms, and the changes from marine to terrestrial environments. From the presence of unconformities, we get a sense of the gaps in the record, when the land stood high, with erosion rather than deposition as the dominant activity, sometimes for hundreds of millions of years. From the Vishnu Schist to the Kaibab, we have traversed over one billion years of geologic time and have seen a great part of Earth revealed. The wonder of it is that the geologist is able to *read* this incredible record and make sense of the changing landscapes created by restless Earth in the ancient and dimly understood past.

1. How is the origin of petroleum related to sedimentary rocks?

The formation of crude oil is a complex process that is not completely understood. But the occurrence of petroleum is almost always associated with marine sedimentary rocks. Organic matter such as marine planktonic plants and animals is the source material, just as land plants are the raw material in the formation of coal. In the sea, the bodies and shells of the organisms accumulate in quiet basins, where the supply of oxygen is low and there are few bottom scavengers. The action of anaerobic bacteria at this stage converts the original substances into simpler organic compounds. Burial of the organic-rich layers, possibly by turbidity currents, may be the next step. Further conversion of the hydrocarbons must take place at considerable depth, on the order of 2 kilometers (about 1.2 miles) or more, where elevated temperatures and pressures occur. Slow cooking under the thick sedimentary blanket for perhaps 2 or 3 million years allows the chemical changes to take place that produce crude oil and natural gas.

The crude oil migrates from its source rock through the pores and fractures in overlying formations into a reservoir layer, such as a porous sandstone, that has abundant spaces in which hydrocarbons can accumulate. Geologic traps such as anticlines or faults plus an overlying impermeable layer prevent further migration of the oil (Refer to Lesson 8).

Petroleum occurs around the world in many sedimentary basins and is also found beneath the sea floor in sedimentary rocks, as in the North Sea, the Gulf of Mexico, Alaska, and southern California. (Review Box 6.1, pages 133-134.)

2. Why are shales the most widespread sedimentary rocks?

Shales are the most widespread sedimentary rocks because they are made of clays, which are the most widespread products of weathering of the rock-forming minerals such as feldspars, ferromagnesians, and micas (Review Lesson 12.) The clays are carried suspended in stream water to lakes, lagoons, or out to sea to settle slowly on the sea bed. Through compaction by overlying layers and cementation by circulating ground waters, the muds and clay particles become shale.

3. What can we infer from a sedimentary rock in which the particles are well rounded and well-sorted?

Rounding means that the particles have undergone abrasion during transport, usually by streams, from a source at some distance from the area of deposition. Sorting – the separation of particles by size – is dependent upon the carrying or transporting power of the agent of deposition. Streams sort their deposits by size, leaving the largest boulders and cobbles upstream since they are the heaviest, and carrying the lighter sands and gravels downstream. Thus sandstones and conglomerates may be deposited within the stream bed or carried to a beach, to be further rounded and sorted by the waves. A shale, made of fine clays and muds, probably was deposited in the quiet waters of a lagoon, a lake, or a delta front or settled offshore on the continental shelf, slope, or even the deep sea floor. Wind also separates particles into fine *dust*, which may be airborne over great distances to form a deposit known as *loess*, and sand- sized particles, which are carried close to the ground and may be deposited in sand dunes. Glaciers do not sort their rocky debris by size, although the meltwaters emerging from the terminus of the ice will sort the particles like any stream.

4. The word clay has two meanings in this course. Differentiate between them.

Clay as a mineral was first introduced to you in Lesson 15 in the section on chemical weathering of feldspar and other minerals. Clay minerals are aluminum silicates with a sheet structure similar to mica, which accounts for the structure of shales and their ability to split into thin layers. In this lesson, *clay* is also used as the size of a particle (Table 14.1, page 304) that is likely to be a clay mineral but may be any mineral with a diameter of less than 1/256 millimeters.

5. Why are most sedimentary rocks of marine origin?

Almost three-quarters of the surface of the continents is blanketed in marine sedimentary rocks of all ages. This is a remarkable occurrence when you think about it. For one thing, this tells volumes about the crustal instability that has allowed marine waters to flood the lands repeatedly all through geologic time. Ice-cap melting and tectonic factors such as formation of broad mid-oceanic rises associated with rapid spreading rates have caused the displaced water to invade vast continental lowlands. Continents have grown, and former continental shelf areas with great thicknesses of marine sedimentary rocks have become incorporated into continental interiors. Deposits made by streams, wind, and glaciers on the continents are not widespread; hence sedimentary rocks formed on land have a relatively small surface extent and are subject to being eroded away. And finally, continents are eroding uplands and have few very large basins for deposition of sediments that will remain intact long enough for thick sedimentary rocks to form.

6. How would sedimentary rocks deposited in a shallow marine environment differ from those deposited in a deep marine basin?

Shallow marine refers to the continental shelf, and from studies of sediments being deposited there today, geologists are able to interpret the environment of deposition of ancient sedimentary layers. Continental shelves are covered for the most part with clastic sediments, with the coarsest material near shore or near river mouths and finer silt and clay farther from the source. Sandstones, conglomerates, and shales are characteristic of shallow marine deposition. Fossils of shallow water creatures are also good indicators of the depth of the water, and limestones containing shells are being deposited today in warm shallow seas. On modern continental shelves, however, the patterns are complex; the lowering of sea level during the Ice Age exposed portions of the shelves as dry land and altered the patterns of deposition. Currents, tides, and strong wave action also affect the deposits.

In the deep ocean, far from land, sediments are generally very fine particles, such as wind-carried dust, volcanic ash, and the shells of microscopic organisms that slowly settle to the sea bed. Local graded bedding may indicate a turbidity current that flowed across the sea floor.

In general, most of the sedimentary rocks that cover the continents are shallow marine deposits and thus consist of shales, limestones, and sandstones. These rocks can be seen from the bottom of the Grand Canyon and in the walls of the Dead Sea (the lowest place on Earth's land surface) to the top of the Himalayas, the Alps, the Andes, and in the ancient folded layers of the Appalachians.

7. What are some of the structures that indicate a sedimentary rock was formed in a continental environment?

Structures that are indicators of deposition in a terrestrial (continental) environment are cross-bedding related to sand dunes, mud cracks left in lake beds following a dry spell, or in muds left by flooding river; the asymmetric ripple marks made by running water; impressions of rain drops in fine-grained sediments; fossil footprints of dinosaurs and other land-living animals; and fossils of plants and animals of the swamps, especially in coal-bearing regions (Figure 14.25, page 314).

8. How does the source area influence the character of a sedimentary rock?

The source area is the locality that was eroded and provided the sediment that eventually became the sedimentary rock. The type of source rock will determine, to a great extent, the resulting sediment. If the source area is a great granite mountain range, the sediment may contain fragments of quartz, feldspar, ferromagnesian minerals, and mica. During the period of transport, however, the particles will undergo various changes, including rounding, sorting, and further mechanical and chemical weathering. If the distance from the source to the basin of deposition is short, the quartz, feldspar, and biotite mica will remain intact and angular, and the deposit will be designated an *arkose*. If the distance is great, only the quartz and clays from the weathered feldspars will remain to reach the basins.

Some of the minerals will be dissolved in the water, and other fine particles will be added to the clay deposits.

On the other hand, if the sedimentary rock consists almost entirely of well-rounded quartz grains, the source was probably a previously deposited sandstone in which the particles were already sorted both by size and mineral content. Many sedimentary rocks are composed of recycled sediments that previously were deposited, lithified, uplifted, and once again eroded. With almost three-quarters of the continental surface covered by sedimentary rock, it is understandable that over geologic time many of the source rocks will be sedimentary.

Some of these pure recycled sandstones are an important source for the manufacture of glass products. Other sedimentary rocks are studied, using current, cross-bedding, thickness of formation, and type of rock, to determine the location of the source, which may be a mineral deposit of commercial value (Figure 14.37, page 322) such as uranium. Diamonds are found in Ice Age glacial debris in the Midwest, but the source rocks have never been located.

Activities

1. Look at the wonderful photograph facing page 302. From the description of the rock, can you name the three formations, starting with the sandstone at river level? (See the "Case Study" for formation names.)

2. Mentally run your fingers over the rocks shown in Figure 14.10, page 308. What should they feel like? (Answer: cemented sand grains.) Many buildings are constructed of sandstone and limestone. Sometimes (but not always), the touch test can help you tell them apart.

3. What is a definitive chemical test for limestone? With the addition of a few drops of dilute hydrochloric acid, the limestone should fizz vigorously from the release of carbonic acid during the reaction. Note: Do NOT use this test on buildings, statues, or any other structures, only on specific lab samples or in the field far from habitation. Do you think *coquina* (Figure 14.17, page 311) would fizz with acid? What about chalk? What does this tell you about the chemical composition of shells of marine creatures? (Right: They are mostly made of $CaCO_3$, or calcite, the same mineral as limestone.)

4. In Figure 14.19, page 312, the term *bioclastic* is used. What does this mean?

5. When you use salt on your french fries, think about the origin of this sedimentary deposit. Why are salt, gypsum, and borates termed *evaporites*?

6. Read Box 14.3, page 324, "Transgressions and Regressions." Carefully look at the sequence of formations deposited under a transgressing sea. Where have you learned about this sequence?

7. The best activity of all is to go to the Grand Canyon or any area where thick sequences of sedimentary rocks occur. Most of the formations will be thin-bedded shales, thick sandstones, or limestones. Look for fossils; study the texture of the rock; try to imagine the environment in which it formed. These are geologic puzzles that are not too difficult but are eminently satisfying to solve.

Self-Test

1. From what you observe standing on the rim of the Grand Canyon, it is apparent that the structures most characteristic of sedimentary rocks are

    a. fossil foot prints.
    b. ripple marks.
    c. mud cracks.
    d. bedding planes.

2. Sediments that are termed *clastic* are classified primarily by

    a. chemical composition.
    b. size of particles.
    c. rounding of particles.
    d. thickness of the deposit.

3. Shales are the most widespread sedimentary rock. This is related to the fact that

    a. quartz is one of the most abundant minerals.
    b. widespread evaporation of sea water results in shale deposits.
    c. clays are the most widespread weathering product of the rock-forming minerals.
    d. the minerals in shale settle out of water quickly and form thick deposits.

4. *Sorting* in sediments refers to separation of particles by

    a. grain size.
    b. rounding.
    c. mineral content.
    d. origin, as environment of deposition.

5. Of the following, which is most likely to have a crystalline texture?

    a. Shale
    b. Arkose
    c. Quartz sandstone
    d. Rock salt

6. Sedimentary structures are important to the geologist as indicators of the

   a. age of the sediment.
   b. water temperature.
   c. minerals in the source area.
   d. depositional environment.

7. Most of the sedimentary rocks that occur on the continents are

   a. wind-deposited sand seas.
   b. shallow water marine deposits.
   c. products of evaporation of vast inland seas.
   d. relics of the continental glaciers of the ice ages.

8. Sediments by definition must

   a. be limited by size to sand and gravel.
   b. be formed in a marine environment.
   c. have well-marked bedding planes.
   d. be unconsolidated.

9. A sedimentary formation exhibiting graded bedding most likely originated as a

   a. turbidity current on a sea floor.
   b. wind deposited sediment on a desert floor.
   c. lagoon deposit transitional between land and sea.
   d. tidal flat alternately covered and exposed by the tides.

10. From your examination of the layers in the Grand Canyon, you have learned that fossils are most abundant in

    a. shale and clay layers.
    b. limestones.
    c. terrestrial sandstones.
    d. conglomerates and breccias.

11. The origin of the Coconino Sandstone is somewhat controversial. It shows cross-bedding and well sorted quartz grains. These features can be characteristic of sediments deposited as both

    a. glacial moraine and glacial outwash.
    b. dune sands and river channel deposits.
    c. tide flats and delta deposits.
    d. deep sea and continental shelf deposits.

12. The lithification of sedimentary rocks is a result of all of the following except

    a. cementation.
    b. compaction.
    c. heating.
    d. crystallization.

13. If you are testing a quartz sandstone with hydrochloric acid and it fizzes, you might assume

    a. there is abundant clay present.
    b. the particles are feldspar and ferromagnesians.
    c. the rock contains plant remains.
    d. the cement must be calcite.

14. Coal is considered a sedimentary rock, and it originates from

    a. cementation of oolites.
    b. compaction of organic-rich plant material.
    c. crystallization of coralline algae.
    d. evaporation of swamps and lagoons.

15. Ripple marks are least likely to form in which of the following environments?

    a. Deep sea
    b. River or stream
    c. Tidal flat
    d. Near shore

18

METAMORPHIC ROCKS

This lesson will help you understand the causes of metamorphism and the kinds of rocks produced.

After reading the textbook assignments, completing the exercises in this study guide, and viewing the lesson's video portion, you will be able to:

1. Recognize how metamorphic rocks form.

2. Describe the roles of temperature, pressure, and fluids in metamorphism.

3. Indicate the factors that control the mineralogical composition and texture of a metamorphic rock.

4. Differentiate between confining pressure and directed pressure, and describe the texture and structural features of the foliated rocks that are formed.

5. Discuss the concerns a builder or engineer might have about building on foliated rock.

6. Recognize how metamorphic rocks are named and classified.

7. Compare the two most common types of metamorphism – contact (thermal) metamorphism and regional (dynamothermal) metamorphism – and the rocks that result from each.

8. Discuss the relationship between regional metamorphism and plate tectonics.

9. Explain the concept of metamorphic facies and how it relates to the classification of metamorphic rocks and their environments of origin.

10. Describe how hydrothermal rocks are formed and recognize their economic importance.

**INTRODUCING
THE LESSON**

Metamorphic rocks, the third great family of rocks, make up an estimated 27 percent of Earth's crust yet have never been observed in the process of formation!   Volcanic rocks, which can be studied during eruptions, tell of igneous processes and the heat within Earth; sedimentary rocks, tied to cycles powered by external sources, are well understood since they all form on Earth's surface. But metamorphic rocks, which are equally important as clues to the dynamics of this planet, are seen only when mountains belts are uplifted, deeply eroded, and their ancient roots exposed to view.

Most metamorphic rocks form at plate boundaries, where the continuing march of the oceanic crust against the moving mass of the continents exerts pressure upon the rocks between, as if they were caught in some gigantic geologic vise. Few of the original minerals emerge unscathed. New minerals form in which the crystal structure is compressed, the minerals are aligned, the spaces between grains are eliminated, and a dense rock results. As these *high pressure-low temperature* rocks are slowly carried downward with the subducting oceanic lithosphere, they are further altered by the heat of Earth's interior, which in the space of about 10 kilometers (6 miles) may increase over 200° C.

Metamorphic rocks also form around igneous intrusions. There, the heat of the crystallizing magma in mountain-size batholiths may reach 1,000° C, and the country rocks become baked, like pottery in a kiln.

The rocks we call *metamorphic* (from the Greek words *meta* meaning "changing" and *morph* meaning "form") have been altered while in the solid state, either in texture or in mineral composition or both. All the rocks you have studied in *Earth Revealed* (igneous rocks, sedimentary rocks, even previously metamorphosed rock) will become unstable and will metamorphose when subjected to a change in environment – in this case, increased pressure or heat or exposure to active chemical solutions.

These changes are attempts to reach a form of equilibrium within new environmental conditions; the process is similar to the one described in the lesson on weathering. Weathering occurs when rocks that formed at high temperatures and pressures deep within Earth are uplifted during mountain building and exposed to surface conditions of low temperatures and pressures. Metamorphism moves in the opposite direction: rocks that form or are exposed on the surface, such as shales and limestones or basalts and granites, become metamorphosed when thrust downward into the high temperatures and pressures encountered at depth. Whereas weathering generally changes rock to disassociated sediments and soil, metamorphism can change a rather dull, even uninteresting stone into a new hard crystalline rock, often with bright and colorful minerals.

In this lesson, you will learn how the factors that cause metamorphism affect various *parent rocks*. Marble, for example, so prized by artists and sculptors through most of human history, is a metamorphic rock composed of calcite crystals whose parent rock is limestone. The slate of blackboards, pool tables, roofs and floors is metamorphosed shale, a sedimentary rock derived from compacted and compressed sea-floor muds. The properties of slate that make it so useful are its hardness and chemical durability and its ability to cleave into thin flat slabs – all resulting from *low-grade metamorphism*.

In some regions, the metamorphosed rock can be traced back to the unchanged parent rock, providing information about processes and conditions deep within the crust. In most cases, however, the study of metamorphic rock is complicated because of almost limitless possible combinations of temperature and pressure as well as a tremendous variety of parent rocks. You will learn how the most common metamorphic rocks are classified and identified; and most important, you will be able to relate plate tectonics to the occurrence and formation of these valuable and, in many cases, beautiful rocks.

Completing the following eight steps will help you master the lesson objectives and achieve the goal for this lesson:

Step 1: Read the TEXT ASSIGNMENT, Chapter 15, "Metamorphism, Metamorphic Rocks, and Hydrothermal Rocks."

Step 2: Study the KEY TERMS AND CONCEPTS as noted in the study guide.

Step 3: Watch the VIDEO using the VIEWING GUIDE in the study guide.

Step 4: Read the study guide's PUTTING IT ALL TOGETHER section, which will help you summarize and integrate all of the information in this lesson.

Step 5: Complete any assigned lab exercises.

Step 6: Complete any assigned ACTIVITIES found in the study guide.

Step 7: Review the material in this lesson and complete the SELF-TEST found in the study guide.

Step 8: Go back to the LEARNING OBJECTIVES and make sure you have learned each one.

LESSON ASSIGNMENT

1. Before reading all of Chapter 15, read the "Introduction" to the chapter to understand the subject matter of this lesson. Look at the rock cycle diagram on page 329. Note that sedimentary rock can undergo metamorphism, but also note that igneous rock has an arrow pointing to metamorphism. Finally, follow the line leading from metamorphic rock to weathering, and from metamorphic rock to magma. You should be able to describe the significance of each part of the rock cycle after completing this lesson.

TEXT ASSIGNMENT

2. Examine the photograph facing page 329 What is the actual width of the object photographed? The thin red plates are composed of what mineral? Look at Figure 15.1, page 330, for a larger view of the outcrop of metamorphic rock.

3. Now read Chapter 15, pages 329-347, paying close attention to the photographs and diagrams. The diagrams of subduction zones are very helpful in understanding the sites of origin of metamorphic rocks.

KEY TERMS AND CONCEPTS

The key terms listed under "Terms to Remember," page 346 in your text, will aid your understanding of the lesson's material. Make sure you look up the definitions of these terms in the glossary or the chapter. All principles and processes in this lesson are important, but the number of metamorphic rocks to learn has been limited to those that are most widespread.

VIEWING GUIDE

Once you have read the text material for the lesson and studied the terms, you are ready to view the lesson's video portion. Review the questions that appear in this section to help you watch for important points and to prepare you for what you will see in the video.

Before viewing, answer the following questions:

1. What are the three most important factors controlling the characteristics of metamorphic rocks?

2. Describe *confining* pressure and tell how it differs from *directed* pressure.

3. Compare the texture of foliated rocks with non-foliated rocks. Name two common metamorphic rocks in each category.

After viewing, answer the following questions:

1. Note Dr. Jean Morrison's work in the stable isotope lab. What role do fluids play in the metamorphic process?

2. What are the major factors that determine the texture and mineral content of a metamorphic rock?

3. What did Dr. Douglas Morton discuss about the origin of metamorphic rocks?

4. How is pressure indicated in the appearance of a metamorphic rock?

5. If you wanted to study metamorphic rocks, where would be the best place to look for them?

6. How are foliated rocks formed? Contrast regional and contact metamorphism and describe some metamorphic rocks characteristic of each environment.

7. Discuss how one parent rock, such as shale, can give rise to different metamorphic rocks.

8. How are metamorphic rocks used as indicators of ancient plate motions and mountain building?

You have read the text assignment and seen the video portion of the lesson. This section, made up of a SUMMARY, a CASE STUDY, the QUESTIONS, suggestions for ACTIVITIES, and a SELF-TEST, will help you pull the information together.

PUTTING IT ALL TOGETHER

Summary

Metamorphic rocks are impressive testimony to the dynamic nature of the planet and to the plate movements that swallow whole ocean floors, shift slabs of crust halfway around Earth, and crunch the rocks between. Equally impressive are the forces that raise rock bodies tens of kilometers from their place of origin on some ancient sea floor, only to have them thrust downward to regions of intense heat and pressure, changing them to a different rock made of new minerals. It is only when these metamorphosed rocks are again uplifted during mountain building and exposed on the surface that geologists can puzzle over them, wondering about the *parent rocks*, where they formed, and what factors contributed to such profound alteration.

Geologists must also consider the relationship between intrusive igneous rocks (Lesson 14) and metamorphism. Metamorphic rocks form at depth, and evidence of the high temperatures that caused their softened or plastic condition is preserved in contorted bands and layers and metamorphic minerals (Figure 15.1, page 330). What would happen if the temperature were raised even further? Some of the minerals would start to melt. The rock would become partially molten, forming pockets of magma and the layers and banding would likely be severely contorted or destroyed. To summarize: metamorphism is recognized as a solid state change, an adjustment to changes in environment. In some cases, the environmental changes come close to the temperature at which igneous rocks form, but they are not high enough for all the rock to become completely molten.

What are the Effects of Temperature, Pressure, and Fluids?

In examining the factors that control the characteristics of metamorphic rocks, we must start with the parent rock. Generally, no new minerals or chemical compounds are added to a rock during metamorphism. The precious marble mentioned earlier is composed of calcite (calcium carbonate), which is likewise the chemical composition of limestone, the parent. Most metamorphic rocks are usually high in silica, because most of the rock-forming minerals present in the parent rocks are silicates.

Temperature is a crucial factor during metamorphism. As you learned in previous lessons, as one goes deeper into Earth, the temperature increases. Deep burial alone can cause a rise in temperature, but most metamorphic rocks form where the temperature rises at a greater rate than the normal geothermal gradient. Minerals are stable only within a given temperature range, and rocks subjected to metamorphism change progressively in their mineralogy as they are heated. The range of stability can be determined from laboratory experiments that provide geologists with convincing evidence of the temperature at which metamorphism occurred. Metamorphism rarely takes place below about 200° C even though some minerals may be unstable at that temperature.

Pressure is another important factor and results from the weight of the overlying rocks and from plate movements. *Confining pressure* occurs at depth within Earth's crust and acts uniformly in all directions, as seen on the deep-sea diver in Figure 15.2, page 331. Minerals have specific stability limits in terms of pressure and will change to new minerals at some pressure limit. The conversion of graphite to diamond at high pressure is a prime example. Pressure forces mineral grains together, eliminating pore space and forming minerals that are denser than low pressure forms. It is almost impossible to imagine the squeezing force of up to 30 metric tons (33 tons) of pressure that are applied to every square centimeter of rock deep in the crust, but it is no wonder that metamorphism takes place under those conditions.

Directed pressure is also commonly present during metamorphism and differs from confining pressure in that it is applied unequally on the surfaces of a body. Directed pressure tends to deform rocks through compression or shearing. Shearing flattens the object parallel to the applied pressure, and the rocks are frequently broken and moved, as in faults (Figure 15.3 A and B, page 331). *Compressive* pressure flattens the object perpendicular to the applied pressure. Both shearing and compressive pressure develop in response to mountain building activity. Stretched pebbles or fossils record such pressure (Figure 15.4, page 332).

The effect of *fluids* is very important in metamorphic reactions. Generally fluids rich in water or carbon dioxide have a tremendous effect on how a rock responds to stress. Water aids in softening rocks so that they become plastic at lower temperature. Even if the temperature and pressure are high enough to cause a metamorphic reaction, some water is generally necessary to allow the chemical changes to take place. Water can bring in or remove elements from the rock and can move ions from one mineral to another, forming a new stable mineral under the existing conditions of temperature and pressure. Metamorphism is a very slow process and without fluids probably would not happen at all or happen only at insignificant rates.

Metamorphic Textures: Foliated and Non-Foliated Rocks

Texture, as you learned in the lessons on igneous and sedimentary rocks, is an important criterion in the identification of rocks. Metamorphic rock textures are classified as *foliated* or *non-foliated* (Figure 15.7, page 334). *Foliation* refers to the banded or layered features resulting from the parallel orientation of the minerals. Incidentally, the term *foliation* has the same origin as the word *foliage*, from the Latin *folia* meaning "leaf," and the term was first used in the geologic sense by Charles Darwin in 1846.

A rock becomes foliated when directed pressure or stress causes newly forming minerals to recrystallize and grow in an oriented fashion parallel to each other. If a flat, *platy* mineral such as mica (Figure 9.18 A and B, page 208) is crystallizing in a rock that is undergoing directed pressure, the mineral grows either parallel to the direction of shearing (Figure 15.5 C, page 332) or perpendicular to the direction of compression (Figures 15.5 B and 15.7 B, pages 332 and 334).

Rocks that split easily along nearly flat and parallel planes show *slaty cleavage* owing to the orientation of the mica flakes. With further exposure to higher temperatures and pressures, larger mica flakes will develop that are approximately parallel to each other and give the rock a shiny irregular foliation called *schistosity* (Figures 15.6 and 15.11, pages 333 and 340). When the minerals have separated into light and dark bands, the rock is described as having a layered or *gneissic* texture (Figure 15.12, page 340). Although the terms slate, schist, and gneiss refer to common metamorphic rocks, the descriptions apply to foliated textures as well and may describe rocks of varying mineral composition.

Non-foliated rocks show no banding or layering, vary from fine- to coarse-grained, are homogeneous, massive, are tough, and are sometimes difficult to identify. Some have been completely recrystallized and have none of their original features; others have been only slightly modified by metamorphic processes. The crystals are equidimensional and, because they have not been subjected to directed pressure, do not grow in any preferred orientation (Figure 15.7 A, page 334). Limestone recrystallizing into marble is one example of an interlocking texture that is non-foliated (Figure 15.8, page 335).

Contact Metamorphism and the Origin of Hornfels, Marble, and Quartzite

The two most common types of metamorphism are called *contact* metamorphism and *regional* metamorphism. Both result from the combined effects of temperature, pressure, and chemically active fluids

Contact metamorphism is sometimes called *thermal* metamorphism. As the name suggests, such metamorphism occurs where high temperatures are dominant and where directed pressure is not a strong factor. The rocks formed in this environment are generally non-foliated. Thinking back to previous lessons, you will recall that the heat from molten magma in an intruding batholith can literally *bake* the country rock, which, in the presence of fluids emanating from the crystallizing magma, recrystallizes into various metamorphic forms.

Depending on the size of the intrusion, the contact zone of altered rock – called an *aureole* or halo – is generally relatively narrow, about 1 to 10 meters (3 to 33 feet). Larger aureoles, up to a kilometer (.6 mile) wide can occur where fluid migrations have been significant. The baked zone surrounding a batholith, of course, will be larger than that around a dike or sill. The aureole itself is zoned, with the high temperature minerals developing close to the margin of the batholith and changing into progressively lower temperature minerals away from the contact. A metamorphic rock commonly occurring in the contact zone is *hornfels*, a dark dense hard rock in which the parent rock is generally shale or basalt but can be other rocks as well.

Marble forms during thermal metamorphism, recrystallizing from limestone and retaining the same chemical composition (calcite) (Figures 15.8 and 15.14 A, pages 335 and 344). Marble can be distinguished from limestone by the large, usually equidimensional crystals and lack of fossils. The color is often pure white, although marbles may have any color due to impurities. Green colors are a result of various iron- or magnesium-bearing silicate minerals. Browns and reds in marble indicate the presence of iron oxide (hematite), while black and gray areas are colored by carbonaceous matter.

Marble is similar to limestone in that it will effervesce in weak hydrochloric acid and is relatively soft (about 3 on Mohs' scale of hardness) making it amenable to carving, cutting, polishing, and use as a building facade. The term comes from the Latin *marmor* meaning "shining stone," and the beautiful warm luster of polished statuary marble is caused by light penetrating into the rock and being reflected from the surfaces and cleavage planes of the deeper-lying crystals.

Some of the fine marbles used by archaic and classical Greek sculptors and in ancient Rome were quarried from islands in the Aegean Sea as well as from the Greek mainland, Anatolia, and Italy. During the Mesozoic era, between 65 and 225 million years ago, this area was submerged under tropical seas, and skeletons of marine creatures formed thick carbonate deposits on the sea floor. Over time, the limy deposits were metamorphosed during plate collisions and later uplifted to form beds of white marble.

In the United States, Vermont is the largest producing state for fine marble, with Tennessee second. One of the finest white marbles comes from a quarry at Marble, in west-central Colorado. It owes its existence to the contact metamorphism that took place when a Tertiary igneous intrusion invaded a Mississippian-age limestone. Marble occurs in many countries but, unfortunately, it is subject to weathering, especially in moist climates and in urban areas, where carbon dioxide, sulfur dioxide, and nitrous oxide are abundant in the atmosphere and are contributing to corrosive acid rain. (Review Lesson 15 on weathering.)

Quartzite is metamorphosed quartz sandstone, and in it the pore spaces once separating the individual sand grains are filled with newly crystallized quartz. Most quartzites form at intermediate to high temperatures and are as hard as a single quartz crystal (7 on Mohs' scale). Quartzite is durable and, being composed of quartz, is not susceptible to chemical weathering. (Refer to Table 15.1, page 335, on non-foliated rocks.)

Regional Metamorphism and the Origin of Slates, Schist, and Gneiss

Most metamorphic rocks are foliated, especially those that are the product of *regional* metamorphism, where pressure and temperature are related to plate convergence and subduction. This is where mountain belts rise, and a great volume of rock below the mountains – the "roots" deep in the crust – experiences extreme changes in temperature and pressure. This is also the zone of magmatic intrusion; intense heat and directed stress account for the widespread areas of foliated rock.

From field studies, it was discovered that a single parent rock can recrystallize into one of several metamorphic rocks, depending on the conditions of pressure and temperature. Slate, as mentioned above, is associated with low temperatures and pressures (only slightly higher than those at which sedimentary rocks form) and is derived from shale, the sedimentary parent rock, but can form from other fine-textured rocks such as volcanic tuff. The microscopic flat clay minerals in the shale recrystallize into equally fine-grained mica flakes, which become aligned by directed pressure (Figure 15.9 A and B, page 339). Slate characteristically exhibits *slaty cleavage*, in which the rock splits easily into thin slabs. The original bedding planes of sedimentary rocks can be preserved during the rather gentle regional low-grade metamorphism, but the slaty cleavage may form at nearly right angles to the bedding (Figure 15.9 A and B, page 339).

During *progressive metamorphism*, the increasing intensity of directed pressure and higher temperature will form a series of different rocks. The slates will first grade into *phyllite* (Figure 15.10, page 339), then into a *schist* in which the mica grains are large enough to be visible to the naked eye. A *schist* (from the Greek word *schistos* meaning "divisible") is characterized by a wavy rock cleavage where the original bedding is almost never identifiable. Almost any rock can be metamorphosed into a schist, but the most likely parent rocks are shale, siltstone, and muddy sandstone (Table 15.2, page 338). The elements in the original rocks are important because they can influence the metamorphic product. A *garnet-mica schist*, for example, will have red garnets of various sizes studded through the silvery mica, a very handsome combination (Figure 15.11, page 340).

At the highest temperatures and pressures, a *gneiss* (pronounced "nice," an old Saxon miners' term for a rock that is rotted or decomposed) will develop, and the minerals will separate into light and dark bands, giving the rock a layered appearance (Figure 15.12, page 340). The dark layers consist of dark micas and hornblende (amphibole), while the light layers are feldspars and quartz. The bands, unlike the cleavage planes of slate, do not make uniformly parallel planes that continue for long distances; more commonly, they are strongly distorted (Figure 15.1, page 330). The rock was deformed in the solid state, but at temperatures high enough to undergo plastic flow slowly. The temperature at which gneiss forms is close to that which causes partial melting and at which granite solidifies. The mineral composition can also be similar to that of granite (Table 15.1, page 335) and, in fact, gneisses can also form from preexisting granites.

If partial melting does take place – as it will at great depth in the crust – the liquid magma, generally of felsic composition, will rise out of the zone of metamorphism, leaving behind the solid metamorphic component of the rock. Generally these are the minerals that remain stable and do not melt at high temperatures. Sometimes the liquid portion will intrude fractures or foliation planes and will solidify, forming a variety of gneiss called *migmatite*, a curious mixture of granite and schist. Migmatites are common throughout the world wherever high-grade metamorphic rocks exist and are testimony to partial melting that once occurred at great depth beneath emerging mountains (Figure 15.15, page 344).

Metamorphism at a Convergent Plate Boundary

Figures 15.13 A and B, page 341, even at a first glance, should be very familiar to you after several lessons involving plate tectonics. The diagram covers a large region about 325 kilometers (195 miles) wide by a depth of about 150 kilometers (90 miles). (Note the scale on the diagram.) The diagrams in the figure illustrate both pressure (Diagram A) and temperature (Diagram B) variations across a converging plate boundary.

Near the slab of ocean floor descending into a deep sea trench is a zone of high pressure-low temperature metamorphism where sediments of the cold sea floor and of the continental shelves are being subducted and undergoing shearing. As the diving, wet, oceanic crust undergoes partial melting at a greater depth and as rising magma underplates the continental crust (Figure 10.21, page 235), a zone of intense compressive directed pressure develops. A different form of metamorphism, one of *high temperature-high pressure*, will take place. From these diagrams, it is apparent that metamorphism can take place over a range of different pressure and temperature combinations and produce minerals that would be unique and stable in each distinct metamorphic environment.

Metamorphic Facies

Basically, a metamorphic *facies* is a group or assemblage of minerals that formed in response to a specific environment. During field studies, rocks of a given composition are traced from areas of little or no metamorphism to areas of intense metamorphism. The minerals in the rock changed as the metamorphic environment changed. Laboratory experiments conducted on metamorphic rocks to determine the temperature-pressure environment in which the facies rocks and minerals were created allow geologists to determine pressures and temperatures experienced by a rock millions or billions of years ago. In this way, conditions deep in Earth's crust where metamorphism takes place can be inferred.

Each facies is named after the predominant minerals or rock type. Box 15.2, Figure 1, page 342, shows the various metamorphic facies with corresponding temperature, pressure, and depth in the crust. Note that hornfels (a rock that is non-foliated and forms in zones of contact metamorphism) is a low pressure rock but forms at a wide range of temperatures. Amphibolite forms over a wide range of pressures but within a narrow temperature range. Depth alone does not determine the temperature. Using Box 15.2, Figure 2, page 343, and referring back to Figure 15.13, page 341, you can see the relationship among plate tectonics, regional metamorphism, and the distribution of metamorphic facies. Blueschists, as an example, form adjacent to subduction zones, where the temperatures are low and the pressures high. The greenschist facies is located closer to the magmatic arc where the temperatures are higher.

Hydrothermal Rocks

Hydrothermal (hydro means "water" and thermal means "heated") rocks are formed by precipitation from solution in hot water, and although water is also important in metamorphic processes, these rocks are difficult to classify as metamorphic rocks (Figure 21.13 A and B, page 486). *Metasomatism* – sometimes called *replacement* – refers to the simultaneous solution and deposition by heated fluids in which minerals are replaced by others of different chemical composition. The reactions are caused by introduction of fluids derived from external sources. During contact metamorphism, the cooling magma is the source of replacement ions. Metasomatism also occurs during regional metamorphism, during which ions and fluids released by metamorphic reactions result in large crystals of feldspar and quartz (Table 15.3, page 343).

A common hydrothermal deposit is vein quartz, which can occur near bodies of cooling magma (Figure 15.16, page 345). The source of the water may be magmatic or, more likely, deep circulating ground waters that become heated and leach metals and other materials from surrounding rocks. Hydrothermal rocks range from high temperature to low temperature occurrences and are the most common source of commercial metal deposits (Figures 15.17 and 21.14, pages 345 and 486).

The combination of minerals found in a metamorphic rock is the basis for determining the metamorphic facies to which the rock belongs. From the elements in the minerals, geologists can interpret the temperature and pressure at which metamorphism took place, a geologic giant step toward understanding the dynamics of Earth. The events leading to metamorphism, we now know, are plate collisions, subduction, and mountain building, or perhaps divergents at boundaries of millions, even

billions of years ago. Very ancient regionally metamorphosed rock and
granite form large masses within continental interiors such as the Canadian Shield. These rocks are an affirmation of ancient plate movements
and now-eroded mountain belts that built the planet we know today.

CASE STUDY:
Marble Sculpture:
Real or Fake?
Ask the Geologist!

For more than 7,000 years, marble sculpture has mirrored the artistic
and spiritual progress of Western civilization. Collectors from the Re-
naissance to the present have lavished princely sums on marble statuary
from the ancient world. For as long as there has a been a market for
marble sculpture, however, there has also been a thriving industry in
copies and forgeries. To tell authentic pieces from fakes, experts have
relied on each object's style and art-historical context. They have also
relied on the appearance of the surface layer, or *patina*, to guess the age
of the carving. Forgers often try to fake a patina by a variety of methods,
from burying the sculpture in cow pastures to applying acid or pastes to
it that mimic an ancient crust.

As prices for such works have skyrocketed, geologists and geochemists have been called in to help settle questions that art historians and
the trained eye of experts cannot resolve. In 1984, a famous museum in
California was offered a chance to buy an archaic Greek *kouros*, which is
a larger-than-life marble figure of a youth. The two-meter-tall statue,
said to be more than 2,500 years old, was well preserved but was unknown to art historians. The asking price was higher than had ever been
paid for an ancient statue. It was reported in newspapers that the sum
was somewhere between $8 million and $12 million! Some of the art
experts declared it was authentic; others, however, considered it "ques-
tionable." Some experts said that ancient marble surfaces fluoresce an
amber color mottled with purple (not proved). The kouros fluoresced an
uneven light purple, a color associated with modern surfaces. Considering the importance of the statue, the doubts raised, and the price asked,
museum officials wanted conclusive scientific evidence of its authentic-
ity before they would consider buying the piece.

The question was: could the statue have been sculpted between 540
and 520 B.C. as suggested by art historians, or had it been carved in more
recent times and then treated artificially to appear ancient?

This is not an easy problem to solve, since stone objects are much
more difficult to authenticate than paintings, ceramics, and other human
artifacts. The stone's absolute age reveals nothing about when it came
under the sculptor's chisel. Methods do exist, however, for determining
when a stone surface was first exposed to the environment. Geologic
weathering takes place over millions of years and can be studied using a
variety of tools such as a polarized light microscope, an electron microscope, and an electron microprobe. Weathering of a piece of ancient
marble sculpture occurs over a few thousand years at most, but the basic
processes are similar.

To conduct the examination of the kouros, a small core was drilled,
very carefully, from the right knee, measuring about one centimeter in

diameter and about two centimeters in length. The initial examination revealed that the marble was composed of relatively pure *dolomite* (calcium magnesium carbonate). (See Table 15.1, page 335.) Dolomite marble is more durable and generally more resistant to weathering than calcite marble, and that could have explained the excellent preservation of the kouros. The kouros surface revealed a thin, tan patina of iron oxides and soil-clay minerals along with manganese oxide encrustations similar to those found on ancient marble outcrops. Further research showed the weathered surface was covered with a continuous layer of calcite. The finding of calcite was significant because dolomite is known to change into calcite through weathering. The calcite did not appear on any of the fresh surfaces inside the small core.

The next question was whether the calcite could have been produced in a modern laboratory, and the answer was only with great difficulty. It was unlikely that the calcite on the kouros was produced artificially in a uniform layer over the two-meter length of the statue. Nor could an artificial calcite layer have been applied to the stone since the trace elements found in the dolomite were also found in the thin calcite patina. The most reasonable explanation of the calcite on the kouros is that it developed through at least several centuries and no more than several millennia of weathering. After 14 months of intensive geological and geochemical study, the museum decided to buy the kouros, and it now stands upright and complete, a wonderful piece, appearing much as it must have in ancient times.

The geochemical techniques used in this study apply only to dolomite marble. Unfortunately, calcite marbles are much more common, and although they also weather, different criteria are required for authentication. Studies of the patina surfaces have revealed some statues to be genuine that were previously in doubt, and others were found to be indeed 20th century copies. Unfortunately, until recently, restorers of ancient marble sculpture cleaned the surface with acid, a practice that removed the weathering crust and made it impossible to apply these geochemical methods of authentication. Thus the validity of many works will remain in question forever because they were acid-cleaned. In the last few years, however, there has been much progress in understanding the natural weathering of ancient marble sculpture and in protecting the artistic legacy of human history.

(This case study is based on an article that appears in *Scientific American*, June 1989, "Authenticating Ancient Marble Sculpture," by Stanley V. Margolis, Vol. 260, Number 6.)

1. If metamorphic rocks make up over one-quarter of Earth's crust, why have they never been seen in the process of formation?

Questions

Metamorphism takes place under conditions of high temperature and high pressure that occur at depth within the crust. Metamorphic rocks are seen only after plate movements and mountain building have uplifted portions of crust, subjected them to deep erosion that removed the overlying rocks, and exposed the changed rock on Earth's surface.

2. If one parent rock can produce several different metamorphic rocks, what influence does the parent really have on the final rock?

The parent rock contributes certain elements and compounds, which in many cases do not change appreciably during metamorphism. For example, in the metamorphism of quartz sandstone to quartzite and limestone to marble, there is no change in the original quartz and calcite mineral content.

Natural rocks, however, have a tremendous diversity and contain many elements that may affect the final metamorphic equivalent. Metasomatic elements (or ions) carried by circulating solutions will also participate in the chemical reactions that change the final metamorphic product.

3. In trying to identify metamorphic rocks, what is the first thing to look for?

Metamorphic rocks are identified by first determining if the rock is foliated or non-foliated. *Metamorphic Rocks* in Appendix B in your text has an excellent key for identifying common metamorphic rocks.

4. What causes foliation in some metamorphic rocks? What are the three variations of foliation?

Foliation refers to the parallel alignment of minerals that accounts for the texture and structure of the rock. Minerals such as micas are *platy*, meaning they have a sheet-like internal structure and appear as microscopic thin plates. (Refer to the photograph facing page 329.) In the absence of directed pressure, the orientation of the plates will be random, with no structure or foliated texture. When subjected to directed pressure, such as compression, or under the influence of shearing, the mineral plates will line up either at right angles to the direction of compression, or parallel to the direction of shearing. Foliation is a characteristic of rocks formed during regional metamorphism and represents the imprint left by conditions achieved deep in Earth's crust millions of years ago. The three types of foliation are slaty, schistose, and gneissic and indicate in that order increasing levels of metamorphism.

5. Why can't valuable metamorphic minerals such as gem stones be duplicated in the laboratory if their chemical content is generally of common elements?

Many metamorphic minerals can be duplicated in the laboratory, but the purity of gem stones is difficult to achieve because of the time element. Reaction times are very slow. The high temperature and pressure of metamorphism would have to be sustained for years to produce the required result. A few tiny diamonds have been fabricated in the laboratory but mostly as a scientific curiosity.

6. What is the effect of heat in causing metamorphism?

Heat is probably the principal factor involved in metamorphism of rock. At high temperatures, chemical reactions are more rapid; the ions in the minerals move and diffuse more quickly and thus activate recrystallization of minerals. Rock tends to decrease in strength at high temperatures and is more likely to deform or even to flow plastically in response to directed pressure. The degree of metamorphism depends upon the stability of various minerals within a certain temperature range as well as the effects of pressure and the presence or absence of chemically active fluids.

7. If metamorphic rocks are characteristically hard, why should builders be hesitant to build on them?

When we say metamorphic rocks are hard, we have to distinguish between foliated and non-foliated rock. It is true that quartzite, a non-foliated rock, is one of the hardest rocks known; hornfels is also hard. Marble, however, another non-foliated rock, is made of calcite, which is relatively soft, about 3 on Mohs' scale of Hardness. The foliated rocks have planes of weakness between the layers, and when the foliation planes are parallel to a sloping surface, such as a hillside, they are prone to landsliding. Water between the layers acts as a lubricant; buildings add weight, but gravity alone could cause instability in tilted layers. Metamorphic rocks are often fractured, another reason to look elsewhere for stable building sites. (See Box 15.1, page 335.)

8. We know that metamorphism takes place at convergent plate boundaries, but does it also occur at divergent boundaries such as the mid-oceanic ridges?

Since the crests of the mid-oceanic ridges are areas of active basalt volcanism, it is not surprising that the nearby rocks are subjected to metamorphic conditions. Water from the volcanic activity and the heat from below that generates the magma are the factors that cause metamorphism. Zeolite facies (low temperature and pressure) and greenschist facies mineral assemblages (Box 15.2, Figure 1, page 342) have been recovered from sea-floor rocks near the spreading centers.

Activities

1. Unless you live in a mountainous region, it is unlikely that you will see metamorphic rocks in your neighborhood. The geology department at your local college will have a collection you might be able to see. A natural history museum will have a display of rocks and

minerals, and you will be able to compare the three great rock families you have studied in this course.

2. A great place to study polished igneous and metamorphic rocks is city buildings where red granite, black gabbro, and beautifully swirled gneiss are commonly used as a building stone.

3. Go to an art museum and study the pieces of sculpture. You may see some fine examples of marble or other kinds of rock used as an art form.

4. If there is a builders supply store near you, take a field trip and look for slate used in roofing and for floor tiles. Decorative slates come in many colors and can be quite attractive. Also, look for marble table tops and counter tops. Why is marble not the best material for coffee tables, bar tops, and other load-bearing surfaces? (As you may have guessed, it is soft and susceptible to dissolution by acid solutions. Orange juice, for example, will leave a ring that may penetrate into the stone and resist removal.)

5. Stone cutters have many kinds of decorative rock and may be willing to give you a small piece to start your own collection.

Self-Test

1. Weathering and metamorphism are both reactions to changes in environment, except that in the case of metamorphism

   a. no new minerals are formed.
   b. water does not play an important part.
   c. elevated temperatures and pressures are required.
   d. the rocks must first become molten.

2. Metamorphic rocks are most likely to be found in all of the following except

   a. around batholiths or large igneous intrusions.
   b. interbedded with thick deposits of sedimentary rocks.
   c. in areas of mountain building.
   d. along spreading centers on the sea floor.

3. Foliation is an important characteristic of regional metamorphic rocks and is caused by

   a. confining pressure.
   b. high water content in the minerals.
   c. location within a contact aureole.
   d. directed pressure.

4. The elements found in most metamorphic rocks are derived from

   a. rocks becoming molten.
   b. the chemical composition of the parent rocks.
   c. intense directed pressure at great depth.
   d. facies change.

5. Each of the following metamorphic rocks could be a product of the metamorphism of shale except

   a. mica schist.
   b. phyllite.
   c. quartzite.
   d. slate.

6. The metamorphic rock that most closely resembles an igneous rock is

   a. gneiss.
   b. slate.
   c. greenschist.
   d. quartzite.

7. If you wanted to learn more about foliated metamorphic rocks, a good place to look for them would be

   a. in continental interiors.
   b. adjacent to volcanic necks, dikes, and sills.
   c. near active faults.
   d. in deformed portions of mountain ranges.

8. The high temperatures of regional metamorphism are related to all of the following except

   a. geothermal gradient.
   b. friction from earth movements.
   c. impacts by meteorites.
   d. heat radiated from nearby magma bodies.

9. During partial melting of metamorphic rocks, the minerals that will be the first to melt are

   a. olivine and calcium plagioclase.
   b. pyroxene and amphibole.
   c. biotite and sodium feldspar.
   d. quartz and potassium feldspar.

10. As a result of partial melting of metamorphic rocks, the molten material will have the chemical composition of a

    a. granite.
    b. zeolite.
    c. shale.
    d. mafic magma.

11. Low temperature-high pressure metamorphic rocks are characteristic of

    a. aureoles around batholiths.
    b. areas adjacent to subduction zones.
    c. magmatic arcs.
    d. a granulite facies.

12. Out in the field, you would recognize a migmatite by its

    a. slaty cleavage.
    b. ores of silver, zinc, gold, and lead.
    c. mixture of schist and granitic layers.
    d. marbles and hornfels.

13. True marble is

    a. a fine-grained quartzite.
    b. a form of gypsum.
    c. ancient stalactites.
    d. metamorphosed limestone.

14. A building constructed of blocks of quartzite is more durable than one made of marble because

    a. marble is foliated and has planes of weakness.
    b. quartzite is one of the hardest rocks, and marble is relatively soft.
    c. quartzite is foliated but resistant to weathering.
    d. marble occurs in migmatites and easily breaks apart.

15. The primary sources of many economically important metallic ores are

    a. blueschists in converging plate boundaries.
    b. gneisses and schists.
    c. hydrothermal veins.
    d. slates and phyllites.

19

RUNNING WATER I: RIVERS, EROSION, AND DEPOSITION

This lesson will help you describe the erosional and depositional characteristics of moving water.

After reading the textbook assignments, completing the exercises in this study guide, and viewing the lesson's video portion, you will be able to:

1. Distinguish between the tectonic and hydrologic cycle.

2. Describe and differentiate between water flow in a channel and sheet flow.

3. Give examples of the techniques farmers use to slow sheet erosion in their fields.

4. Indicate the factors that influence the velocity of a stream.

5. Recognize the relationship between stream velocity and the erosion, transportation, and deposition of sediment.

6. Compare three mechanisms by which a stream erodes the rock and sediment: hydraulic action, solution, and abrasion.

7. Describe how sediment is transported as bed load, suspended load, and dissolved load.

8. Describe the characteristic longitudinal profile of a stream.

9. Describe the various features created through repeated cycles of river erosion and deposition, including potholes, bars, braided streams, meandering streams, flood plains, deltas, and alluvial fans.

10. Recognize the role human activity can play in intensifying or reducing the potential danger of natural flooding.

INTRODUCING THE LESSON

From "Old Man River" to "Row, Row, Row Your Boat," countless composers have been inspired by the sparkling music of flowing streams. If we add the work of painters, sculptors, and writers enchanted by the Seine in Paris, the Thames in London, the Nile, or the Mississippi, we get a sense of the fascination and importance of rivers to human life and activities.

It is no coincidence that the greatest cities on Earth are located on river banks: Paris, London, Cairo, New York, New Orleans, to mention but a few. Cities were founded near river landings, where ships and barges could bring in goods and carry away raw materials, where pioneers and explorers could follow routes from the sea to the continental interior, and, in desert regions, where rivers provided fresh water for irrigation. Rivers were also natural barriers, and the crossing of the Danube by the barbarians is cited as one of the events leading to the fall of the Roman Empire.

While rivers can be life lines, they can also be ferocious agents of destruction. Most river cities have been victims of severe flooding, especially in the days before dams and flood control projects partially tamed the rampant streams. Inundation of cities and erosion of valuable agricultural lands are still intermittent problems that have not been completely solved.

To the geologist, streams are the greatest sculptors of the landscape, shaping the hills and valleys, eroding the highest mountains in a continuing battle with tectonic forces, and transporting the worn particles to a distant flood plain, basin, or ocean. In this lesson, you will learn the nature of stream flow, the factors that affect velocity, and the mechanics of *stream erosion, transportation,* and *deposition.* The landforms that result from stream action, such as *meandering streams, oxbow lakes,* and *deltas,* will be discussed, and the role that human activities play in causing or modifying the dangers of erosion and flooding will be examined.

LESSON ASSIGNMENT

Completing the following eight steps will help you master the lesson objectives and achieve the goal for this lesson:

Step 1: Read the TEXT ASSIGNMENT, Chapter 16, "Streams and Landscapes."

Step 2: Study the KEY TERMS AND CONCEPTS as noted in the study guide.

Step 3: Watch the VIDEO using the VIEWING GUIDE in the study guide.

Step 4: Read the study guide's PUTTING IT ALL TOGETHER section, which will help you summarize and integrate all of the information in this lesson.

Step 5: Complete any assigned lab exercises.

Step 6: Complete any assigned ACTIVITIES found in the study guide.

Step 7: Review the material in this lesson and complete the SELF-TEST found in the study guide.

Step 8: Go back to the LEARNING OBJECTIVES and make sure you have learned each one.

1. Before reading all of Chapter 16, pages 349-381, read the "Introduction" and study the photograph facing page 349. Note the little waterfalls, the turbulence, the water-worn boulders, and the mass wasting on the far side of the stream.

2. Read the chapter as far as "Valley Development" on page 368, carefully noting the diagrams and photographs. (The second part of the chapter will be studied in Lesson 20.)

TEXT ASSIGNMENT

The key terms listed under "Terms to Remember," page 380 in your text, will aid your understanding of the material in this part of the lesson. From the list, record the terms that appear in this lesson. At the end of Lesson 20, you should have learned any terms remaining in the list. Be sure to use the glossary and the text for any unfamiliar terms. Here is an additional term for this lesson:

KEY TERMS AND CONCEPTS

Alluvium: A general term for all stream deposits of comparatively recent time. The term *alluvial* pertains to alluvium and is used to describe fans, *flood plains*, slopes, *terraces*, and other features originating from stream deposition. "Alluvial diamonds" are diamonds found in river gravels, and "alluvial gold" is gold found in association with water-worn materials.

Once you have read the text material for the lesson and studied the terms, you are ready to view the lesson's video portion. Review the questions that appear in this section to help you watch for important points and to prepare you for what you will see in the video.

VIEWING GUIDE

Before viewing, answer the following questions:

1. What is meant by the hydrologic cycle? (See Figures 1.5 and 17.1, pages 6 and 384.)

2. What factors control stream velocity?

3. Compare stream capacity and stream competence.

4. Compare the movement of the bed load with that of the suspended load.

After viewing, answer the following questions:

1. Note the physical model of Red Eye Crossing and the ship channel at Baton Rouge, Louisiana, prepared by the U.S. Army Corps of Engineers. What is the special problem encountered by shipping using Red Eye Crossing?

2. How does the Army Corps propose to rectify the ship channel problem? Why is the solution less than perfect?

3. Under what conditions will a stream erode or deposit within its channel?

4. How do changes in base level influence a stream profile?

5. What are the three types of stream load?

6. How does the capacity of a stream relate to channel slope and discharge?

PUTTING IT ALL TOGETHER

You have read the text assignment and seen the video portion of the lesson. This section, made up of a SUMMARY, a CASE STUDY, the QUESTIONS, suggestions for ACTIVITIES, and a SELF-TEST, will help you pull the information together.

Summary

Streams, aided by mass wasting, are the most important geologic agents both in shaping the landscape and in eroding, transporting, and depositing sediment. The effect of streams can be seen in every terrestrial environment, be it tropical, polar, desert, mountain, or shore. Human activities have been concentrated on the banks of streams since prehistoric times because of the benefits of transportation and communication. Rivers, however, rise from time to time and flood adjacent cities and fields, and despite many flood control projects, major floods are still very damaging.

The Great Cycles: Tectonic and Hydrologic

If streams are such powerful agents of erosion and have acted since the first drops of water splashed on barren rocky lands, we might wonder why any continents remain above sea level. Tectonic movements, as you have learned in previous lessons, are driven by internal heat flowing from Earth's interior and result in plate collisions that raise land, build mountains, and extend continents. While erosion lowers the land, continents are renewed and stand in mute testimony to the tremendous forces unleashed by the internal heat engine.

The hydrologic cycle is essentially solar powered and equally impressive because it involves all of the possible paths of water in the atmosphere, the sea, and on land. Most water in the hydrologic cycle – about 97 percent – is in the oceans. On land, it includes the water that seeps into the ground to become ground water, the water in glaciers and lakes, and the small percentage that becomes "run-off" in streams and rivers, the subject of this lesson (Figure 17.1, page 384).

Under the influence of the hydrologic and tectonic cycles, the landscape is never static but is constantly evolving. While the process generally operates slowly (flooding can be an exception), given geologic time nothing on land is immune to the work of streams.

What is a Stream?

A stream (or river) is a narrow body of water that flows in its channel between the stream banks and over the stream bed (Figure 16.2, page 351). Streams have a *head* (near their source in the mountains) and a *mouth* (at the lower end, be it delta, alluvial fan, lake, or ocean). The cross section of a stream valley changes, from the head towards the mouth, from a narrow V-shaped canyon to a broad open valley and flood plain (Figure 16.1 B and C, page 350). The stream also has a longitudinal profile as if viewed from the side; the profile is a long concave upward sweep, steeper in the mountains and flatter near the sea (Figure 16.1 A, page 350). (Remember: these diagrams in the text are just that, *diagrams*. Streams are variable, complex, dynamic, ever changing; the diagrams just illustrate the general principles.)

Not all water flowing over the land is confined to channels. After heavy rains, the water may flow downhill as a thin surface layer called *sheetwash*, moving considerable amounts of soil and other weathered rock debris. The water may become concentrated in small irregularities in the ground surface, forming gullies and rills. The rills (small flows cut into the surface that are parallel to each other) coalesce into small streams, which join to form larger streams and create a tree-like network of flowing water (Figure 16.3, page 352).

Stream Velocity

The work done by streams, such as erosion, transportation, and deposition, is dependent on the stream's velocity, which is defined as the direction and speed of the water in the stream bed. It is measured in meters or feet per second to miles or kilometers per hour. The average rate is about 5 or 6 kilometers (3 or 4 miles) per hour but can vary from about 1 kilometer (0.6 mile) per hour to the maximum recorded rate of about 24 or 31 kilometers (15 to 19 miles) per hour during flood time.

Measurements have shown that the velocity of a stream is not consistent from one bank to the other but that the greatest velocity is near the middle, where the friction against the banks and stream bed will be the least (Figure 16.6 B, page 354). When a stream goes around a bend, however, the water is thrown to the outside of the curve, shifting the region of maximum velocity toward that bank and causing it to erode. On the inside of the curve, where the water moves more slowly, sediment is deposited (Figure 16.6 A and C, page 354). (Think forward in time: what will happen to the shape of the stream channel if erosion continues on the outside of each bend, and sediments are deposited on the inside? Peek at Figure 16.23, page 362).

What Controls a Stream's Velocity?

Gradient – the downhill slope of the stream bed – is the most important factor controlling stream velocity. Generally, the steeper the gradient, the greater the rate of stream flow. In a high waterfall, where the drop is almost vertical, the rate may approach free fall, as in Niagara Falls, some of the falls in Yosemite, or Victoria Falls in southern Zimbabwe. In a steep mountain stream, the water may drop 10 to 55 meters per kilometer (50 to 200 feet per mile) and will leap and tumble within its banks as it rushes downstream. (The increased turbulence and friction within the rocky channel, however, may moderate the velocity despite the steep gradient. The water may spend a good part of its time moving up, down, sideways, or even backward before it makes its way downstream. See the photograph facing page 349.) Toward the mouth, where the gradient has dropped to about 0.15 meters per kilometer (0.5 feet per mile), the velocity generally has decreased as the river wends its way to the sea. (Increased discharge or volume of water at the lower end may actually increase the velocity slightly, over-riding the effects of the reduced gradient.) "Old Man River" isn't just rolling along; he's actually going pretty fast but in a smooth, efficient way. (Refer to Figures 16.7 A and B, page 354, and Figure 16.8, page 355).

The *shape of the channel* in cross-section is another factor that influences stream velocity. If the channel is wide and the river shallow, or if the channel is rough and filled with boulders, the friction between the river bed and the water is at its greatest, and the velocity will decrease (Figure 16.10 B and C, page 355). If, on the other hand, the stream encounters a layer of hard rock, resistant to erosion, the valley will become narrow, and the water will surge with increased velocity through the restricted opening. Obstructions in the channel such as slump, landsliding, or even bridge abutments will also cause a local increase in velocity (Figure 16.11, page 356). The channel shape that resembles a semi-circle offers the least frictional resistance and allows for the greatest velocity (Figure 16.10 A, page 355). Because all the factors that affect velocity also affect the erosive power of the stream, you can see that in a stream such as the Amazon or the Mississippi the delicate balance between erosion, transportation, and deposition is constantly changing from place to place as the long rivers make their way to the sea.

Discharge is an important factor that influences stream velocity. Discharge is defined as the quantity of water passing a point in a given interval of time. It is stated in cubic feet or cubic meters of water per second. (Refer to page 356.) The discharge of most rivers is far from constant and is strongly affected by location and seasonal changes. In northern or high altitude rivers, discharge fluctuates with the melting and freezing of snow and ice. Streams of the arid southwestern United States also show a great range of discharge, from completely dry to a raging torrent following a sudden cloudburst. The discharge of the Amazon, the world's largest river, is estimated to be about 150,000 cubic meters per second; one day's volume would satisfy the water needs of New York City for about 5 years. This is almost 10 times the discharge of the largest river in the United States, the Mississippi.

With an increase in the volume of water (discharge), the velocity will increase as the river flows down valley. In most streams, the discharge increases downstream, primarily because entering tributaries add their water to the flow (Figure 16.3, page 352). Ground water may also enter the river through the stream bed and contribute to the discharge (Figures 17.8 and 17.11, page 389). To accommodate the increased discharge, stream valleys become wider and deeper downstream. During floods, both discharge and velocity may increase enormously, and erosion and transportation of sediment will increase proportionately. When the flood waters abate, the power of the stream to erode and transport drops, and deposition of sediment, usually mud, will occur, adding to the despair of families living on flood plains.

Streams Cut Their Own Valleys

When you look down from the rim of the Grand Canyon, it doesn't seem possible that the narrow, muddy Colorado River could be the agent that excavated such a huge gash in the Colorado Plateau. But that is exactly what happened and is still going on through a variety of geologic processes. The next question is: where *is* all the rock that the river removed during downcutting and valley widening?

Stream erosion is accomplished in three ways: hydraulic action, solution, and abrasion. *Hydraulic action* refers to the ability of turbulent water to pick up and move rock and sediment in the river channel. It is also a quarrying process, during which the water, under pressure, will be forced into cracks in the river bed and banks, forcing out rock fragments and moving them downstream (Figure 16.12, page 357). *Solution* is an effective process of weathering and erosion especially in limestone rocks. As you remember from Lesson 15 on weathering, limestone – composed of calcite or calcium carbonate – is a relatively soluble sedimentary rock. Thus, stream channels in limestone regions will be deepened and widened by solution in addition to other river processes; the removed calcium carbonate will be carried away in solution by the stream.

The third and most effective erosional process is *abrasion*. The loosened particles of sand, gravel, and pebbles, moved by hydraulic action, are the tools of the stream that grind away at the river banks and the stream bed, and by friction and impact wear them away. The tools themselves are abraded, accounting for the rounded pebbles and boulders common in streams. All of these processes work together with mass wasting along the banks to deepen and slowly widen the valleys of the land.

Stream Transport

The sediments eroded from the banks and stream bed, and the rocky rubble delivered into the stream by slump and other forms of mass wasting are termed the *load*. Load includes *dissolved* load, *suspended* load, and lesser amounts of *bed* load.

The part of the river's burden called the *bed load* consists of the larger, heavier particles that stay near or on the stream bed. Only during the flood stage will the velocity and water pressure be sufficient to transport these larger rocks actively (Figures 16.5 A and B, 16.15 and 16.16, pages 353 and 358). In terms of work accomplished by a stream in cutting down or widening its channel, the bed load is by far the most important. The boulders, cobbles, pebbles, and gravel roll, slide, or drag along the bottom of the stream in a motion called *traction*, vigorously abrading the stream bed and each other.

Particles the size of sand grains, also part of the bed load, move downstream by *saltation* (from the Latin word *saltare* meaning "to jump, leap, or dance"). A small local increase in the velocity of the water caused by an eddy or other turbulence will lift the sand above the stream bed; while the sand grain is suspended in the water, it will be transported downstream a short distance before it settles back to the bottom of the stream (Figure 16.15, page 358). In rapidly moving streams, much of the sand in the bed load will saltate, so that each particle becomes part of the suspended load for a short time; as the velocity drops, the particles will sink to the bottom, dependent on another stream eddy for their leaping, bounding journey downstream.

The *suspended load* can be seen when comparing a clear mountain stream in the high mountains with a muddy flow in a desert wash after a heavy rain. The fine sediment that clouds the water consists mostly of clay-size particles that will remain suspended in the water even after the velocity drops (Figure 16.31 D, page 367). How long the clays stay afloat depends on the size, shape, and density of the particles, the velocity of the current, and the degree of turbulence. Flat particles like clays and mica follow a complex path downstream; lifted by the water, they start to settle, then are swept up and swirl erratically in the next eddy, suspended above the bed but not eroding it, just muddying the water (Figure 16.15, page 358). While clay and silt particles are easily suspended once picked up, it is hard for the water to lift them because molecular forces tend to bind them in a cohesive mass. Only high water velocities dislodge them, but they stay suspended until the water almost stops moving (Box 16.2 Figure 1, page 359).

The *dissolved load* is invisible but transports more material than either the bed load or the suspended load! The source materials are the soluble products of weathering leached out of rock and soil; they will go wherever the river goes and will precipitate only under special conditions. When the water in an arid region evaporates, for example, a white residue will be left that can cover the ground and look like snow. These are the crystallized salts from the dissolved load, which, incidentally, in sufficient concentration can be fatal to crops. Since the dissolved load varies from stream to stream, the water of each river has its own distinctive taste (faint, but distinctive). The dissolved load is estimated to be millions of tons per year, a figure that makes one think the continents are literally dissolving away.

Capacity and Competence

Capacity and competence refer to a river's ability to carry sediment. *Capacity* is the term for the potential load of sediment that a stream can carry and depends upon the velocity and discharge of the current. A sluggish stream meandering across a swamp is capable of moving very little material compared to a fast moving stream in flood stage. A small tributary with a low volume, moving at about the same velocity as the two kilometer-wide Mississippi, has a capacity that is clearly a fraction of that mighty river.

Competence is a measure of the largest particle size that the stream can carry and, as you have learned, depends primarily upon the velocity. At low velocities, many streams run clear, and the sediments rest undisturbed on their beds. With increasing velocity, the turbulence increases, and larger and larger particles are picked up. There are many examples of streams at flood stage moving automobiles caught in their path. During the famous Johnstown (Pennsylvania) flood of 1889, several railroad locomotives were swept from the roundhouse and disappeared downstream. A single flood may cause more erosion than 50 to 100 years of normal stream flow.

The relation between stream competence and stream velocity is not a simple one. From an examination of Box 16.2, Figure 1, page 359, it can be seen that it takes a higher velocity to pick up (erode) sediment than to transport it. Also, higher velocities are needed to erode larger particles, as would be expected, but also to pick up fine clays. Note that heavy particles are quickly dropped as the stream velocity starts to lessen but that the clays can stay suspended even after the velocity is zero.

Stream Deposits

Because stream velocity and discharge are ever changing, sediments that were picked up at higher altitudes and deposited where the gradient is lower are only temporarily in place and will be moved downstream again when the volume or velocity increases. Stream deposits called *bars* form in the middle of the channel as ridges of sediment or along the stream banks. Later, flood waters may sweep the channel clean of old bars, but when the water level subsides, new bars will form as boulders and cobbles drop out first, followed by gravels and sands (Figures 16.5 B, 16.16, 16.17, and 16.18, pages 353, 358, and 360). The fine muds will stay in suspension and float downstream.

Streams that are heavily laden with mostly bed-load material will develop a series of bars on a broad valley floor while the water flows around them in a network of interconnected rivulets. The stream has no main channel and is referred to as a *braided stream* (Figures 16.19 and 16.20, page 361). Meltwaters from glaciers carry heavy loads of debris and cause the relatively small discharge to flow in a series of shallow braided channels. The channels can change their course hourly, daily, or seasonally, responding to changes in discharge, slope, and channel depth. Braided streams also occur in desert regions, where the volume of water is relatively small and the amount of coarse debris is large, as on the surface of an alluvial fan.

If the stream load is primarily fine-grained silt and clay, the stream – particularly at the lower end – will tend to develop a series of curves called *meanders*. The word comes from the name of a river in Turkey fabled in ancient times for its twisting, winding course of looping bends. (The term also comes from the Latin word *maendere* meaning "to wander" (Figures 16.6, 16.9, and 16.21, pages 354, 355, and 361). Meandering and braiding are major river patterns.

You have learned that because a river's velocity is higher on the outside of a bend, the river will be deeper there and the bank undermined by deep scouring. The bank that fails by slumping into the stream is sometimes called the *cut bank*. Much of the sediment eroded from the cut bank is carried downstream to be deposited as a sand bar (also called a *point bar*) on the inside of the next bend, where the water velocity is low (Figure 16.23, page 362).

Meanders wander, as the name suggests, migrating in the direction of the cut bank (Figures 16.22 and 16.23, page 362). During flood time, the neck between two bends in the river becomes narrow, and with the increased discharge the river will erode a new, more direct channel called a *cutoff* (Figure 16.25, page 363). The abandoned meander will become a crescent-shaped *oxbow lake* that eventually fills with sediment (Figures 16.24 and 16.26, pages 362 and 363). Meandering rivers have caused a variety of problems, especially where a county line or state line is located in the center of the river or where homes, commercial buildings, and bridges line the river. Because lateral migration of the stream may affect valuable property, remedial measures, including lining the outside of the bends with concrete, broken rock (riprap), or junked cars, have been tried to control wayward streams.

Flood Plains

The flood plain is a broad area that results from lateral erosion by migrating meandering streams (Figure 16.39, page 371). Repeated floods blanket some flood plains with fine muds and silts, while others are covered by sediments deposited in point bars (Figure 16.28, page 364). The lower Mississippi River, from near its junction with the Ohio River (near southern Illinois) down to the Gulf of Mexico, is flowing on the deposits of its own flood plain. Meanders are features of flood plains, as are point bar deposits, oxbow lakes, and especially *natural levees* (Figures 16.27 and 16.28, pages 363 and 364).

Levees form when a flooding river overtops its banks and the velocity of the water is suddenly decreased when the thin film of water spreads widely over the flood plain (Figure 16.2 A and B, page 351). As with any stream, a decrease in velocity means deposition of sediment, and in this case, it is adjacent to the main channel (Figure 16.29, page 366). After many episodes of flooding and deposition, a *levee* or raised ridge of sediment, which will contain the water during low flood stages, forms on either side of the channel. Sometimes boats can be seen moving down the river 3 or more meters (10 feet) higher than the cars on roads in lowland areas. Exceptional floods can overtop levees – despite efforts to raise the levees with sandbags or other materials – and inundate the flood plain with water, mud, and silt.

Deltas

Now back to an earlier question: where is all the rocky material that was eroded out of the Grand Canyon? In fact, what happens to all stream loads? The dissolved load, of course, ends up in the ocean or in large lakes and contributes to the salinity and other ions in the water. Some streams will evaporate in playa desert lakes, and the dissolved minerals will form a salt crust such as in the famous Devil's Golf Course in Death Valley, California. The bed load, consisting of boulders and cobbles, is dropped upstream along the banks or on the bed, to be abraded into smaller rounded particles that will slowly be transported downstream. Some of the sands and silts and the suspended load will be deposited as the velocity is decreased when the stream enters a body of still water at the mouth. The sediments form a generally triangular body called a *delta*, named by Herodotus in the 5th century B.C. for the Nile delta that resembled the Greek letter "delta" (Figures 16.31 A and B, page 366). The name is so appropriate that it has been adopted by most languages throughout the western world.

The river entering the delta breaks into many small shifting channels called *distributaries* that carry the water and sediments away from the main stream. Not every stream forms a classic delta shape, however. The form of the deposit depends on such variables as the amount of sediment carried and the erosional force of waves and currents in the sea or lake. The delta of the Mississippi River is known as a *birdfoot delta* because it resembles a claw or talon. The shape is a result of heavy sediment carried by the river and the relatively quiet waters into which the sediment is deposited (Figures 16.30 and 16.31 D, pages 366 and 367). The structure of a delta is extremely complex and constantly changing. The internal construction of a small delta is shown in Figure 16.32, page 367, where the surface distributaries, the horizontal topset beds, the tilted foreset beds, and the fine clays of the bottomset beds are illustrated.

Geologically, the best known delta in the United states is that of the Mississippi River, partly because about 90,000 oil wells have been drilled in its sediments and also because repeated surveys have been made by hydraulic engineers responsible for the operations of the waterway. The delta of the Mississippi has been growing for many millions of years. It started out around Cairo, Illinois, in Cretaceous times and has advanced about 1,600 kilometers (960 miles) since then. But like all deltas, the Mississippi delta grew in one direction, then shifted to another, seeking a shorter path to the sea. This formed a complex mass of sediment, as one delta piled on top of another. Dredging has kept the major distributary channels, called "passes," fixed in position for many decades, to keep the waterways open to navigation.

This strange world of the delta – half water and half land – is the home for a great variety of water birds and many plants and animals, as well as the site of many cities. Some of the greatest deltas, such as those at the mouth of the Ganges, the Indus, the Tigris-Euphrates, the Nile, and the Yangtze, are also some of the densest population centers in the world. The advantage of living on a delta is the network of communication among landlubbers, seafarers, and those who make their living along the banks of rivers. Such cities, however, are at constant risk of being inundated and of subsiding as the sediments become compacted. Venice, Italy, is an excellent example of what differential settling can do to a city built on delta mud.

Sediments eroded by the Colorado River are deposited in a broad delta at the head of the Gulf of California. At one time, the Gulf extended far north of its present site, but the growing delta of the Colorado separated the northern part of the Gulf from its connection with the Pacific Ocean. The area behind the delta is now a rich valley where dates, citrus, and other crops grow in abundance, watered by irrigation from the Colorado. With construction of Hoover and other dams on the Colorado, the sediments are now settling out in the lakes, and the old delta receives little sediment. Similar but more severe problems face the delta region of the Nile, where lack of annual flooding, encroaching salt water, and subsidence cast doubt on the wisdom of building the great Aswan Dam.

Not all of the world's rivers form deltas where they enter the sea. Two large rivers that do not are the Columbia and the St. Lawrence – the St. Lawrence because it does not pick up much sediment in the short run between Lake Ontario and the Gulf of St. Lawrence; the Columbia because of the way it discharges directly into the open sea, where powerful waves and strong longshore currents quickly redistribute the stream's load of sand and silt.

Human Activities and Streams

Sheet erosion was mentioned as a water flow unrestricted by channels or banks. In tilled fields, where plowing has removed the anchoring cover of vegetation, sheet erosion can remove as much as 100 tons of precious topsoil per acre in a single year. As you learned in Lesson 15, soil is very slow to form and must be protected as a natural resource.

Several methods have been devised to prevent or moderate sheet erosion. Plowing of fields parallel to contours rather than up and down retards the flow of water, prevents gullying, and moderates the rate of sheet erosion. Another method is called *strip planting*, in which alternate rows of different crops are planted along the contours. *Terracing* parallel to the contours also decreases soil erosion. (You learned about terracing in Lesson 16 on mass wasting.) Mulching with organic matter is also used as a way to mitigate sheet erosion. (Read Box 16.1, page 351, and note Figure 1.)

Major flooding by streams is not necessarily a seasonal event but may occur every 10, 50, or even 100 years. Two 50-year floods can also occur in consecutive years! However, studies of flows in natural channels (without dams or flood control projects) show that, on average, streams overtop their levees about every 1 to 2 years. In addition, in cities, where paving prevents water from being absorbed naturally by the soil, increased run-off into streets and storm drains may overload the drainage system during wet weather. Thus the nature of cities themselves can cause local flooding. Bridges, docks, piers, and buildings built on river banks or flood plains can constrict water flow, raising water height and velocity and increasing local erosion. While upstream dams and flood control projects can partially reduce the dangers of flooding, it cannot prevent destruction by the 100-year flood, for example. Constructing artificial levees or building up natural levees can help, but the best method for preventing loss of life is to avoid building on flood plains (Figures 16.4, 16.5 A and B, page 353; and 16.27 A, B, and C, page 363 Also, read Box 16.3, pages 364-365).

CASE STUDY: The Nile

Any mention of the Nile brings to mind the glorious antiquities of ancient Egypt, the golden treasures of King Tut's tomb, the soaring Pyramids, and the enigmatic Sphinx. The Nile is tied to the long span of human history, but the river also has a geologic past – not as well-known perhaps, but filled with wondrous events!

Since the Nile is the longest river on Earth, about 6,695 kilometers (4,160 miles), and has a remarkable history that only recently is being revealed by satellite photographs and geophysical investigations, it is well worth our study. The second largest river system, by the way, is the Mississippi-Missouri, at about 6,300 kilometers (3,890 miles).

When viewed from the air, the Nile appears a narrow green belt that runs not only the length of Egypt on its northward journey to the Mediterranean Sea but also flows through the Sudan, Kenya, Tanzania, Ethiopia, Zaire, and Uganda. In Egypt, beyond the thin green belt and the 4,000 year-old cities, temples, and tombs of the kings that line the river, there is only barren desert, spanning the length and breadth of this nearly rainless country. The narrow lush valley represents 4 percent of Egypt's land but holds 96 percent of its 44 million people. Without the Nile to nourish the land, no grand civilizations would have flourished here, and no viable modern nation could have endured.

The Nile is actually the confluence of three principal streams. The Blue Nile, receives most of its water from the heavy summer rains on the highlands of Ethiopia and contributes about four-sevenths of the water. The White Nile supplies about two-sevenths of the water and has its sources in the headwaters of Lake Victoria, a region of heavy year-round rainfall. And the Atabara also has its headwaters in the rainy highlands of Ethiopia and contributes the remaining one-seventh. The Atabara joins the main stream about 200 miles north of Khartoum, where the Blue and White Niles meet and is the last tributary on the long descent to the sea.

The geologic history of the Nile is a part of the tectonic evolution of North Africa. During the Eocene, about 50 million years ago, part of the ancient Tethys Sea covered the area that is now Cairo and the delta of the Nile and left thick deposits of limestone on an unstable continental shelf. At the end of the Eocene, about 40 million years ago, the sea retreated from northeast Africa, and streams eroded the uplifted land.

It was in the late Miocene time that the amazing history of the Nile began. Tectonic uplift of North Africa raised the floor of the shallow narrow channel at Gibraltar above the level of the Atlantic Ocean and cut off the supply of sea water to the Mediterranean. The water in the Mediterranean slowly evaporated, leaving a series of brine pools on a deep sea floor that descended over 12,000 feet below sea level.

At the same time, a river that geologists call the "Eonile" or "Dawn Nile" formed that was near but not exactly in the same location as the present Nile. As the level of the sea dropped, the river, with a lowered base level and renewed gradient, cut a majestic canyon from the vicinity of the present Nile Delta to Aswan, about 600 miles to the south. (Refer to Figure 16.8, page 355.) In depth and width this valley was comparable to the Grand Canyon of the Colorado, but the Eonile canyon was twice as long!

The depth to bed rock was determined using seismic data and by drilling a series of test holes through the river sediments. At Aswan, bed rock was reached at a depth of about 170 meters (560 feet), deepened to about 2,500 meters (8,250 feet) near Cairo, and reached about 4,000 meters (13,200 feet) under the north part of the Delta. (This last figure may have partly resulted from faulting or downwarping of the old continental shelf.)

The channel at Gibraltar subsided about 3 to 5 million years ago, and once again, the Mediterranean was filled with Atlantic seawater. The great chasm of the Eonile became a long thin arm of the sea as the Mediterranean waters rose and inundated the valley. Marine sediments were deposited, filling about one-third of the depth of the canyon. During the Pliocene, about 2 million years ago, the climate changed and became cool and wet. New streams flowed into river system, and their deposits filled in more of the canyon. In the early Pleistocene, there was yet another climatic change: aridity set in. Egypt became a desert, and this river, called the "Paleonile," apparently stopped flowing or became an intermittent stream!

Later in the Pleistocene, another river formed in a valley parallel to the modern Nile but 10 to 15 kilometers (6 to 9 miles) to the west. Through a series of adjustments and stream captures in the Ethiopian highlands, a new course was cut separate from those of the older Niles. This modern Nile formed only about 10,000 years ago, probably with the capture of the waters of the White Nile. And so the Nile completed the filling of its deep hidden canyon with its own sediments brought in from its distant headwaters. Compared to the discharge of the Amazon, the Congo, or the Mississippi, the discharge of the Nile is small. Nevertheless, the integrated river became the longest river in the world, extending from the Mediterranean to Lake Victoria, which is about 600 kilometers (1,000 miles) farther than from San Francisco to New York City!

For many centuries, it was the annual flooding of the Nile that brought new silt, soil, and nutrients from the heavily laden Blue and White Niles and the Atabara to the early agricultural civilizations along the river bank. The great river was unpredictable, however, and led the ancient Egyptians to acquire mathematical, engineering, and astronomic skills far in advance of other civilizations in order to cope with lean years and floods. Basins were constructed, with feeder canals to save the water for later irrigation. With completion of the Aswan High Dam in 1971, the river was controlled and, with older existing dams, has a storage capacity to hold back the entire flood for later use. The dam's hydro-electric plant has tripled Egypt's power output. It has expanded the cultivated area by over 1.8 million acres and protected the country against high floods while providing a dependable source of water for irrigation even in years when the river is low.

But do benefits outweigh costs? The new (and old) lands under irrigation need large amounts of artificial fertilizer since natural nutrients no longer arrive on the now tamed river. In the Delta region, salt water encroaches because of reduced fresh-water flow, affecting agriculture in this heavily cultivated region. The delta (typical of most deltas) has subsided because of compaction of sediment. No new sediments are being brought in because they are all being deposited above the Aswan Dam. The Delta front has been considerably eroded and modified by coastal currents and is retreating. Fishing near the Delta has declined, and animal pests have increased in the fields since flooding no longer keeps their populations down. The marvelous antiquities of Egypt, such as the Sphinx, are threatened by increased moisture in the air and by deterioration of ancient stone. Even more serious is the spread of disease, caused by bilharzia, a protozoan carried by snails that live in the quiet waters of basins and ponds and that infect humans. Despite the dam's benefits, poor people still live in mud huts and wash their clothes and animals and dump their wastes in the Nile. (But at night, as you sail quietly downstream, you can see the eerie lights of television screens beaming through windows of those huts . . . soap operas are very popular.)

Questions

1. Where does the water come from that fills the rivers of Earth?

Although many rivers such as the Colorado and the Nile flow through arid regions, the source of the water is rain and melting snow that flows into the river systems from highlands, mountains, or other regions of abundant rainfall. Despite water loss through evaporation and seepage into the ground, most large rivers have enough water to carry them across wide arid regions into the sea. It is true, however, that some smaller streams, especially in the southwestern United States, are intermittent or seasonal and vanish into desert basins.

2. What is meant by a longitudinal profile, and what does it illustrate about stream mechanics?

The longitudinal profile is a hypothetical cut through a stream from the headwaters to the mouth and is used primarily to illustrate the changes in elevation and gradient along the stream course. Changes in stream velocity and discharge can be inferred from this profile. Most streams or their tributaries originate on mountain slopes or highlands where there is abundant water or melting snow. The gradient is steep, and the rivers are characterized by V-shaped valleys, waterfalls, and rapids. The energy of the stream is spent in eroding the channel, generally during storms or after the spring thaw. Mass wasting commonly takes place where there are steep slopes, adding rock debris to the stream load.

Downstream, the gradient is less steep, but the discharge increases due to entering tributaries. The valleys become broader, develop meanders, flood plains, and other features such as levees and oxbow lakes. The flood plains can reach enormous widths, extending far beyond the river's banks. These lowlands may be flooded only once in 50 to 100 years; because of rich sediments, they are used for farmland and pastures despite the danger of severe flooding. The construction of homes, towns, and cities continues in these threatened areas. Unfortunately, humans seem to have short memories about past disasters.

3. How does sediment load affect stream velocity, and what is the principal source of stream load?

When the sediment load increases drastically, there can be a corresponding decrease in the percentage of flowing water. The more sediment a stream carries, the muddier it becomes, until finally it may change into a mud flow.

Much of a river's work is to carve downward, steepening slopes and undercutting banks that fail by mass wasting. This is the source of most of the debris that enters the stream's transport system. Thus running water and mass wasting work together to carve and widen stream valleys, particularly in areas high on the longitudinal profile.

4. What are potholes?

The force of moving water is well known to anyone who has tried to swim in the ocean or in a swiftly moving stream. Abrasion, the most important erosional factor, is a common phenomenon and known to anyone who has used sandpaper, cleaning powder, or a nail file. The friction of cobbles and pebbles against the rocky side and bed of a stream channel and against each other, tumbled or dragged along by the rapidly moving stream, becomes a potent force in wearing away the walls of even a mighty canyon. Potholes are smooth cylindrical hollows in a rocky stream bed, where sediments, usually gravel or pebbles, in swirling water will scour out rounded holes in the rocks. An assortment of pebbles may be inside a pothole, all well rounded from abrasion with the side of the pothole and with each other (Figures 16.13 and 16.14, page 357).

Potholes, valleys, and great canyons are not carved in a day, however, or eroded with the same force all year around. The greatest erosion takes place during flood stages, which may occur with variable frequency from once per year to perhaps once per hundred years. There is also geologic time to be considered, time for streams to do their work in shaping the landscape.

5. Since rocks are heavier than water, how does water, even at high velocities, carry its load downstream?

Most rocks are three times heavier than water, so rock particles larger than small sand grains are generally transported only during floods. Boulders and pebbles are dragged or rolled as part of the bed load. Smaller particles, including those that are sand-sized, are also part of the bed load and move downstream by *saltation*. The suspended load is the only solid portion of the load that is continuously carried in the flowing stream. The particles are of small, usually clay-sized minerals, typically clay or mica. While spherical particles will fall directly through the water, flat particles stay in the water or slowly drift down. Specific gravity is important because denser substances, such as gold nuggets or flakes, with a density of 16 to 19 times that of water, will be deposited more rapidly than feldspar flakes, with a density of about 7. (Incidentally, panning for gold uses the specific gravity of gold to separate it from the lighter rock particles in *placer* – that is *river* or *glacial* – *deposits*.)

6. Why do rivers meander?

Meanders are normal for rivers flowing on low gradients in plains or lowlands where the sediments are fine sands, silts, or muds. In almost all kinds of terrain, however, meanders are more common than long straight stretches of river. Meanders occur even in streams that do not carry sediments, such as meltwater streams flowing on top of glaciers. The Gulf Stream in the Atlantic Ocean meanders, and jet streams in the stratosphere appear to have a meander pattern. Meanders even occur within cave streams where erosion occurs primarily by solution.

The meander is a stable form, and a river that meanders will not change its pattern and become straight. The meandering stream continually excavates and redeposits the sediments as the meanders change position. The net effect is to widen the valley and gradually move the sediments downstream. The looping curves have been studied and are considered to be the most efficient means of dissipating the river's energy. The meanders can start when the water strikes some obstruction in the stream bed; once started, meanders become self-perpetuating. Actually all flowing water has some degree of turbulence and does not flow with the same velocity from bank to bank. An eddy or jet of faster moving water could start the meanders, which increase in size as the water swings from one bank to another.

It has also been suggested that meandering is a course of the least resistance – a mechanism to achieve the equilibrium required between a river's velocity and the downhill gradient. By meandering, a river lengthens its passage, thus decreasing its velocity in much the same way skateboarders or skiers learn to control their downhill velocity. Although imperfectly understood, meanders are the most significant cause of lateral erosion and valley widening.

7. What is an alluvial fan, and how does it differ from a delta?

The basic conditions that lead to formation of a delta also apply to alluvial fans, except that the fans form where the stream does not reach the sea but flows from mountains to a flat plain. The velocity of the stream may be high in the steep mountains, and the stream may carry a sizeable load of rock debris. At the mountain front, the channel breaks up into a series of shallow distributaries, which start to deposit the water's coarse sediment load. The deposit is generally cone shaped with the apex at the point where the stream leaves the mountain canyon. Alluvial fans can be formed in any environment but are commonly seen in arid regions (Figures 16.33 and 16.34, page 368). Deposition on the fan is intermittent in arid regions since sediment is carried from the mountains only during rare rainstorms or flash floods.

Activities

1. Think about the Grand Canyon of the Colorado and the great canyon that lies under the Nile Valley. How are they alike, and how are they different in origin? (In the case of the Colorado, the land surface was uplifted, which renewed the stream gradient. In the case of the Eonile – what happened to increase the stream gradient?)

2. On Figure 4.26, page 81, find Lake Victoria and Lake Albert and trace the Nile River to its delta in the Mediterranean Sea. Using the map scale of about ¾ inch for each 500 miles, determine the length

of the Nile. It is interesting to contemplate the fact that some of the sediments on the delta originated in the African rift valley.

3. Using a United States atlas or Figure 16.3, page 352, trace the course of the Mississippi River and the Missouri River from their headwaters to their confluence (the place where they join). What major city is found near there? What major river joins the Mississippi where Illinois, Kentucky, and Missouri come together? (Look for Cairo!)

4. If you live near a river, take a field trip. Look for the high water mark from the last flood. If the stream is small, look for some of the features described in this lesson: potholes, meanders, flood damage, mass wasting along the banks, erosion of fields, and so on. You will learn much about river mechanics from on-site study. Even a tiny stream can tell many things.

5. If you enjoy visiting art museums or have access to books of art reproductions, plan to spend a little time looking at 18th and 19th century paintings of landscapes. Many will realistically depict river scenes, sometimes so accurately that you can see the influence of stream erosion on the landscape.

1. Most of the water in the hydrologic cycle is contained in   **Self-Test**

   a. streams and rivers.
   b. glaciers and ice sheets.
   c. ground water.
   d. oceans.

2. Methods of controlling sheet flow include all of the following except

   a. contour plowing.
   b. planting alternating rows of different crops.
   c. constructing levees.
   d. terracing of hillsides.

3. The most important factor controlling stream erosion, transportation, and deposition is the river's

   a. width.
   b. velocity.
   c. length.
   d. depth.

4. The velocity of a stream toward the mouth tends to decrease, but this effect may be overridden by

   a. increased load.
   b. widening of the channel.
   c. the presence of a body of still water at the mouth.
   d. increased discharge.

5. In a normal flowing stream, it will take a much higher stream velocity to pick up sediments from the stream bed than to transport them. This is especially true of

   a. clays and silts.
   b. sands.
   c. gravels.
   d. boulders.

6. The dissolved load of a stream depends upon the rock in the bed and banks. The greatest dissolved load would be obtained from which of the following rock types?

   a. Shale
   b. Basalt
   c. Limestone
   d. Gneiss

7. The rate of flow of a stream is not constant across the river from one bank to the other and from top to bottom. It tends to be slowest

   a. on the top.
   b. on the sides and bottom.
   c. in the center of a straight stretch of water.
   d. on the outside of a bend.

8. The process of saltation changes to suspension when

   a. the velocity is high enough to pick up sand grains and transport them.
   b. the velocity decreases and suspended muds settle to the stream bed.
   c. strong currents cause the gravels and pebbles to jump into the water.
   d. rocks of the bed dissolve and the particles are transported by the water.

9. In terms of erosion by streams, the most effective agent(s)

   a. are chemicals in the dissolved load.
   b. are muds and silts carried in the suspended load.
   c. is hydraulic action of clear water.
   d. is abrasion by large particles in the bed load.

10. Of the following features, which is the most unlikely to be found on a flood plain?

   a. Natural levees
   b. Cut banks and point bars
   c. Potholes
   d. Oxbow lakes

11. When engineers try to compute the capacity of a stream, they are discussing the

    a. maximum velocity ever reached by that stream.
    b. total load of sediment the stream can carry.
    c. rate of erosion of the stream valley.
    d. total volume of water that will stay within the banks.

12. An examination of the longitudinal profile of a stream will show that the steepest gradients of a stream bed are most likely to be found

    a. in the delta region near the mouth of a stream.
    b. on the outside bend of a meander.
    c. in a meander neck cutoff.
    d. in the headwaters of a mountain stream.

13. Of the following, the greatest discharge of a stream is most likely to occur in

    a. the headwaters following spring melting of snow.
    b. the upper waters above the confluence of the tributaries.
    c. the downstream section below the confluence of the tributaries.
    d. a distributary on a delta.

14. If a stream is heavily loaded with sediment and has banks that are easily eroded, the

    a. stream will tend to develop a braided pattern.
    b. stream will develop strong meanders.
    c. velocity will tend to decrease.
    d. stream pattern will stabilize over long periods.

15. All of the following statements are correct about alluvial fans except

    a. they form from streams that do not reach a sea or other body of water.
    b. they are limited to arid regions.
    c. they form where a stream's velocity decreases as it flows out of a mountain valley.
    d. the flow of water branches as it flows over the sediments.

20

RUNNING WATER II: LANDFORM EVOLUTION

This lesson will help you understand the critical role of running water in sculpting Earth's surface.

After reading the textbook assignments, completing the exercises in this study guide, and viewing the lesson's video portion, you will be able to:

1. Describe how the tectonic and hydrologic cycle work together to shape the land.

2. Explain what a drainage basin is, what a drainage divide is, and how drainage divides change with time.

3. Indicate how streams shape their valleys by downcutting, lateral erosion, headward erosion, and sediment deposition.

4. Describe how mass wasting affects the development of stream valleys.

5. Relate the concepts of base level and graded stream to the process of valley development.

6. Summarize how vertical crustal movements influence landscape evolution.

7. Recognize the factors that affect slope erosion.

8. Describe the relationship between the drainage pattern that a river and its tributaries develop and the nature and structure of underlying rocks.

9. Recognize the ways in which stream terraces and incised meanders form.

10. Describe the origin of superposed and antecedent streams.

INTRODUCING THE LESSON

When flying over any continent on a clear night, we can see the lights of cities, towns, and villages sparkle through the blackness to relieve the dark emptiness below. By day, we would see that nearly every community is on or near a river. Even away from urban areas, streams of various sizes, from small brooks to large rivers, represent the principal agent of erosion responsible for shaping the land surface we see. The diversity of features created by streams is striking and ranges from rugged mountainous landscapes to arid badland topography to broad fertile flood plains with quiet crescent-shaped oxbow lakes. There are deeply incised meandering streams in Utah, sharp peaks and steep walled cliffs in the Rockies, streams flowing off the Cascade volcanoes like spokes in a wheel, and curved ridges carved out of the sedimentary formations of the ancient wooded Appalachians.

The processes of erosion, transportation, and deposition by running water have contributed to the vast array of landforms and are major contributors to the on-going evolution of continental surfaces. In Lesson 19, you learned that running water is an important part of the hydrological cycle; in this lesson, we will discuss the influence of the tectonic cycle as we interpret the origin of landforms created by running water.

For example, why is it that most streams cannot erode their channels below sea level but that streams flowing into Death Valley carve their canyons more than 250 feet deeper than that? Do streams extend their length through growth at the mouth or at the head . . . or both? Some streams meander over broad, low-gradient flood plains, while others cut deep meanders and have no flood plain at all. Consider streams flowing outward from a volcanic dome: what kind of pattern will they make? Why do some stream patterns resemble a trellis, while most others are tree-like or *dendritic*? Admittedly, it is sometimes difficult, if you live in a major metropolis, to recognize the effects of streams. This lesson will help you understand and discover how stream action shapes landscapes everywhere.

LESSON ASSIGNMENT

Completing the following eight steps will help you master the lesson objectives and achieve the goal for this lesson:

Step 1: Read the TEXT ASSIGNMENT, Chapter 16, "Streams and Landscapes." Pay special attention from page 368 to the end of the chapter.

Step 2: Study the KEY TERMS AND CONCEPTS as noted in the study guide.

Step 3: Watch the VIDEO using the VIEWING GUIDE in the study guide.

Step 4: Read the study guide's PUTTING IT ALL TOGETHER section, which will help you summarize and integrate all of the information in this lesson.

Step 5: Complete any assigned lab exercises.

Step 6: Complete any assigned ACTIVITIES found in the study guide.

Step 7: Review the material in this lesson and complete the SELF-TEST found in the study guide.

Step 8: Go back to the LEARNING OBJECTIVES and make sure you have learned each one.

1. In Chapter 16, pages 349-381, review the first part of the chapter, and then study from page 368 to the end, starting with the article on "Valley Development."

2. Re-read and study the Chapter 16 "Summary" as a review of Lessons 19 and 20.

3. Try the "Questions for Review" and "Questions for Thought." They are not difficult and will help you prepare for the SELF-TEST.

**TEXT
ASSIGNMENT**

The key terms listed under "Terms to Remember," page 380 in your text, will aid your understanding of the material in this part of the lesson. Admittedly, it is a long list – but not difficult. By the end of this lesson, all the words should be familiar. If not, use the glossary, text, and study guide for explanation. In addition, learning the following terms will be useful for this lesson: *turbulent flow*, *stream piracy*, and *stream rejuvenation*.

**KEY TERMS
AND
CONCEPTS**

Once you have read the text material for the lesson and studied the terms, you are ready to view the lesson's video portion. Review the questions that appear in this section to help you watch for important points and to prepare you for the video.

**VIEWING
GUIDE**

Before viewing, answer the following questions:

1. In what ways does a stream widen its valley?

2. What are some of the features of a youthful stream?

3. What are some of the features of the old age stage of valley development?

4. In what ways may stream patterns demonstrate that regional uplift has occurred?

After viewing, answer the following questions:

1. Why have the Atchafalaya and Old River water control systems in Louisiana been constructed?

2. Why does the development of hydroelectric power near these water systems have to be carefully regulated?

3. What special geologic problems face the Atchafalaya and Old River water control systems?

4. Why did the concept of uniformitarianism inhibit acceptance of the present theory of origin of the Channeled Scablands?

5. What evidence was cited in the video to support a flood origin of the Scablands?

6. How are youthful, mature, and old age landscapes distinguished?

7. What geologic evidence exists for stream rejuvenation or changes in base level?

PUTTING IT ALL TOGETHER

You have read the text assignment and seen the video portion of the lesson. This section, made up of a SUMMARY, a CASE STUDY, the QUESTIONS, suggestions for ACTIVITIES, and a SELF-TEST, will help you pull the information together.

Summary

Streams cut their own valleys, which they deepen and widen by removing rock material from their banks and stream bed, especially in the upper and middle reaches of the stream. The valleys also grow longer by headward erosion and by deposition at the mouth in alluvial fans and deltas. Streams are *pirates* under certain conditions; by headward growth, the waters of one stream may capture those of a lesser stream, add to its discharge, and increase the area of its drainage basin (Box 16.4, pages 375-376). The drainage basin of the Mississippi River, for example, drains half the land area of the continental U.S. (Figure 16.3, page 352). A *drainage divide* is a ridge or mountain range that separates one drainage basin from another. The Continental Divide in the United States, for example, separates streams that flow to the Pacific Ocean from those that flow to the Atlantic and the Gulf of Mexico.

Valley Development by Downcutting and Widening

You learned in Lesson 19 that high stream velocities aided and abetted by solution, hydraulic action, and abrasion are causes of stream erosion. Some streams cut deeply into bed rock to form *slot canyons* (Figure 16.35 A, page 369), relatively uncommon forms usually seen where erosion is intermittent but rapid. Slot canyons occur in soft rocks in arid regions where, after extended rainfall and flash flooding, great volumes of abrasive rock and debris flow rapidly through the canyons.

Most canyons are V-shaped because, as the stream is cutting downward, sheetwash plus mass wasting widen the valley, causing the sides to flare out (Figures 16.35 B and 16.37, pages 369 and 370). Think of the Grand Canyon (photograph facing page 175); the Colorado River is only about 90 to 120 meters (300 to 400 feet) wide, but the top of the Canyon is about 16 to 21 kilometers (10 to 13 miles) across. Was the water ever that wide when it was at the top of the Canyon? No. The widening of the Canyon is primarily a result of sheetwash, undercutting of the banks, and mass wasting, aided by differential weathering of the exposed formations (Figure 16.44, page 373). The Colorado River is like a giant conveyor belt, carrying out the sands, red silts, and clays that entered the stream from the cliffs above.

In addition to sheetwash and mass wasting, a meandering stream is responsible for lateral erosion. As the stream swings from one side of the valley to the other, the banks along the outside of the bend are undercut. The valley walls collapse, and debris falls into the stream and is carried away by running water (Figure 16.22, page 362). Thus, while the width of the stream channel stays about the same, the loops of the meanders broaden and thus widen the valley and flood plain (Figures 16.21, 16.24, 16.26, 16.27, and 16.39, pages 361, 362, 363, and 371).

Extending Stream Length

Gullying, mass wasting, and sheet erosion at the head of a stream increase its length in the headward direction (Figure 16.40, page 371). Gullying, a serious erosional problem in agricultural lands, is difficult to stop, especially if there is widespread sheet erosion across tilled fields. Streams also increase their length through deposition of sediments in deltas, some of which extend hundreds of kilometers beyond the end of the main channel (Figures 16.31 and 16.32, pages 366-367). These are the processes whereby streams extend their tributaries headward, increase the area of the drainage basin, deepen and widen their valleys, and lengthen those valleys through deposition at the lower end.

Base Level

If streams could continue downcutting indefinitely, how deep would they get? The ability of a stream to erode and transport its sediments depends on its velocity, which in turn depends primarily on the gradient. Where a stream flows into a body of standing water or into the sea, the velocity drops almost to zero (but not quite, since the river must preserve its forward motion). The stream loses its competence and cannot erode below sea level, the universal limit of downcutting, which is called *base level*.

Base level is not a stable feature, however, and is sensitive to local tectonic changes such as uplift or subsidence. World-wide changes in base level also occurred, when, for example, during the glacial ages, water was removed from the sea and frozen into the glaciers, causing sea level to be lowered drastically. During the last glacial age, the base level dropped at least 100 meters (330 feet), and with the renewed gradient, Pleistocene rivers cut deeply into their channels. As the ice melted, sea level rose, raising the base level, and streams began to deposit their sediments in their glacial age channels. This cycle of erosion and deposition was repeated many times, resulting in complex depositional patterns in deltas and at many river mouths. Sea level is apparently rising at present; how will this affect streams, world-wide? Hint: look at a map showing Chesapeake Bay on our Atlantic Coast.

Base level, however, is not always sea level. The Dead Sea, 392 meters (1,292 feet) below sea level, is the lowest base level on Earth. Death Valley, lying about 85 meters (282 feet) below sea level, is always cited as an example of an unusually low base level. (Both Death Valley and the Dead Sea are not stream-cut valleys, but have been down-dropped by tectonic movements along faults.) Streams flowing from the mountains that surround Death Valley carry their sediments into this dry enclosed basin (Figure 16.36 B, page 369). Much of the sediment load is deposited in alluvial fans, where the streams leave the steep narrow mountain canyons and spread out on the flat valley floor (Figures 16.33 and 16.34 A and B, page 368). Other features such as high mountain lakes and reservoirs behind dams that interrupt the longitudinal profile of a stream may act as an elevated, local, and often temporary base level (Figure 16.36 C, page 369). Is Death Valley also a temporary base level?

The Graded Stream

After a region undergoes tectonic uplift and a stream begins its journey to the sea, the longitudinal profile is anything but smooth. Waterfalls and rapids occur where the water encounters hard rock layers that resist erosion and project upward in the stream bed (Figure 16.37, page 370). The valleys usually are narrow and V-shaped, and the stream energy is concentrated on downcutting to smooth out irregularities in the stream bed.

Eventually, the waterfalls retreat upstream or are worn away, the rapids are abraded flat, and the long profile becomes smoothly concave upward (Figure 16.1 A and 16.38, pages 350 and 370). This is a profile of equilibrium in which all factors – erosion, transportation, and deposition – are balanced. In this condition, the stream is said to be *graded* and is just able to transport the available sediment load without further erosion or deposition in the stream bed. If any changes occur, the stream will adjust to restore equilibrium. If the gradient is increased (by uplift or drop in sea level), the velocity of the stream and its ability to erode and transport will increase until the long profile is again smoothed (Figure 16.8, page 355). Lessening the gradient causes the sediment load to be deposited, raising the stream level (Figure 16.7, page 354). When a river is dammed to make a reservoir, a local base level is established, upsetting the natural balance achieved over eons of time. Sediment normally in transport begins to fill the reservoir. Erosion occurs below the dam and proceeds unchecked because no sediment is allowed to replace what has been taken away.

It is unlikely that a condition of perfect equilibrium is ever completely achieved in natural stream systems. A cloudburst may disturb the balance by increasing the discharge. The collapse of a large river bank adds excess sediment to the channel. Tectonic movements vary from slow to rapid, and, like variations in climatic regimes, affect the graded stream's equilibrium. Thus, adjustments continually take place.

Stages of Valley Development

Streams go through stages of *youth, maturity, and old age* and one stream, from head to mouth, usually shows youthful characteristics at its head and tributaries, merging into old age at the mouth. The landscape also evolves through stages of youth, maturity, and old age, ending as an almost flat plain, called a *peneplain*, with most of the hills, valleys, and other topographic features lowered by running water. While this is a useful way of interpreting the changes that take place during valley development, they are not necessarily related to any particular time span.

Youthful streams have narrow valleys that are V-shaped in cross-section and are found where the streams are actively downcutting. The gradients are usually steep, and rapids and waterfalls are common. Flood plains are lacking or minimal, and there are no meanders (Figure 16.37, page 370). On many rivers, these conditions exist only near the headwaters, although narrow valleys may form where resistant rocks slow the downcutting. Lakes may form in narrow valleys, but like falls and rapids they are temporary features, geologically speaking.

Maturity is reached when valley widening begins after the stream has reached grade and is no longer rapidly downcutting. Widening generally starts near the mouth of the stream, where the increased discharge promotes more efficient erosion. Lateral erosion exceeds downcutting; streams start to meander and to widen the flood plains. A few oxbow lakes form from meander cut-offs. The wide valleys are usually fairly flat in cross-section.

During *old age*, a stream has a very broad flood plain produced by lateral erosion and deposition during flooding. The flood plain is usually blanketed with thick alluvial sediments. Stream meanders are extensive and continually shift. Oxbow lakes, natural levees, and *backswamps* are common. The stream flows on a very low gradient, but increased discharge from tributary streams keeps stream velocity relatively high.

Stages of Landscape Development

Just as river valleys progress through a series of changes, so do landscapes. Is there some orderly sequence by which mountains are worn down to plains? From studies of youthful actively-rising mountains such as the Himalayas to old, mostly eroded mountains such as the Appalachians, a natural progression of landscape evolution comes into view. One of the first theories on landscape development, which dates back to the end of the 19th century, used the terms *youth, maturity*, and *old age* to characterize the stages in the evolution of a stream-cut landscape.

The story starts with tectonic uplift of a broad flat plain produced during a previous erosional cycle. Streams will have a new, increased gradient as the land surface is now high above base level. During the youthful stage, the landscape will be mostly flat upland; streams will be actively downcutting, producing narrow straight valleys, and tributary development will be minimal (Figure 16.41 A, page 372).

As the dissection of the landscape proceeds into maturity, the flat upland between streams is eroded away. This is a time of high relief, high peaks, and deep valleys, with rugged topography and maximum dissection of the land (Figure 16.41 B, page 372). The primary streams are mature and widen their valleys, but the many tributaries are still youthful.

As the valley sides are worn away – mostly by rain wash and mass wasting – the steep slopes become gentle, the land surface is lowered, and the stream gradients are lessened (Figure 16.41 C, page 372). Erosion continues until the region is lowered almost to base level. A flat plain is left, with possibly a few resistant erosional remnants to break the monotony (Figure 16.41 D, page 372). This is called a *peneplain* ("almost a plain") and represents the old age of the landscape. Peneplains are developing today along the flanks of the Appalachian Mountains. Ancient peneplains can be seen in the flat surfaces of unconformities, such as the one at the top of the Precambrian near the bottom of the Grand Canyon. (Review Figures 8.5 to 8.11, and 8.14, pages 179, 180, and 181.) There are many flat unconformities throughout the geologic record, attesting to the repeated rise and fall of mountains in this geologic "tug of war" between plate tectonics and the erosive power of the hydrologic cycle.

While the theory is useful in understanding landscape evolution, it is flawed by its simplicity. Mountain building is an extremely long, complex process that is intermittent but, over geologic time, continuous; mountains do not stand still, waiting to be eroded. Mountains are worn down, yet may rise again as a consequence of changes in plate motion. Base levels change, and uplift or folding and faulting renew stream gradients or alter stream patterns. Climates also change, sometimes from moist to arid or temperate to arctic, affecting the amount of water in the streams. All rocks do not weather or erode in the same way, as so vividly demonstrated in the Grand Canyon (Figure 16.44, page 373). The underlying rock structure, also the result of crustal movements, strongly affects the landscape (Figures 16.44, 16.45, and 16.46, page 373). Thus, it is difficult to predict the future of a particular landscape. Its future may be very different from its present, or, conversely, the landscape may persist in much the same form for a long time.

Drainage Patterns

The arrangement of a river and its tributaries, as seen in high altitude photographs or on maps, is called the *drainage pattern*. The most familiar pattern, *dendritic*, resembles branches of a tree; tributaries join the main stream at an acute angle, which forms a V pointing downstream. (The word *dendritic* is from the Greek word *dendron* meaning "tree.") Dendritic drainage is typical of a land surface floored by massive igneous or metamorphic rocks or by horizontal sedimentary rocks having a uniform resistance. The drainage basin of the Mississippi River (Figure 16.3, page 352) is a famous example of dendritic drainage. Another pattern is called *radial drainage* and develops on a volcanic cone or dome, where the stream flow is away from the center in all directions, arranged like the spokes of a wheel. (See Figures 16.48 A and 11.19, pages 374 and 252, showing drainage on Mount Fuji.)

Streams frequently follow faults or fractures where the rock is crushed and easily eroded; they also frequently follow or formations that are softer and more susceptible to erosion. A *rectangular pattern* of drainage develops where the underlying rock has a network of joints or fractures in a right angle system. The tributaries enter the main stream at angles of approximately 90 degrees following the fracture zones (Figure 16.48 B, page 374). Another pattern, called *trellis* drainage (like vines trained on a trellis), forms over tilted rock layers of differing hardness. The main streams are parallel to each other, flowing in the valleys eroded out of less resistant formations. The short tributaries flow down harder parallel ridges and enter the main stream at high angles. In Figure 16.48 C, page 374, note the synclinal structure of the underlying formations and how it affects the drainage patterns.

Thus we can see that drainage patterns are influenced by rock types and structures, climate, runoff, even vegetation, and by the geologic history of the areas in which they occur. The experienced geologist is able to learn much about the geology of an area just by studying the stream patterns.

Landscape Development and Vertical Crustal Movements

You have learned some of the effects of crustal movements and renewed gradients on stream erosion. One of the most interesting stream patterns involves *entrenched meanders* (Figures 16.53 and 16.54 A and B, page 378). If a stream with well-marked meanders is within a region undergoing slow uplift, the increased gradient will cause the stream to cut deep canyons in its meanders and become *entrenched* in narrow V-shaped valleys. A lowering of base level, as you learned in the case study on the Nile River, could also have a similar effect on a meandering stream.

Terraces, or step-like platforms above a river on the flood plain, are remnants of former valley floors and are also indications of crustal instability and, most likely, uplift. Streams flowing on broad flood plains can build up thick deposits of sands and silts. Sometimes the depositional cycle will change to a period of erosion, and the stream will cut downward and form a new flood plain. The remnants of the original flood plain will be left as *paired terraces*, raised flat surfaces on either side of the river. A new flood plain will eventually form at a lower level (Figures 16.51, page 377).

River terraces can result from several factors that influence drainage patterns, including regional uplift, climatic change with increased rainfall, and discharge and a lowering of base level because of an glacial age. In middle-latitude countries, some of the terraces lie on debris that flooded the streams from meltwaters of glaciers from the last glacial age. By counting the number of paired terraces, it is possible to decipher how many events of uplift or other changes have taken place in the history of a stream.

Sometimes it take a little geologic detective work to figure out why a stream will cut through a mountain, especially when it seems more logical to take another path, such as around the obstacle. Let us assume the stream is flowing on a sediment-covered plain and as it cuts down through the sediments it encounters a buried ridge of granite. The river is *let down* or *superposed* on the ridge, and its energy is concentrated on downcutting its valley. Usually, a narrow V-shaped canyon will be cut through the resistant formations, forming what is sometimes called a *water gap* (Box 16.4, Figure 2, page 375).

In another instance, the river is *antecedent* and was flowing in its course before folding or uplift of the mountain. The ridge, or anticline, rose across its path, but the river was able to erode its valley at about the same rate as uplift occurred. This is not uncommon in areas where the streams are large, have well developed drainage basins, and the regions are tectonically active (Box 16.4, Figures 1 and 3, pages 375-376).

In all of the above examples, we see the effects of crustal movement on evolution of landforms. We can also understand why, on a dynamic planet such as Earth, it is difficult to interpret all the geologic events of the past and even harder to visualize the continental landscape of the future.

**CASE STUDY:
The Channeled
Scablands**

Since the time of James Hutton in the 18th century, the principle of uniformitarianism (Chapter 1 in the text), has played a vital role in the validation of geology as a science by substituting plausible or observed processes for the mythical or fanciful explanations that had so long dominated natural science. In fact, many geologists, dedicated to strict uniformitarianism, were reluctant to accept the possibility of a catastrophic flood as the cause of the remarkable landscape in East Central Washington that came to be called the *Channeled Scablands*. (Read "Astrogeology" Box 16.1, page 379. Take special note of Figure 1 A and B.) The region is part of the Columbia basalt plateau (Lesson 13 and Figures 11.24 and 11.37, pages 254-263) where loess (wind-blown silt and clay) covers much of the basalt landscape (Figure 19.22, page 445).

The origin of the Channeled Scablands must be traced to the Pleistocene glacial age, when the mountains surrounding the Columbia Plateau were glaciated. U-shaped valleys and hills of glacial debris, evidence of glaciation, show that tongues of ice moved toward the basin from several directions.

Along the northern rim, a large lobe of ice reached about 50 kilometers (30 miles) on to the ancient lavas. The rest of the basalt surface is streaked with a gigantic system of abandoned channels trending southwestward toward the low point of the basin. The widest of the channels is more than 32 kilometers (20 miles) across, and the deepest cut is 273 meters (900 feet) below the surface. The surface of the basalt, stripped of the thin loess covering, has been scoured in a way never seen before, hence the name *Channeled Scablands*. It has been estimated that about 7,252 square kilometers (2,800 square miles) of the Columbia basin have this distinctive appearance.

This landscape, unlike any produced by known rivers – even in flood time – was named and intensively studied by J Harlen Bretz of the University of Chicago from the 1920s to the late 1960s. His investigation led to some startling conclusions and embroiled him in a heated controversy that lasted for many decades. From his detailed study of the array of erosional and depositional landforms, Bretz concluded that they could only be accounted for by a catastrophic event – a truly gigantic flood. Needless to say, this theory was not well accepted by the geologic community firmly entrenched in uniformitarianism.

Bretz found that the patterns of the huge channels suggested their origin was the result of one or more immense floods of short duration rather than erosion by rivers of normal size. The channels divide and reunite in a braided network that drains generally southwestward. In some places, the current spilled over low divides from one channel to another. Huge potholes and rock basins occur along the channel bottoms and even on the divides. Unlike most braided streams, these channels have very little sediment. The long profile is not smoothed, and the basins have not been filled in, characteristic of a sudden event rather than a long period of erosion and deposition.

Other evidence of a flood includes dry coulees (canyons) with abrupt cliffs marking former huge cataract systems (waterfalls) and plunge pools, all carved in basalt. Associated with the now-dry channels are enormous gravel bars containing granite boulders as large as 6 meters (20 feet) across that must have been transported by flood waters. Gravel deposits formed gigantic current ripples up to 6 meters (20 feet) high and about 91 meters (300 feet) from crest to crest.

One difficulty other geologists had with Bretz's theory was that the source of the flood waters was not immediately apparent. The problem was resolved with the discovery that the continental ice sheet that covered western Canada during the last glaciation dammed a huge lake in the vicinity of Missoula, Montana. Lake Missoula, as it came to be known, contained between 2,000 and 2,500 cubic kilometers (349 to 512 cubic miles) of water at its maximum and covered an extensive area including what is today the University of Montana campus. The lake remained in existence only so long as the ice dam was stable. When the lake got deep enough, it floated the ice dam free, and water emptied rapidly from the lake as if a plug had been pulled. The only possible exit route was across the Columbia Plateau and down the Columbia River canyon to the sea.

Bretz's idea slowly gained support as additional study continued throughout the Columbia basin. It is now generally agreed that not one but many floods crossed the plateau. From the size of the giant ripples, hydraulic engineers estimate that the water depth was between 12 and 150 meters (40 to 500 feet) and that the average flow velocity was 32 to 65 kilometers (19 to 40 miles) per hour. Such investigations and the discovery of other examples of flood-sculptured terrain have led to a more complete understanding of the dynamics of catastrophic flooding and the manner in which it can drastically alter a landscape.

Nearly 50 years after publication of Bretz's original paper on the Channeled Scablands, a group of scientists representing 19 countries visited the Lake Missoula shorelines, the gigantic current ripples, Grand Coulee, the great gravel deposits, and other features of the Columbia Plateau. The next day, Bretz (who was then about 90 years old) to his great satisfaction received a telegram of "greetings and salutations" from the assembled scientists, which closed with the sentence, "We are now all catastrophists."

Questions

1. What is meant by a 50-year flood?

A 50-year flood refers to a level of precipitation and resulting discharge thought to occur on an average of once in 50 years. There is no certain way to predict when a major flood will occur, and a 50-year flood may occur at longer or shorter intervals. It is possible to have two 50-year floods in successive years, but that doesn't mean the third one won't happen for another 49 years! By analyzing the frequency of occurrence of past floods of different sizes, scientists can establish the *probable* interval between floods of a given magnitude.

Flood maps, which show to what extent floods of various sizes might inundate a flood plain, have been prepared for major river cities. Such maps are an important part of wise land-use planning and should be consulted if construction close to a river is being considered. Flood maps have to be updated fairly often because each flood provides new data, which improve flood-recurrence statistics. In addition, urban growth, cultivation, road-building, and other paving – all of which prevent rain from infiltrating into the ground – increase storm runoff and the incidence of larger floods.

Annual flooding of most streams is expected, and in most natural streams, bank-full discharge occurs about every 1 to 2 years. The recurrence of major flooding is also a natural event that should be anticipated. While upstream dams and flood control projects can partially reduce the dangers of flooding, they cannot prevent the destruction that accompany the 50-year flood. This big flood is considered a catastrophe only because homes and cities have been built on the banks and on the flood plains. On the other hand, flooding rivers add soil and nutrients to their broad valleys, valuable agricultural resources we must learn to understand and appreciate. (See Figures 16.2, 16.5, 16.27, pages 350, 353, and 363, and read Box 16.3, pages 364-365.)

2. What is stream rejuvenation, and why does it occur?

The phrase to *rejuvenate*, when applied to streams, has several meanings: to render young again; to stimulate, as by uplift; to renew erosive action; to develop youthful features of topography in an area previously worn down to a base level. Streams that are *rejuvenated* have increased velocities because of uplift (or a drop in sea level) that starts a new erosion cycle.

The term was first used in 1906 and is no longer in common usage, although the concept is still valid and useful. Plate tectonics was totally unknown in 1906, and the reasons for uplift were poorly understood, although the effects on the landscape and on stream behavior were well described. Today, tectonic movements are known to be among the principal causes of changes in stream behavior and in landscape development. Other factors include climatic changes that increase or lessen the discharge, changes in base level, such as the drying up of the Mediterranean Sea (Lesson 19), or lowering and raising of sea level several times during the glacial ages, even the downdropping of a basin as in Death Valley.

A classic example of tectonically-induced rejuvenation is the Colorado River cutting into a plateau that was uplifted about 2 kilometers in recent geologic time. Thus all streams should reach old age unless tectonic movements occur and restart the cycle, which is what frequently happens. Changes are always occurring someplace on Earth, and existing streams may have experienced a number of cycles of rejuvenation. These can be read in the shape of raised stream terraces, entrenched meanders, and *superimposed* streams (See Box 16.4, pages 375-376).

3. Why do streams have so many different drainage patterns?

Drainage patterns vary because there are so many different rock structures that influence valley development. Streams flowing freely over uniform rocks, such as horizontal sedimentary layers, are not restricted by underlying rock structures and will develop dendritic drainage. Seen from the air or on a map, the pattern will resemble a branching tree, hence the name. If the underlying rock structure is jointed, rectangular patterns will result. Radial drainage forms on volcanic cones, and streams flowing over dipping sedimentary strata of differing hardness will develop trellis drainage.

4. What are stream terraces and why are they considered significant indicators of tectonic instability?

Stream terraces are flat areas that rise step-like in matched or unmatched pairs on either side of a stream. They are erosional features that may be cut in rock and have a thin veneer of sediments or, more often, they are the abandoned remains of a former flood plain that has been uplifted and is now being eroded by a rejuvenated stream. Some terraces are not paired and occur at different elevations on either side of the stream. These will form when a stream is downcutting the valley and lateral erosion from meandering is occurring at the same time (Figure 16.52 B, page 378). Unless eliminated by erosion from major floods, each terrace level is evidence of regional or local uplift (or lowered base level) and a cycle of renewed stream energy. If the uplift is considerable, the stream will run more rapidly and will cut through the flood plain while following its meandering path. It will deeply erode a looping but youthful V-shaped valley, forming *incised meanders*.

5. What is meant by the Continental Divide, and where is it?

A divide is a ridge or mountainous area dividing one drainage basin from another. A drainage basin is the total area drained by one stream and its tributaries. (See Figure 16.3, page 352, which shows the drainage basin of the Mississippi River and the Continental Divide in the United States.)

About 80 kilometers (50 miles) west of Denver, U.S. Highway 40 crosses Berthoud Pass at an altitude of 3,450 meters (11,314 feet) above sea level. At the crest of the pass is the *Continental Divide*, part of an almost continuous ridge and mountain crest system that runs generally north and south the length of North America. The streams to the east of the Divide flow into the Atlantic or the Gulf of Mexico; those draining to the west flow into the Pacific Ocean. Some of the mountain peaks along the Divide, especially in Colorado, reach heights over 4,242 meters (14,000 feet), rivaling Mount Whitney in California, the tallest peak in the continental United States.

Another continental divide exists in northern Wisconsin, about 80 kilometers (50 miles) southeast of Duluth, near the small town of Cable, where U.S. Highway 63 crosses a low east-west rise only a few hundred meters above sea level. This divide separates all of the water that flows to the north, to Lake Superior and the other Great Lakes and to the St. Lawrence River, from water that flows to the south through the Mississippi and its tributaries.

6. What is stream piracy?

As impressive as the Continental Divide might be, divides change with time. Streams, by headward erosion, extend their tributaries into the mountain divides, deepen the valleys and transport away the rocky debris. With time, and barring tectonic uplift, the divide will be lowered. A strong stream, whether through greater discharge, increased slope, or other factors, may erode back through the divide and cut into the valley of another stream on the opposite side. The drainage of the lesser stream will be taken, or captured, and its drainage added to that of the larger stream, thus enlarging its drainage basin. This is called *stream piracy*, a term applied by 19th century geologists for stream capture. Piracy explains narrow valleys and gorges that are strangely lacking in flowing water; the once-present stream was captured by a "pirate" stream.

7. **How do streams get started, and how does mass wasting contribute to their growth?**

The start of a stream is usually *slope wash*, a shallow runoff of rainwater on the surface of the ground. It may form little rills of water that gradually come together, forming larger rivulets that will enter a stream that leads to a river. If the rainfall is sufficiently heavy, it will produce *sheetwash*, a continuous sheet of flowing water. Both slope wash and sheetwash running over a hillside move small particles of silt, sand, and even small pebbles down slope. Sheetwash causes headward erosion, eating back at the head of a gully, where slump is enlarging the little valley (Figure 16.40, page 371). As a valley develops, various forms of mass wasting bring sediments and rock debris to the valley floor. Soil creep, landslides, rock falls, and weathered rock particles all move downward by gravity. The slopes retreat, although the width of the stream at the bottom may be narrow. (See Figure 16.37, page 370. Note evidence of rockslides and rockfalls).

8. **What happens when a river comes to a large body of standing water?**

The ocean, the largest body of water, is the ultimate base level of all streams. (For streams that flow into a lake, the elevation of the lake is also a base level. For tributary streams, the base level is the elevation at which they join the main stream.) A stream cannot erode below base level because as the powerful current enters the body of standing water, its energy and forward momentum are dissipated. The current loses momentum gradually, depending on its volume and velocity. For small streams entering a wave-swept coast, the waters mix so rapidly that the current disappears almost immediately. The discharge of the Amazon, on the other hand, is so enormous (about 4 billion gallons per minute) that the river maintains its integrity many kilometers away from shore. Seafarers, if they had known their geology, could have had a drink of fresh water from the Amazon far out at sea.

Activities

1. The best suggestion for *seeing* what you've learned in this lesson is to visit a river, if one is a reasonable distance from your home. Look for a V-shaped valley and signs of active erosion if the stream is *youthful*. In an *old age* stream, expect meanders, a flood plain, natural levees, oxbow lakes, and river terraces.

2. *Wild rivers* and the efforts to dam them have generated a lot of public debate. Can you think of good reasons to leave the wild rivers alone? What are the stated reasons for damming wild rivers?

Self-Test

1. A stream valley becomes wider toward the mouth of the stream. The process most important in accomplishing this is

   a. meandering of the stream.
   b. deposition of sediments at the point bar.
   c. formation of the delta.
   d. channelizing of the stream.

2. During development of a stream valley, the first event is most likely to be

   a. erosion by sheetwash.
   b. abrasion of the stream bed.
   c. deposition of river bars.
   d. slump and rockfalls.

3. Despite its distinctive stair-step, cross-canyon profile, the Colorado River flowing through the Grand Canyon would be considered a

   a. mature stream.
   b. youthful stream.
   c. meandering stream.
   d. superposed stream.

4. The drainage pattern observed most often usually forms on

   a. tilted sedimentary layers.
   b. fractured intrusive rocks.
   c. parallel folds of unequal hardness.
   d. horizontal uniform sedimentary layers.

5. The lengthening of a stream is an on-going process and is accomplished by all of the following except

   a. headward erosion in the tributaries.
   b. deposition of sediments in a delta or alluvial fan.
   c. cutting off meanders.
   d. stream piracy.

6. Rejuvenation of a stream might be accomplished by all of the following except

   a. melting of glaciers after an glacial age.
   b. regional uplift.
   c. increased discharge caused by climatic change.
   d. down-faulting of a basin of deposition.

7. The term *base level* could conceivably apply to all of the following except

   a. a reservoir formed behind a dam.
   b. a desert lake surrounded by mountains.
   c. the point where a tributary joins the main stream.
   d. the place where the main stream starts to meander.

8. When geologists say a condition of equilibrium has been reached in which a stream is neither eroding nor depositing its sediment, they are referring to a

   a. mature stream.
   b. graded stream.
   c. meandering stream.
   d. braided stream.

9. A youthful stream usually will lack

   a. waterfalls and rapids.
   b. a narrow V-shaped valley.
   c. a flood plain.
   d. a high gradient.

10. High altitude photographs often show crescent or U-shaped features on river flood plains. These are caused by

    a. rivers over-topping their banks during flood times.
    b. winding tributaries to the main stream.
    c. resistant layers of folded rocks.
    d. cut-off meanders that form lakes.

11. The final stage in valley development, theoretically, would be

    a. a graded stream.
    b. a peneplain.
    c. a mature landscape.
    d. sheet erosion.

12. In the present landscape of the continents, which of the following is the least common landform?

    a. V-shaped valleys
    b. Meanders
    c. Stream terraces
    d. Broad peneplains

13. Entrenched meanders are spectacular landforms and are good indications of

    a. the old age of valley development.
    b. stream rejuvenation.
    c. easily eroded rock underlying the valleys.
    d. transition from a youthful to mature stage in valley development.

14. The term *25-year flood* would refer to

    a. the flood level that is reached about once in 25 years, on average.
    b. the cumulative level of all floods during a 25-year period.
    c. an unnatural condition that occurs during a 25-year span owing to human intervention.
    d. an annual flood level that follows a 25-year cycle.

15. Some stream valleys show several levels of paired terraces. In interpreting the history of the stream, each pair would indicate

    a. several cycles of lateral erosion.
    b. periodic increased sedimentation on the flood plain.
    c. approaching old age in landscape development.
    d. a cycle of stream rejuvenation.

MODULE V

CARVING THE LANDSCAPE

21

GROUND WATER

This lesson will help you understand the distribution of ground water and how ground water is important to life.

After reading the textbook assignments, completing the exercises in this study guide, and viewing the lesson's video portion, you will be able to:

1. Explain why ground water is a critical natural resource.

2. Distinguish between porosity and permeability and indicate how these factors influence ground water.

3. Describe how ground water forms and describe its rate of migration.

4. Sketch the relative positions of the saturated and unsaturated zones, the water table, and perched water tables.

5. Indicate how permeability and the slope of the water table control the velocity of ground-water flow.

6. Recognize the relationships among springs, streams (gaining and losing), and ground water.

7. Explain why certain rock types make good aquifers.

8. Recognize the relationship between wells and the water table and how a pumped well and the heavy use of ground water affect the water table.

9. Discuss several ways in which ground water can become contaminated.

10. Relate the effects of ground-water action to the formation of caverns and depositional cave features and karst topography.

11. Explain how hot springs and geysers are formed.

12. Recognize the potential uses of geothermal energy.

13. Account for the origin of pressure in an artesian well.

Ground water that lies beneath Earth's surface is not only a geologic agent of erosion and deposition, it is also one of our greatest natural resources. It is generally hidden from view, however, and its presence is not apparent to the casual observer. Then how do we know it even exists? The perceptive observer sees ground water emerging in lakes and ponds, in hot and cold springs, in swamps, marshes and bogs, in some rivers and streams, and in gushing artesian wells and the millions of water wells that have been dug or drilled throughout human history.

You can see the work of ground water most dramatically in the eruptions of the famous Old Faithful Geyser in Yellowstone Park in Wyoming. Or you can descend underground with a light to explore the many miles of passages and glistening caverns at Carlsbad Caverns, New Mexico, or Mammoth Cave, Kentucky, or in the Dordogne Valley in France, where early man made wonderful cave paintings. You will marvel how the slow-moving ground water removed huge quantities of limestone bed rock, molecule by molecule, and carried it away in solution or returned it in part to the caves as *stalactites* and *stalagmites*.

The source of ground water is rain or melting snow that percolates into the ground as described in Figure 17.1, page 384. About 99 percent of the water on the surface of Earth is contained in the world's oceans and frozen in the ice sheets of Antarctica and Greenland. Only about 0.6 percent of the water on Earth is ground water, but this is about 35 times greater than the volume of all the fresh water contained in lakes or flowing in streams!

Ground water is actually present nearly everywhere in the pores and fractures of the bed rock, but it appears at different depths and in varying amounts. In a moist climate, a well may have to penetrate only a few meters to reach water; in arid regions, it may be hundreds of meters deep. The target of the well driller is the water table, the upper surface of the saturated zone below which the pores of the rock are filled with water. This is a significant boundary because it represents the upper limits of all readily usable ground water. Below the water table, the ground water percolates laterally, usually a few centimeters per day, twisting its way through the openings in the rock. Comparing the rate of motion of ground water with that of the well-known garden snail, the snail is estimated to be about 10 times faster!

In this lesson, you will learn how the limited amount of water on the planet, including ground water, is recycled over and over again. Some of the water molecules used by the dinosaurs 100 million years ago may be in your tea kettle today! You will read about the vast underground aquifers that store water for farm crops and for urban uses. You will also learn about sources of pollution of ground water and the results of excess removal of ground water from aquifers. You will read about geothermal energy, its potential and its problems, as we move toward an energy-hungry 21st century.

Completing the following eight steps will help you master the lesson objectives and achieve the goal for this lesson:

Step 1: Read the TEXT ASSIGNMENT, Chapter 17, "Ground Water."

Step 2: Study the KEY TERMS AND CONCEPTS as noted in the study guide.

Step 3: Watch the VIDEO using the VIEWING GUIDE in the study guide.

Step 4: Read the study guide's PUTTING IT ALL TOGETHER section, which will help you summarize and integrate all of the information in this lesson.

Step 5: Complete any assigned lab exercises.

Step 6: Complete any assigned ACTIVITIES found in the study guide.

Step 7: Review the material in this lesson and complete the SELF-TEST found in the study guide.

Step 8: Go back to the LEARNING OBJECTIVES and make sure you have learned each one.

1. Read Chapter 17, pages 383-403.
2. Re-read the "Introduction" and "Summary" in the chapter for a good review.

The key terms listed under "Terms to Remember," page 402 in your text, are important and will supplement your knowledge of water learned from the chapter on ground water. Use the glossary and the summary in the text to help you understand the definitions.

Once you have read the text material for the lesson and studied the terms, you are ready to view the lesson's video portion. Review the questions that appear in this section to help you watch for the important points and to prepare for what you will see in the video.

Before viewing, answer the following questions:

1. List some examples of the geologic work of ground water.
2. What kind of rocks make good aquifers, and why?
3. List potential sources of contamination of ground water.

After viewing, answer the following questions:

1. Explain why "dowsers" for ground water seem to have some success.

2. What are the sources of ground water, and which is the most important?

3. What are the properties that make a rock a good aquifer? What kind of rocks form an aquiclude?

4. Most of the great caves of the world are found in limestone bed rock. Is this just a coincidence, or is there a good reason, geologically?

5. If you start to dig a well in your backyard, how will you know when you reach the water table? Is the water table really flat like a table?

6. From the animation shown in the video, what are the requirements for artesian flow from water wells? If you were prospecting for artesian water, what geologic indicators would you look for?

7. What is the source of water for the Orange County, California water district? How do they recharge their ground water supplies?

8. What is Orange County doing to keep out salt water? How are land fills being constructed to prevent pollution of ground water?

9. What are some suggestions for dealing with changing ground water supplies?

PUTTING IT ALL TOGETHER

You have read the text assignment and seen the video portion of the lesson. This section, made up of a SUMMARY, two CASE STUDIES, the QUESTIONS, suggestions for ACTIVITIES, and a SELF-TEST, will help you pull the information together.

Summary

Although ground water itself is seldom visible, the geologic work of ground water can be quite spectacular. The vast network of caves, passages, and chambers that riddle underground limestone bed rock result from solution by acid-bearing ground water. The colorful displays of *dripstone*, such as stalactites and stalagmites, and the sheet-like *flowstone* are remarkable in that they were formed drop by drop by ground water carrying calcium and bicarbonate ions dissolved from the surrounding limestone. If the water table drops and the cave is filled with air, the dripping water will lose a little carbon dioxide, and calcite (calcium carbonate) will precipitate as a solid and slowly form stalactites and other dripstone features.

The *solution pits, sinkholes,* and *disappearing streams* that characterize karst topography are clues to the presence of soluble rock layers below the surface. Above ground, the great roaring explosions of water and steam from geysers such as Old Faithful in Wyoming are breathtaking views that attract thousands of visitors every year. Geysers are a relatively rare phenomenon and are caused by special underground "plumbing" of hot springs in areas of recent volcanic activity. Ground water also plays an essential role in converting unconsolidated sediments into sedimentary rocks by depositing mineral cements around the grains and in the pores. Common natural cements are calcite and silica.

Occurrence of Ground Water

Ground water is that part of the hydrologic cycle that infiltrates the ground and is stored in *aquifers*, which are rock layers – usually sandstone or conglomerate – that are both porous and permeable. About half of the United States is underlain by one or more aquifers. The vast amount of water in most aquifers has accumulated over thousands of years. Both the rate of recharging an aquifer and the movement of the water underground is extremely slow. The water in some aquifers is believed to be in about the same place today that it was 25,000 years ago!

At varying depths below the surface lies the saturated zone, where all the pores in the rock are filled with water. The upper boundary of this zone is the water table, the goal of water-well drillers. Above the water table lies the unsaturated zone, where the pores of the rock are filled partly with water and partly with air. The water table generally follows the surface contours of the land, being higher under the hills and lower in the valleys. The level of the water table changes, rising after a rainy cycle but flattening almost to the valley level after a prolonged drought. The slope of the water table influences the rate of flow, which may vary from a few centimeters to a few meters per day.

Springs form where the water table intersects the land surface. In many localities, faults control the location of springs by moving impermeable rock against the permeable layers. Water will rise along the fault plane, and a line of trees or other vegetation will mark the location of the springs.

The amount of water in streams is also related to the level of the water table. Some streams gain water from the saturated zone, and the stream will flow at the level of the water table. Other streams will lose water to the water table, which may be far below the river bed. The Platte River in Nebraska and the Nile River in Egypt are examples of streams that flow from mountains having substantial rainfall into much drier regions in which the water table lies deep beneath the surface. Water from these rivers leaks downward and recharges the ground water below. Since construction of the Aswan Dam and filling of Lake Nasser in Egypt, the water table in the surrounding area has risen because of seepage from the lake.

Ground Water as a Crucial Natural Resource

Because ground water is a crucial resource, the search for plentiful, good-quality water has gone on all through human history. Geologists today use many techniques to find new high-producing aquifers. They may create computer-generated models or drill small test wells before drilling more expensive supply wells. But despite the many scientific advances in the field of hydrogeology, some people still search ("dowse") for water using a forked stick. Since ground water exists in many locations, the exact placement of the well is not critical; thus dowsers appear to have some success. Statistically, however, people would do better to hire a geologist. (See Box 17.1, page 390.)

Artesian wells tap into tilted confined aquifers at places where the water is under pressure from the weight of the water above. The first wells drilled in South Dakota into the huge High Plains aquifer spouted water hundreds of feet into the air. Continued drilling of thousands of wells lowered the artesian pressure so much that the water no longer reached the land surface. Water in natural aquifers cannot be replenished easily and in some instances seems to be a non-renewable resource. Various methods of *artificial recharge* are being tried such as filling specially constructed surface-infiltration ponds with natural flood water or, in some areas, using clean reclaimed water from water treatment plants.

Serious Problems Affecting Ground Water

All the water that seeps into Earth is contaminated or polluted to some degree at the start of its journey at the ground surface. For example, rainfall and melting snow pick up minerals and bacteria from the soil. But waste water seeping into the ground from an individual septic-tank drain field is *heavily* polluted. Fertilizers, pesticides, liquids that leach from city dumps and toxic waste disposal sites, liquid and solid wastes from industry, and radioactive wastes are all sources of pollution of our ground water resources (Figures 17.15 through 17.19, pages 391-395).

In some areas, the water is naturally purified as it percolates through sand and other fine-grained materials in the aquifer. This natural purification is of great economic importance and has made possible the direct use of ground water by millions of people in the United States. However, many public water supplies are chlorinated as an additional safety factor.

CASE STUDY I:
The High Plains Aquifer: "The Water Carrier"

Many years ago, the United States was divided into 10 ground water regions based primarily on aquifer type without regard to surface-water drainage. Those ground water regions still exist today. One aquifer of particular interest is in the High Plains ground water region, which includes semi-arid to sub-humid plains and plateaus east of the Rockies. The ground water region extends from South Dakota to western Texas and eastern New Mexico and includes parts of Wyoming, Nebraska, Colorado, Kansas, and Oklahoma. This one aquifer system supplies water for more than 20 percent of the irrigated land of this country.

As you have learned in this course, the surface of Earth has undergone many changes, some due to erosion and deposition, and others due to tectonic changes. The High Plains are no exception. Eastward-flowing streams eroded the rising Rocky Mountains and deposited a blanket of alluvium over a vast area covering much of the Great Plains. The edges of the alluvial apron were also uplifted, and streams whittled away the deposit both along its western margin near the base of the mountains and along the eastern limits. The original formation is now isolated from the mountains that were the source of the sand and gravel. The streams that flowed eastward out of the mountains kept the alluvium full of water, but now the aquifers must be replenished primarily by rain and snow within the region itself.

The bulk of the alluvial deposits that form the aquifer in this vast plain is known as the Ogallala formation. The aquifer lies at depths of less than 350 meters (1,155 feet) and covers the older Tertiary and upper Cretaceous bed rock with an average thickness of about 65 meters (215 feet) of sandy and gravelly deposits. The flow of water within the aquifer is generally to the east. Since the formation was originally a giant alluvial fan, the characteristics of the aquifer are those of a typical fan deposit. Particle sizes range from very fine to coarse with abrupt variations from silt and clay to coarse sand and gravel within short distances, both vertically and horizontally. For some reason, the sand and gravel formations of the Ogallala do not take on recharge water as readily as the aquifers in some of the eastern states.

When the formation was first tapped about 100 years ago, it produced about 1,000 gallons of artesian water per minute. Since that time, more than 170,000 wells have been drilled, resulting in a drop of water pressure and a lowering of the water table by as much as 30 meters (100 feet). Further development of ground water resources was spurred by a severe regional drought in the 1930s and again in the 1950s. A dramatic increase in pumping rates led to further decline of the water level. In parts of Kansas, New Mexico, and Texas, the thickness of the saturated zone in aquifer has declined by more than 50 percent. It is estimated that in the dry western plains of Texas, the Ogallala is being drawn down as much as 18 times faster than it is being replenished. The decreased water yield and increased pumping costs have led to major concern about the future of irrigated farming on the High Plains.

Many of the great cities of the world are sinking because of removal of the fluids beneath them, such as ground water or oil and gas. Differential subsidence has dislocated utilities, buckled pavements, and damaged the structures of some of our most famous historical buildings. Most of the cities having problems with land subsidence are located on unconsolidated sediments, either on a river flood plain as New Orleans or London are, on a coastal marsh or delta as Venice, Italy, Houston-Galveston, or Tokyo are, or on ancient lake beds as Mexico City is. In addition, some of these urban areas are located on coasts where long-term tectonic subsidence is occurring at the same time the sea level is rising. (See Lesson 24 on waves, coasts, and beaches.) The most severe problems of sinking have been accelerating during the past 50 years, and human-

**CASE STUDY II:
Why are So Many
of the Great
Cities Sinking?**

induced subsidence may be more than 50 centimeters (20 inches) per year in some places.

The weight of a city is supported partly by the soil or rock on which it is built and partly by the pressures generated by the fluids that occupy the pores in the underground aquifers. In confined aquifers, the fluid pressures are sometimes higher than the pressures generated by the weight of the overburden. When a city is built on recent sediment consisting of interbedded sands, clays, and peats that have not been well compacted and contain considerable moisture, a lot of subsidence occurs when the fluids are withdrawn.

One of the most spectacular occurrences of land subsidence has taken place in the harbor area of Los Angeles-Long Beach, California. Offsetting its effects and arresting its action have cost more than $100 million. Production of oil and gas and drilling for the contained formation water from sands interbedded with clays and siltstones at depths about 350 to 600 meters (1,150 to 2,000 feet) below the surface began in 1936. By 1941, the subsidence was about 9 meters (30 feet) at the center of the area. Since the land originally was only about 2 to 3 meters (7 to 10 feet) above mean sea level, an extensive dike system had to be built around the port facilities, power plants, and industries to prevent flooding.

Re-pressuring of the producing zones was undertaken on a large scale in 1958, and the sinking was stopped by 1962, but only about 33 centimeters (13 inches) of rebound has occurred. The area remains at considerable risk since it is earthquake-prone, and a breach in the dikes could result in wide-scale flooding.

Although the rapid subsidence of the last half century now appears slowed in many cities, there has been almost no recovery in ground elevation, and many of the world's greatest cities remain vulnerable to flooding from the rising sea. Whatever is done to protect these threatened areas may be the largest effort, the greatest challenge, and the most expensive undertaking yet faced by civilized people.

Questions

1. **What is the difference between porosity and permeability, and how do they affect ground water storage and water movement in rock layers?**

Porosity refers to the proportion of the total volume of a rock that is open space. This is a measurement of the ability of the rock to hold water. *Permeability* is the ability of a rock to transmit fluids and depends on the connections between the openings or pore spaces.

Sandstones and conglomerates are both porous and permeable and are important in storing and moving ground water. Shale may hold a great quantity of water but is relatively impermeable. Granite, schists, and some limestones are not actually porous but when jointed and fractured may hold and transmit quantities of ground water. The Hawaiian Islands, for example, are built of massive piles of thin-bedded porous lava flows that hold huge natural reservoirs of ground water recharged by the abundant rainfall.

It is important to understand that the ground water is held in and moves through the pores of the rock in an aquifer. Some of the ancient myths that told of great subterranean rivers are erroneous notions that unfortunately persist today. Small underground streams flow in some caves, but they are uncommon and, in terms of volume of water, insignificant. (Petroleum also exists in the pores of the rock, but the term "oil pool" still reflects the wrong concept of an underground lake or oil-filled cavern.)

2. What is meant by the water table?

The water table is the upper surface of the saturated zone. The slope of the water table generally parallels the slope of the land surface. This is not a static level, however, but responds to short-term changes in weather and climate and to longer cycles such as tectonic movement (Figures 17.2 and 17.11, pages 385 and 389). Note that in the unsaturated zone above the water table the pores of the rock contain both water and air.

3. Describe a confined aquifer.

A confined aquifer consists of a layer of permeable rock such as sandstone sandwiched between layers of impermeable rock. Typically, the aquifer may be exposed at the surface in a mountainous area where there is sufficient rainfall to recharge the permeable layer. The aquifer usually dips downward below the adjacent plain and flattens as it extends away from the mountain (Figure 17.13, page 391). The water in the aquifer can move "downhill" underground, sometimes for hundreds of miles. As the rate of recharge is extremely slow, for all practical purposes, the water in confined aquifers is an exhaustible resource that once gone is not likely to be replenished.

4. Describe an artesian well and the conditions that produce artesian flow.

An artesian well taps into a tilted confined aquifer in which the water is under pressure from the weight of the water behind and above it (See Figures 17.13 and 17.14, page 391). The water will rise above the level of the aquifer and usually above the ground surface, depending on the elevation of the recharge zone and the location of the well. When some artesian wells in the Great Plains were first drilled, the water rose over 33 meters (100 feet) above the land and produced an enormous volume of water (Figure 17.14, page 391).

5. What is a spring, and under what conditions do springs form?

A spring is a flow of ground water emerging naturally at the surface. Examine Figures 17.3, 17.5, 17.6, and 17.7, pages 380, 387, and

388, for examples of springs and the geologic conditions that produce them.

6. How does a geyser differ from the usual hot spring? Explain the interior plumbing that leads to the cyclic eruptions.

A *geyser* is a type of hot spring that periodically erupts hot water and steam. Geysers, the most spectacular manifestations of ground water, are rare phenomena, found only in Yellowstone National Park in Wyoming, in New Zealand, and in Iceland – all areas of recent volcanism.

The internal plumbing of a geyser usually consists of a hot spring with a constricted opening to the surface that permits heated ground water, under pressure in the underground geyser chambers, to reach temperatures near boiling. Bubbles of steam surge through the opening, and some of the water spills over the top, lowering the pressure on the water in the lower chambers. The water then becomes superheated and flashes into steam that blasts the water and vapors out of the chambers. The chambers are then empty, but will fill again with ground water, become heated, and the cycle will be repeated. (See Figures 17.28 and 17.29, page 399, for an explanation of the geyser cycle.)

7. What is karst topography?

Karst topography is a land surface underlain by limestone or other soluble rock characterized by features related to solution by ground water. These include caves, disappearing streams, and bowl-shaped depressions called *sinkholes* formed either from collapse of the roof of a cave or as solution pits in the limestone bed rock. (Review "Solution Weathering" in Chapter 12 for information about chemical weathering of limestone.) The name "karst" is from the Karst re- gion in Yugoslavia, but karst topography can be seen in Pennsylvania, Maryland, Virginia, Indiana, Kentucky, Tennessee, Florida, Texas, and other states, as well as Puerto Rico. The famed *cenotes* of Mexico's Yucatan Peninsula are deep sinkholes.

Sinkhole formation is still an active process, and in one small area of about 25 square kilometers (9 square miles) in Florida more than 1,000 collapses have occurred in recent years. The sinkhole problem in Florida and parts of Texas is related to excessive pumping of ground water that lowered the water table. The supporting effect of ground water (Case Study II) is also seen in California's San Joaquin Valley, the surface of which subsided more than 9 meters (30 feet) from extensive pumping and lowering of the water table (Figures 17.23 and 17.24, pages 397 and 398).

8. How do caves and caverns form?

Caves are naturally-formed underground chambers, frequently connected by long passages. As you remember from Lesson 15 on weathering, limestone is particularly susceptible to solution in slightly acidic ground water. Carbon dioxide, CO_2, which is abundant in the atmosphere and in the soil, combines with the ground water to form weak carbonic acid, H_2CO_3, which reacts with the calcium carbonate of the limestone. The limestone decomposes into calcium ions and bicarbonate ions, which are soluble in water. (Study the chemical reaction on page 246 in the text for both solution of limestone and the development of flowstone and dripstone.) Most caves were probably formed below the water table by ground water moving along the fractures and bedding planes (Figure 17.20 A and B, page 396). If the land is uplifted as a response to tectonic movements, the water table will drop, draining away the water and allowing air to enter the cave system. At this point, stalactites and stalagmites begin to form.

9. What happens to all the mineral matter dissolved from the bed rock?

The minerals such as calcite (from limestone) and dissolved silica will enter the ground water and percolate through the sands and gravels of the aquifer to eventually be precipitated in the pores as a natural cement between the grains of the sediment. This will reduce the porosity and eventually convert the sediment to sedimentary rock, one of the most important geologic aspects of the work of ground water.

It is interesting to note that great thicknesses of limestone were deposited almost without exception in warm shallow seas and that at least 26 of the U.S. states are underlain by these sedimentary rocks. Some of the deposits contain ancient coral reefs, further evidence of the widespread seas that covered much of North America in Paleozoic and Mesozoic times. Ancient limestones occur on every continent, an indication of the great changes that have taken place on the surface of Earth as a result of plate movements, both horizontal and vertical.

10. What are some of the advantages and disadvantages of geothermal energy?

Geothermal energy is obtained from naturally occurring steam and heated ground water. Wells drilled into geothermal sites tap superheated water that is piped to a powerhouse to run a generator. The advantages are that no fossil fuels have been burned, air pollution problems are minimal, and the radiation hazards of nuclear plants are eliminated. But there are also problems and environmental effects. The hot water brought to the surface often contains minerals dissolved from the rocks, such as mercury and lead, and salts such as sodium chloride that corrode machinery and present a difficult disposal problem. But geothermal energy may still contribute toward meeting the world's future energy needs (Figures 17.32 and 17.33, page 401).

11. Why is ground water considered one of Earth's greatest natural resources?

Aside from the water frozen in glaciers and ice sheets, most of the fresh water on Earth occurs in the ground rather than on the surface. Rivers and lakes account for only a small fraction of the total fresh water. But like surface water supplies, ground water is also vulnerable to pollution, and a concerted effort must be made to protect and renew this remarkable resource.

Activities

1. Fill a glass with sand. Now pour in half a glass of water dyed red with food coloring. Can you see the water table? The saturated and unsaturated zones should be easy to distinguish. This little bit of kitchen geology clearly demonstrates the presence of pore spaces between the sand grains.

2. Go to a rock-hound shop and look for specimens of *petrified wood* (Figure 17.25, page 398). The buried wood or organic material has been replaced by silica carried in by ground water. The original material in fossil shells and bones can also be replaced by silica or calcite. Also, look for *concretions*, the strangely-shaped objects formed when ground water deposits excess cementing material around some buried object such as a leaf or fossil (Figure 17.26, page 398).

   The rock-hound shop may also have *geodes*, which are most interesting. They are spheres of rock partly or wholly filled with quartz or calcite crystals. Their exact origin is not clear, but they are most likely the work of circulating ground water. Geodes are usually sawed in half and polished to better show their unusual colors and beautiful perfect crystals (Figure 17.27, page 399).

Self-Test

1. The principal source(s) of ground water is/are

   a. pools of water in caves and caverns.
   b. inland flow from the oceans.
   c. rain or melted snow that percolates into the ground.
   d. rivers, streams, and marshes.

2. Shale tends to have a high porosity but is not suitable for aquifers because it

   a. is well-cemented.
   b. lacks joints and fractures.
   c. is soluble and dissolves in water.
   d. is not permeable and does not allow water to flow.

3. In general, ground water tends to be most abundant

    a. in poorly cemented sandstones and gravels.
    b. the deeper the well is drilled.
    c. in granites and metamorphic schists.
    d. at higher elevations on ridge tops.

4. Underground streams

    a. exist under about half of the United States.
    b. are the most important sources of ground water.
    c. are only found in some caves or caverns.
    d. supply water to most major rivers and streams on the surface.

5. The water table may be described by all of the following except the

    a. level at which water will stand in a well.
    b. top of the saturated zone.
    c. base of the unsaturated zone.
    d. top of the crystalline basement.

6. Karst topography may be identified by all of the following except

    a. sinkholes.
    b. geysers.
    c. disappearing streams.
    d. springs.

7. Bed rock in karst areas is most commonly

    a. shale.
    b. granite.
    c. limestone.
    d. gravels and sands.

8. A good artesian system requires a(n)

    a. tilted confined aquifer.
    b. shallow aquifer open to the surface.
    c. deep aquifer close to the basement rocks.
    d. aquifer in cavernous limestone.

9. Plant roots obtain their water from the

    a. water table.
    b. zone of aeration.
    c. zone of saturation.
    d. zone of soil moisture.

10. Most of the hot springs in the United States occur in the west because

    a. the climate there is hotter.
    b. the region is arid and ground water is closer to the surface, where solar heating can be effective.
    c. this is an area of the most recent igneous activity.
    d. this is an area of greater radioactivity in the rock layers.

11. Purification of contaminated ground water can best be accomplished by percolation through

    a. cavernous limestone.
    b. sandstone or sandy loam soil.
    c. highly fractured granite.
    d. shales or schists.

12. The chemical composition of a stalactite is most likely to be

    a. sodium bicarbonate.
    b. iron magnesium silicate.
    c. calcium carbonate.
    d. iron oxide.

13. Springs may form in all of the following situations except

    a. where fractures in crystalline rock intersect the land surface.
    b. at the mouth of caves.
    c. along faults where permeable rock has been moved against impermeable rock layers.
    d. where confined aquifers are exposed in mountainous areas.

14. When logs are buried, the organic material may be replaced by

    a. silica to form petrified wood.
    b. calcite to form concretions.
    c. both silica and calcite to form geodes.
    d. calcium carbonate to form alkali soils.

15. At present, geothermal energy is

    a. a major source of electrical power all over the world.
    b. most successfully used for space heating and water heating.
    c. most valuable because it has no adverse environmental effects.
    d. an inexhaustible source of energy.

22

WIND, DUST, AND DESERTS

This lesson will help you understand the formation and location of deserts and their geologic features.

After reading the textbook assignments, completing the exercises in this study guide, and viewing the lesson's video portion, you will be able to:

1. Indicate where deserts are typically located and give some reasons for their geographic distribution.

2. Contrast the characteristics of arid regions to humid regions in terms of drainage, base level, and landscape.

3. Describe several landforms unique to arid regions.

4. Explain the origin of deserts within the context of plate tectonics.

5. Describe the role of water and drainage patterns in shaping the landforms of the deserts.

6. Compare wind and water as agents of erosion and deposition.

7. Describe the behavior of wind-blown dust and sand.

8. Indicate how deflation causes blowouts and desert pavement to form.

9. Recognize how sand dunes move, and understand the factors that favor the formation of various dune types.

10. Recognize the origin, distribution, and importance of loess.

11. Explain desertification and its relationship to human activity.

INTRODUCING THE LESSON

Earth is known as the "Water Planet." Seen from space, it is as blue as a sapphire with almost three-quarters of its surface covered by oceans. With such an abundance of water, it is hard to imagine why there are some regions so arid, so lacking in moisture that all living things, including humans, have difficulty surviving there.

These are the deserts – generally regions that receive less than 25 centimeters (10 inches) of rainfall per year. Deserts are widespread; arid and semi-arid regions cover almost one-third of the total surface of the land. These areas occur both north and south of the Equator, chiefly around latitudes 30 degrees north and 30 degrees south. Some deserts occur inshore from regions with cold oceanic currents, such as the deserts of Peru. Deserts also occur in polar regions, contrary to the image of deserts as limited to the hottest places on Earth.

Deserts are not always the shimmering seas of sand we see in films, either, but are more likely to be broad expanses of thin gravelly soils or even barren exposed rock with sparse scattered vegetation. Some deserts are mountainous like the Basin and Range of North America, but others, like much of the Australian desert, are nearly featureless plains.

Sand dunes do occur, and their sinuous shapes soften the stark aspect of many deserts. The dunes are ever changing as they migrate under the force of the wind. Except along sea coasts, wind is a much more effective geologic agent in arid regions than in humid regions. Wind not only moves sand but can pick up and carry vast amounts of dust, as Americans learned during the Dust Bowl disaster of the 1930s.

Streams in the desert, unlike streams in humid regions, almost never reach the sea but die out from evaporation or seepage into the dry terrain. Some desert streams end in playa lakes that are usually flat dry basins covered with mud cracks or are glittering salt flats like the infamous Devil's Golf Course in Death Valley. But despite the pervasive aridity, running water is still the major sculptor of the desert landscape, with occasional flash floods creating most of the unique landforms.

In this lesson, you will learn where global deserts form, and you will be able to relate their location to Earth's atmospheric circulation and to tectonic movements. You will apply geologic processes to decipher the origins of this unique landscape and to unravel the evolving geologic history of the wind, dust, and deserts.

LESSON ASSIGNMENT

Completing the following eight steps will help you master the lesson objectives and achieve the goal for this lesson:

Step 1: Read the TEXT ASSIGNMENT, Chapter 19, "Deserts and Wind Action."

Step 2: Study the KEY TERMS AND CONCEPTS as noted in the study guide.

Step 3: Watch the VIDEO using the VIEWING GUIDE in the study guide.

Step 4: Read the study guide's PUTTING IT ALL TOGETHER section, which will help you summarize and integrate all of the information in this lesson.

Step 5: Complete any assigned lab exercises.

Step 6: Complete any assigned ACTIVITIES found in the study guide.

Step 7: Review the material in this lesson and complete the SELF-TEST found in the study guide.

Step 8: Go back to the LEARNING OBJECTIVES and make sure you have learned each one.

1. Read Chapter 19, pages 433-451, making sure to include the "Introduction" and "Summary" in your reading.

2. Carefully study the photographs and diagrams throughout the chapter.

TEXT ASSIGNMENT

The key terms listed under "Terms to Remember," page 451 in your text, are important and will supplement your knowledge of deserts learned from the chapter on Deserts and Wind Action. Use the glossary and the summary in the text to help you understand the definitions. Here are some additional terms to help you in this lesson:

KEY TERMS AND CONCEPTS

Base level and internal drainage: *Base level* is the lowest point to which most streams flow, usually the ocean. *Sea level* is the theoretical level to which streams can erode. Desert drainage patterns differ in that streams drain toward the local base level which will be the floor of an enclosed basin. The Colorado River in the southwestern United States and the Nile River in Egypt are famous exceptions, as they rise in humid mountains but flow through desert areas. There may be many base levels in arid regions. Some, like the Dead Sea in Israel and Death Valley in eastern California, lie below sea level. As the individual basins fill with sediment, base level will rise.

Badland topography: Bare slopes of shale or ancient lake deposits that are scored by countless gullies resulting from the occasional rains on soft, impermeable rock layers.

Oasis: In north Africa, a green, fertile, and well-watered area surrounded by a sandy desert. The origin is believed to be a result of wind deflation down to the water table. Also, any permanent spring in a desert area.

Loess: A deposit of wind-blown clay and silt-sized particles of unweathered grains of quartz and feldspar. Loess covers much of the north central plains of the United States and the middle latitudes of Europe, and provides the parent material for the fertile soils of the American grain belt. The origin of loess is from wind erosion of the glacial age river valleys and outwash plains lying south of the limits of the continental ice sheets. The large area of loess in central China is, on the other hand, believed to be derived from the Gobi Desert to the north (see Figure 19.22, page 445).

VIEWING GUIDE

Once you have read the text material for the lesson and studied the terms, you are ready to view the lesson's video portion. Review the questions that appear in this section to help you watch for important points and to prepare you for what you will see in the video.

Before viewing, answer the following questions:

1. What is the geological definition of a desert?

2. Geographically, where are most deserts located? List at least three regions.

3. What is the most effective agent in sculpting the desert landscape?

After viewing, answer the following questions:

1. From the video animation, describe how the atmospheric circulation leads to formation of deserts.

2. What landforms develop as a result of erosion in the desert?

3. How does wind influence the desert landscape and what features develop as a result of wind erosion and deposition?

4. What were the factors that contributed to the "Dust Bowl" of the 1930s?

5. Tell the story of the Sahel and why it is important to understand desertification and its causes.

PUTTING IT ALL TOGETHER

You have read the text assignment and seen the video portion of the lesson. This section, made up of a SUMMARY, a CASE STUDY, the QUESTIONS, suggestions for ACTIVITIES, and a SELF-TEST, will help you pull the information together.

Summary

Deserts are regions generally having less than 22 centimeters (10 inches) of rainfall per year. Arid regions cover 33 percent of the land surface, and this area is increasing yearly. Despite the lack of abundant water, desert landforms result primarily from stream erosion and deposition. The strong winds of the desert also make a considerable impact upon the landscape. The landforms in the Colorado Plateau and in the Basin and Range province are determined by their individual geologic history and by the underlying rock structure rather than by climate alone. (Compare Figure 19.7 A and B, page 438, showing landforms of the Colorado Plateau with Figure 19.11, page 440, showing the origin of some Basin and Range topography.)

Changing climates today and improper use of the land are causing *desertification*, the spread of the barren deserts into once populated regions. Deserts evolve as climate, geology, and human activities change. Be sure to look at Box 19.1, Figure 1, page 436, showing the Sahel, south of the Sahara Desert, which explains desertification.

Where are Deserts Located?

Most deserts are located within the belts of dry descending air that occur around latitudes 30 degrees north and 30 degrees south of the Equator. The aridity is a response to the prevailing wind systems of Earth that cause intense evaporation at these latitudes. Deserts also occur in the rain shadow or lee side of high mountain ranges. Moist air blowing in from the ocean will rise, cool, and precipitate moisture on the windward side. The descending dry air on the downwind or lee side will be compressed and heated, leading to intense evaporation and arid conditions.

Some deserts, such as the interior of Asia and China, are located at great distances from a source of water and receive very little atmospheric moisture. Study Figure 19.3, page 435, to become familiar with the location of global deserts. Relate this diagram to Figure 19.2, page 434, showing global air circulation; note particularly the belts of descending air.

A young rising mountain range on the edge of a continent may interact with wind patterns to create a rain shadow, and deserts will form where before there was a humid coastal plain. Other continental plates may have drifted from moist regions into the 30 degrees wind belt of high evaporation, changing from humid to arid conditions. Various combinations of these causes may produce a specific desert.

What are the Effects of Wind in Deserts?

The greater range of temperatures on arid lands causes stronger winds to blow in the desert compared to humid regions. Winds are effective as an erosion agent because of the sparse vegetation and the dry sediments on the desert floor. Wind-blown sand does not rise high into the air but erodes close to the ground. The erosion forms pedestal rocks and cuts away the bases of telephone poles and fence posts. Dust, being finer, is picked up by the winds and carried high in the air over vast distances. The stronger the wind, the more material in transit, and the material moved will be coarser. Very strong winds may even transport gravel.

Winds move sand and heap it up into depositional features called sand dunes (see Figures 19.23 to 19.28, pages 446-448). The type of dune that will form depends on the strength and direction of the wind, the source and amount of sand available, and the vegetation that tends to anchor the sand. Winds can also remove fine particles and create a depression in the land surface called a blowout (see Figure 19.17 to 19.19, pages 444-445).

Anza-Borrego Desert State Park in Southern California changed from being part of an ocean floor to a verdant grassland to its present desert condition as a result of dynamic forces acting along the plate boundary between North America and the Pacific plate. The presence within the park of thick Pliocene beds of oyster shells and coral reefs tell the story of marine life that flourished in ancient seas about 5 million years ago. This episode ended when the Colorado River built its delta across the basin and barred the ocean waters that had entered from the Gulf of California to the south.

During the Pleistocene Epoch that followed, beginning about two million years ago, a freshwater lake developed that was replenished from repeated overflows of the Colorado River and from the humid climate that developed during the glacial age. There were streams, grasslands, and wooded areas around the lake. Sabre-toothed cats, camels, turkeys, ducks, giant turtles, and an enormous condor-like vulture thrived in the moist environment. The old lake deposits can be seen today forming the Borrego Badlands.

Plate movements that had started before the Pleistocene slowly pushed up a series of ridges that formed a mountainous geographic barrier between the coast and the inland Salton Trough. Eventually, the cool moist ocean breezes no longer reached the lowlands. The resulting rain shadow effect, coupled with the drying of the California climate following the glacial age, led to desert conditions. The Horses, camels, and other grazing animals moved to greener fields.

About 1,000 years ago, the most recent freshwater lake, Lake Cahuilla, formed when the Colorado River temporarily shifted its course. The high-water mark can still be seen on the surrounding hills about 12 meters (40 feet) above the present land surface. This lake, which covered about 450 square kilometers (2,000 square miles), became fringed with tule, arrow-weed, mesquite, and palms. The Yuman Indians lived here until the lake disappeared about 500 years ago; then they used the area only for seasonal activities.

Today's Salton Sea was formed in 1905-1907, when the Colorado River broke through irrigation floodgates near Yuma and filled the basin, another episode in the evolution of a desert. The story will go on, however, as Anza-Borrego is within a very active seismic zone adjacent to the great San Andreas fault.

Questions

1. Describe the atmospheric conditions at latitudes 30 degrees north and 30 degrees south? Why are so many deserts located in this belt?

The atmospheric circulation of Earth is determined by the tilt of the axis and the sun's radiation. The Equator receives more of the sun's heat than any other part of the planet. As a result, the air over the Equator is warmed and forms a continually rising moist air mass. When the rising air reaches the higher altitudes, it cools and loses its moisture in the form of equatorial rains. The now dry air moves north and south from the Equator, cools and descends near 30 degrees north and 30 degrees south, in the so-called "Horse Latitudes." The sinking dry air mass, compressed by the weight of the air above, warms and is able to pick up and hold huge quantities of moisture in the form of water vapor. The rate of evaporation is so great that rain rarely falls to Earth. The greatest deserts of Africa, Australia, and the southwestern United States are within these atmospheric belts of drying, descending air.

2. **What is the rain shadow effect? How does it contribute to the aridity of the southwestern United States?**

The *rain shadow* effect occurs when high mountain ranges block air blowing in from the ocean. When the moist air encounters a mountain, it is forced upward, and as it rises, it expands and cools. The moisture in the air condenses and precipitates on the windward side of the mountain. The now dry air descends on the lee side, warming as it is compressed. This process causes intense evaporation and limited rainfall. The deserts in Nevada and Arizona are largely a result of the rain-shadow effect of the Sierra Nevada Mountains.

Death Valley is separated from the oceans by the coast ranges, the Sierra, and the high Panamints. The very dry air is further heated by descending to over 280 feet below sea level to the floor of the Valley. Temperatures as high as 57° C have been recorded there, and the average annual rainfall is about 4 centimeters (1.6 inches) per year.

3. **In what other geographic localities do deserts form? About how much of Earth's land surface is arid or semi-arid?**

Deserts occur in regions far from sources of water, such as the interiors of Asia and China. Deserts also develop on tropical coasts next to cold ocean currents, such as the Pacific coast of South America and the Atlantic coast of southwest Africa. Arid and semi-arid regions cover about one-third of the global land surface.

4. **What are the major differences between drainage in arid regions and drainage in humid regions? Why is base level for most rivers at sea level but in certain deserts it is 61 meters (200 feet) or more below sea level? Why did the early pioneers in the Basin and Range province get into major trouble following the stream beds on their trek westward?**

In humid regions, major streams flow through the land to reach the sea. Sea level is base level, the depth below which streams cannot erode. In arid regions, streams flow toward a land-locked basin, and base level will be the floor of that basin. The floor in Death Valley is actually over 85 meters (280 feet) below sea level!

The early pioneers traveling westward would usually follow stream channels, assuming the river would eventually lead them to the sea. Following the desert streams and dry washes proved to be a disaster. The streams would frequently sink into the sand and vanish, or would lead to an interior dry playa lake or to a badwater pond laden with bitter salts.

5. **What underlying rock structure lends itself to formation of plateaus, mesas, and buttes?**

Flat-topped plateaus, mesas, and buttes form as a result of erosion and parallel retreat of their steep sides. The underlying rocks are horizontal sedimentary layers of differing resistance in the arid climate. These landforms are characteristic of certain regions in the Colorado Plateau such as beautiful Monument Valley and the Grand Canyon in Arizona.

6. **What are the major depositional landforms you would expect to see in Death Valley? What is the underlying geologic structure there?**

Depositional features seen in Death Valley include large alluvial fans, bajadas, and playas. The topography consists of high rugged mountain ranges separated by long flat-bottomed valleys. The steep straight mountain fronts are faults along which the valleys are subsiding and the mountains are rising. Faulting in Death Valley is active and can be seen in the recent fault scarps that cut across alluvial fans.

7. **What is the most important agent of erosion in arid regions? Describe some of the characteristic landforms attributed to this form of erosion.**

The most important agent of erosion in arid regions is running water, though the rains may be infrequent. Thunderstorms are often violent, and one or two inches of rain may fall within an hour and create flash floods. The normally dry river beds become heavily laden with moving rocks and debris eroded from the dry land. The erosive power of the sediment-laden floodwaters tends to produce narrow canyons with steep walls and flat, sediment-covered floors.

The pediment, a major erosional landform in arid regions, is a bed rock surface stripped bare or left with a thin veneer of gravel. The pediment surface slopes gently away from the mountain front toward the interior basin. The origin of pediments is something of a geologic puzzle, but many geologists think the pediment forms when streams sweep back and forth over the rock surface. The materials carried by the streams abrade away the bed rock. A definitive answer to the origin of pediments is still not certain. There is little opportunity to observe the actual formation of the pediment by running water in an area with such infrequent precipitation.

8. **Describe the origin of blowouts, desert pavement, pedestal rocks, and ventifacts.**

Blowouts, desert pavement, oases, and ventifacts are all features of wind erosion. Wind is very effective at transporting particles of a small size. Sand grains are moved along the ground in a series of small hops called *saltation* and rarely rise over 1 meter (3 feet) above the surface. Smaller particles such as silt or clay remain aloft much longer than sand. In fact, wind is the only agent that can transport any material beyond the edges of a typical desert basin.

Blowouts are areas excavated or deflated by wind erosion. Some blowouts may be very large, as in Egypt, where the oasis of Kharga is 189 kilometers (118 miles) long, about 80 kilometers (50 miles) wide, and about 179 to 292 meters (590 to 984 feet) deep! The origin of these large blowouts may be more complex than a simple blowout, but wind certainly was a factor in enlarging and deepening it. The lower limit of a blowout is the water table, where the grains are damp and cohesive.

Desert pavement is a rocky surface formed when the wind removes the fine particles and leaves the gravel behind. *Ventifacts* are pebbles and cobbles that have flat polished surfaces from abrasion by sand and smaller particles. *Pedestal rocks* are strange mushroom-shaped erosional remnants formed by the sand-blast effect of wind erosion. These peculiar landforms illustrate the effects of sand eroding just above the ground surface.

9. **What kind of minerals make up the sand in sand dunes? What factors cause the different types of sand dunes to form?**

The mineral composition of sand in sand dunes depends on the original sand source and the degree of chemical weathering. Dunes near beaches in humid regions are composed largely of quartz grains because all other minerals in the original rock have been decomposed by chemical weathering. Desert dunes often contain feldspar and rock fragments in addition to quartz. Some dunes are formed of calcite. At White Sands, New Mexico, the dunes are made of grains of gypsum. But no matter what the chemical composition, most dunes consist of sand-sized particles, testimony to the extraordinary sorting ability of the wind. The type of dune that forms depends on the amount and source of sand available and the strength and direction of the wind.

10. **Describe loess as seen under a microscope. Where is it found, and how did it get there? What is the importance of loess?**

Loess is a yellowish-tan sediment consisting of mechanically pulverized silt and clay-sized fragments – predominantly quartz but with some feldspar, mica, hornblende, and pyroxene. Carbonate minerals may act as a weak cementing agent. Loess originates as fine particles that were lofted out of deserts or regions of glacial outwash and transported scores of miles before settling down to accumulate on the land. Loess is widely distributed in northeast China, in central Europe, and in the central United States. Loess forms the parent material of some of the richest agricultural soils on earth (see Figure 19.22, page 445).

11. What are the symptoms and causes of desertification? Is this problem found only in under-developed countries?

Desertification, defined as the invasion of desert into a non-desert area or into once-populated regions, is a global problem difficult to reverse. The major symptoms of desertification are declining ground water tables, increasing saltiness of water and topsoil, reduction of supplies of surface water, high rates of soil erosion, and destruction of native vegetation. There are many causes of desertification, including changing global climates, but most involve overuse of land by livestock and humans, excessive withdrawal of ground water, unsound water-use practices allied with population increase, and expanded agricultural production. This condition is not limited to underdeveloped countries. Within the United States, about 10 percent of the land area – an area approximately the size of the original 13 states – has been severely affected by desertification.

Activities

1. Suppose you were appointed to an international committee on desertification. Where would you go to see it happening? What solutions could you present to some of the problems? (Realistic answers are hard to come by.) This is a life-threatening disaster that is taxing the expertise of scientists all over the world. See *Scientific American*, "Special Issue: Managing Earth," Vol. 261, No. 3, September 1989.

2. What are the effects of off-road vehicles on desert regions? What would be a realistic solution to this problem? (Or is it really a problem?)

3. What advice would you give to a first-time desert visitor who is planning to camp there? Where would be a good place to set up camp, and where would be the worst? (No fair setting up camp in a local motel!)

4. Using the diagrams of sand dunes, Figure 19.26, page 448, make a sketch of a barchan, a transverse dune, a parabolic dune, and longitudinal dunes. Label the slip face in each and indicate which way the wind is blowing.

1. All deserts are characterized by

   a. sand dunes.
   b. very high temperatures.
   c. very low rainfall.
   d. alluvial fans.

2. Deserts most frequently occur around latitudes 30 degrees north and 30 degrees south because

   a. lands here are far from oceans.
   b. these are the belts of drying descending air.
   c. these are the belts of rising moist air.
   d. these are the belts of cold ocean currents.

3. Pedestal rocks and eroded telephone poles are good indicators of

   a. erosion by wind-blown sand.
   b. erosion by angular fragments of rock carried in flash floods.
   c. deflation and lowering of the land surface.
   d. previous moist climates.

4. Base level of desert streams is usually

   a. sea level.
   b. the top of the alluvial fan.
   c. the base of the alluvial fan.
   d. the floor of the land-locked basin.

5. The desert conditions of the southwest United States are most likely related to the

   a. cold California current offshore.
   b. location near the Equator.
   c. rain shadow effect of the Sierra Nevada.
   d. horizontal sedimentary layers.

6. Deserts in the Basin and Range province do NOT have

   a. mesas.
   b. alluvial fans.
   c. desert varnish.
   d. pedestal rocks.

7. Features of the Colorado Plateau consist of all of the following except

   a. buttes.
   b. bajadas.
   c. monoclines.
   d. a flow-through stream.

8. The landforms of the Colorado Plateau result from

    a. the recent block faulting and uplift of the mountains.
    b. erosion of the flat-lying beds of sedimentary rock.
    c. deposition in the down-faulted basins.
    d. rising base levels.

9. Of the following, which is NOT characteristic of deserts?

    a. Internal drainage
    b. Flash floods
    c. Thick soils
    d. Thunderstorms

10. Strong winds on the desert are usually caused by

    a. a wide range in daily air temperatures.
    b. moisture evaporating into the air.
    c. strong greenhouse effect.
    d. warm days and warm nights.

11. All of the following are types of sand dunes except

    a. longitudinal dune.
    b. barchan dune.
    c. parabolic dune.
    d. pediment dune.

12. The presence of desert pavement may indicate

    a. frequent flash flooding.
    b. recent deposition of gravels and pebbles.
    c. a former playa lake bed.
    d. an area of deflation.

13. For humans, the most important property of loess is its

    a. ability to stand as vertical cliffs without slumping.
    b. ability to form highly fertile soils.
    c. valuable mineral content.
    d. potential to provide low-cost housing for many people.

14. The sand dunes that form where the supply of sand is limited

    a. always have the horns of the crescent pointing downwind.
    b. are elongate dunes parallel to the direction of the wind.
    c. form around blowouts and pile up damp sand.
    d. are more likely to be covered with dense vegetation.

15. Decreasing ground water supply is usually a symptom of

    a. cooling global temperatures and more water held in glaciers.
    b. decreasing rain shadow effect.
    c. increasing desertification.
    d. increasing use of water by native plants.

23

GLACIERS

This lesson will help you understand the properties of glaciers and the importance of glaciers in the sculpting of Earth's surface.

GOAL

After reading the textbook assignments, completing the exercises in this study guide, and viewing the lesson's video portion, you will be able to:

LEARNING OBJECTIVES

1. Recognize what a glacier is and differentiate between alpine glaciation and continental glaciation.

2. Discuss how glaciation has influenced landscapes in many parts of the world.

3. Indicate how glaciers form, grow, move, and shrink.

4. Compare the movement of a valley glacier to that of an ice sheet.

5. Contrast the erosional and depositional features of both continental and alpine glaciers.

6. Compare the effectiveness of stream, wind, and glacial erosion.

7. Distinguish among various types of moraines and explain what they indicate.

8. Recognize the characteristics of outwash deposits and the landforms built of outwash.

9. Discuss the current theories for the causes of the glacial ages.

10. Describe how past glacial ages affected the distribution of life forms.

11. Indicate the indirect effects of the glacial ages, including the formation of pluvial lakes, changes in sea level, and crustal rebound.

INTRODUCING THE LESSON

The great agents of erosion – streams, wind, and waves – each leave its distinctive imprint upon the landscape, but none is more impressive or scenic than the work of glaciers. The jagged peaks and carved U-shaped valleys of the Alps, the Himalayas, and the Andes, the grandeur of Yosemite Valley in the Sierra Nevada and Lauterbrunnen Valley in Switzerland, the deep fiords of Norway, British Columbia, Alaska, and New Zealand are but a few of the truly spectacular effects of moving ice.

During the glacial age that ended a mere 10,000 years ago, great continental ice sheets covered about 33 percent of the land. At least four times during the past million years, changing climatic conditions allowed snow to remain in the mountains and northern latitudes where previously summer heat had melted it away. Over tens of thousands of years, the snow compacted into ice and the ice built up into ice sheets, some of which reached a thickness of several thousand feet. Then each glacial cycle came to an abrupt end. Within a few thousand years, the ice sheets receded and the climate returned to a warm interglacial regime.

Changes in sea level that reflected the waxing and waning of the ice sheets profoundly altered world geography. During the last maximum extent of the ice about 18,000 years ago, sea level was lowered as much as 100 to 125 meters (330 to 400 feet). Land areas such as Siberia and Alaska were connected and permitted migrations of people and animals to the New World. Today Siberia and Alaska are separated by the Bering Strait, as much as 55 meters (180 feet) deep. During the last glacial age, large lakes such as the Great Lakes formed where none existed before. Even Death Valley – today the driest place in the United States – was occupied by a deep lake.

Modern glaciers cover about 10 percent of the land area of Earth. Most of the ice is locked up in two great *continental ice sheets*: Antarctica, which accounts for about 85 percent of the ice in the world, and Greenland, which accounts for about 10 percent. The rest of the ice is scattered around the world in the small ice caps and valley glaciers found in almost every major mountain range. Most of the world's fresh water is locked up in these glaciers. If all the present ice were to melt, sea level would rise about 100 meters (330 feet). Some inland cities would become seaports, but New York, Tokyo, London, and all of Florida would be inundated with sea water.

A *valley glacier* is a river of ice, born high on a mountain slope, where the winter snowfall lingers throughout the summer. Over the years, more snow will accumulate than is melted away or evaporates. Eventually, a field of stratified ice and recrystallized snow is formed. The ice will move slowly downslope under its own weight, usually following and modifying an older river valley.

Glaciers are individualistic; no two are exactly alike. Some are dirty and rubbly, so covered with rock and soil that they support plants and dense pine forests near the *terminus* (lower end of the glacier). Some are sparkling clean, showing beautiful green and blue ice, and iridescent colors on sunny days. Some are wide, slow moving, even stagnant. A few suddenly "gallop" or surge at many times their normal rate.

Many glaciers around the world are receding, which is not unexpected because the 1980s ended as the warmest decade in more than a century. Nevertheless, the Greenland ice sheet seems to have been growing thicker at a rate of about 2 centimeters (0.9 inches) per year since 1975. It will be interesting to observe the behavior of the world's glaciers and ice sheets as they respond to what appears to be a warming trend.

In this lesson you will learn how glaciers and ice sheets sculpt the landscape and create their unmistakable landforms. You will discover that repeated glacial episodes are part of Earth's dynamic evolution; there is evidence of extensive glaciation almost a billion years ago. Some of the most exciting scientific investigations today are the efforts to correlate the effects of plate tectonics, the variations in Earth's orbit, and the tilt and orientation of Earth's axis with the recurring periods of glaciation. An understanding of the long-term climatic cycles may lead to better weather forecasting, important for agriculture and for feeding Earth's burgeoning humanity.

Have we reached the end of the glacial age, or are we enjoying just another interglacial interval?

LESSON ASSIGNMENT

Completing the following eight steps will help you master the lesson objectives and achieve the goal for this lesson:

Step 1: Read the TEXT ASSIGNMENT, Chapter 18, "Glaciers and Glaciation."

Step 2: Study the KEY TERMS AND CONCEPTS as noted in the study guide.

Step 3: Watch the VIDEO using the VIEWING GUIDE in the study guide.

Step 4: Read the study guide's PUTTING IT ALL TOGETHER section, which will help you summarize and integrate all of the information in this lesson.

Step 5: Complete any assigned lab exercises.

Step 6: Complete any assigned ACTIVITIES found in the study guide.

Step 7: Review the material in this lesson and complete the SELF-TEST found in the study guide.

Step 8: Go back to the LEARNING OBJECTIVES and make sure you have learned each one.

TEXT ASSIGNMENT

1. Read Chapter 18, pages 405-431.

2. Re-read "Introduction" and "Summary" for a good review of the material in this chapter.

KEY TERMS AND CONCEPTS

The key terms listed under "Terms to Remember," page 430 in your text, are important and will supplement your knowledge of glaciation learned from the chapter on glaciers. Use the glossary and the summary in the text to help you understand the definitions.

There are many terms for glacial landforms, and you will become more familiar with them if you divide the long list into at least two: one for features of glacial erosion, such as *arete, cirque, fiord, horn,* etc., and one for features of glacial deposition such as *drumlin, moraine, esker, outwash,* etc. Another list could be of words that refer to glacial processes, such as *alpine glaciation, basal sliding, plastic flow,* etc. Here are some additional terms to help you in this lesson:

Zone of accumulation: The upper part of the glacier where snow lingers from year to year (Figure 18.6, page 410). The zone of accumulation can make up about two-thirds of the total surface of a valley glacier. In equatorial mountains such as Kilimanjaro in Africa and the summits of the Andes in South America, the ice fields and snow line lie at altitudes of 4,800 to 5,400 meters (15,744 to 17,712 feet). In mid-latitudes, as in the Sierra Nevada of California and the Swiss Alps, they lie close to 3,000 meters (9,840 feet). Nearly all of the Antarctic ice sheet down to sea level is within a zone of accumulation! Glaciers develop where there is a combination of sufficient winter snowfall and low summer temperatures that limit melting.

Zone of wastage or ablation: Ablation refers to all the processes by which snow or ice are lost from a glacier. The zone of ablation is the lower part of the glacier, where the snow from last year and sometimes the underlying ice *wastes,* or melts, evaporates, or in the case of tidewater glaciers that flow into the sea, *calves* and breaks away into icebergs. When the amount of wastage of the ice is greater than the amount of snow in the zone of accumulation, the front of the glacier will become thinner and will recede, although the net movement of the ice will continue to be forward or downslope. Many of Earth's glaciers seem to be receding at present, but they are not all acting in unison (Figure 18.7, page 410, and Figure 18.8, page 411).

Continental ice sheet: A mass of ice, irregular in shape, that covers more than 50,000 square kilometers (18,000 square miles) of land and whose flow patterns are not necessarily guided by the underlying topography (Figure 18.3, page 408). Only two ice sheets exist today, in Greenland and in Antarctica. Ice sheets flow down and out from a central high point where the ice is thickest. The center of Antarctica is buried under ice to a depth of several thousand meters! In the recent geologic past, ice sheets planed and scoured much of North America and as far south as central Europe. The last major glacial interval, which began more than 100,000 years ago, ended not only recently (10,000 years ago) but also suddenly – at least on a geologic scale of time.

Glacial surge: Glacial motion, popularly called "galloping," in which a glacier, either stagnant or flowing normally for years, speeds up for a relatively short time to as much as a hundred times the normal rate. Flow rates usually vary from a few millimeters to a meter or so per day but can increase to an astounding 50 to 65 meters (165 to 215 feet) per day during a surge. (On an average slope, with gravity as the only moving force, water can flow about 100,000 times as fast as ice.) A surge may last for two or three years, and the cycle of repeated surging is in the range of 10 to 100 years. Some experts believe surging is the result of a buildup of pressure in the meltwater at the base of the glacier; the pressure reduces friction and increases the rate of basal sliding. Theoretically, if the water pressure were to reach the pressure of the overlying ice, the glacier would be floated off its bed.

Other theories about glacial surge include the buildup of a thick block of stagnant ice within the glacier while the lower part is melting. The block acts as a dam until the pressure of the flowing ice behind it forces it to give way suddenly. Actually, surging is still one of the unsolved problems of glacial mechanics.

Once you have read the text material for the lesson and studied the terms, you are ready to view the lesson's video portion. Review the questions that appear in this section to help you watch for important points and to prepare you for what you will see in the video.

VIEWING GUIDE

Before viewing, answer the following questions:

1. Under what conditions will glaciers form?

2. What landforms indicate that a region has been glaciated?

After viewing, answer the following questions:

1. According to Dr. James Zumberg, what are the requirements for glaciers to form?

2. What is the process by which snow is transformed into glacial ice? (Watch the graphics and animation for helpful hints.)

3. How did Dr. Zumberg describe how a glacier moves?

4. Most of the glaciers of the world are flowing downslope, but the ends are receding. Can you explain this in terms of the glacial budget?

5. What is the current theory about why some glaciers surge?

6. The causes of the glacial ages are still not resolved. What theories were advanced by Dr. Lawford Anderson? What theories were presented by Dr. Chester Langway?

7. Note the maps in the video showing the Bering land bridge. During the last glacial age, what were some of the effects on the land and on the sea in this area?

8. Recent research in Greenland and Antarctica has involved drilling and extracting cores of ice from the ice sheets. What information can be gleaned from analyzing the ice?

PUTTING IT ALL TOGETHER

You have read the text assignment and seen the video portion of the lesson. This section, made up of a SUMMARY, a CASE STUDY, the QUESTIONS, suggestions for ACTIVITIES, and a SELF-TEST, will help you pull the information together.

Summary

Glacial erosion has produced some of the most scenic mountainous terrain on earth. The great cirques at the valley heads, the broad U-shaped valleys, the gouged out bed rock depressions that became sparkling glacial lakes called *tarns*, and the fiords of Norway and other places attract scores of visitors each year. Even the vast remote Antarctic ice sheet, surrounded by towering icebergs, has its own almost mysterious fascination that attracts scientists and hardy tourists (Figure 18.5, page 409 and photograph facing page 405).

Glaciers form where there is adequate winter snowfall and cool summer temperatures. The snow does not melt away completely each summer, but accumulates year by year. The snowflakes become converted to ice, and when thick enough, the glacier moves slowly downhill under its own weight. Glaciers move by slippage on meltwater along a bed rock base, and by internal deformation or plastic flow of ice crystals under pressure (Figure 18.8, page 411).

The *glacial budget* is determined by the annual accumulation of snow and ice and the annual loss due to ablation. If the two are balanced, the glacier will maintain a stationary front, although the ice within the glacier is still flowing forward.

Glacial Landforms

The landforms produced by glacial erosion include the U-shaped valleys that have been deepened and straightened by the moving ice. In some of the most striking instances, as in Yosemite Valley in California and Lauterbrunnen Valley in Switzerland (called the "finest glaciated valley in all of Europe"), the sides are steepened until they appear almost vertical. The head of a glaciated valley is usually a *cirque*, which is a steep-walled horseshoe-shaped erosional amphitheater. *Hanging valleys*, rock-basin lakes called *tarns*, steep jagged divides between valleys called *aretes* (from the French word for ridge or fishbone), and great triangular pinnacles called *horns* are all part of the erosional landscape (Figures 18.16 through 18.24, pages 415-418).

Glacial deposits called *moraines* consist of the rocky unsorted debris from glacial erosion that comes to rest beneath or along the edges of the glacier (Figure 18.13, page 414). The *terminal moraine* at the end of the glacier may form a hummocky, crescent-shaped ridge of glacial till. When the ice melts at the lower end of a glacier, the coarser rock material carried by the moving ice is left behind, and some of the finer particles will be carried away by the meltwater streams (Figures 18.26 through 18.29, pages 419-420).

Causes of Glacial Cycles

Evidence of at least four major episodes of glaciation during the last glacial age is seen in the multiple layers of *till* and the soil layers that developed during the warm interglacial periods. Pollen from terrestrial plants, fossils of certain animal species, even beetle species tell of glacial episodes separated by periods of warm climates.

The causes of the glacial ages are still being debated. (See "What Drives Glacial Cycles," by Broecker and Denton in *Scientific American*, January 1990.) The most likely explanation is related to changes in Earth's orbit and in the tilt and orientation of its axis of rotation. (Read Box 18.3, pages 424-425.) But there are many theories, and eventually the best of them will be brought together to form a coherent explanation.

The glacial ages affected life on earth in many different ways. The alternation of cold glacial periods with warm interglacial events caused the climatic belts and the plants and animals that occupied them to shift hundreds of kilometers. The fossils of certain mammals such as the cold-climate muskrat reveal that climates in Florida were cool when glaciers pushed southward into the northern United States. Other fossil occurrences, such as those of hippopotamuses in Britain, show that during some interglacial intervals climates were warmer than they are today.

Some animal species adapted to the changing climate. The woolly mammoth and the woolly rhinoceros were much hairier than living relatives, presumably as a response to the cold. We know about these animals from cave drawings made by early humans. The caves at Lascaux in France and at Altamira in Spain are magnificent records of late Pleistocene life and give us some idea of what today's animals looked like hundreds of thousands of years ago.

Other species became extinct. In Alaska and Siberia there are huge bone beds, frozen into the permafrost and recovered during gold mining, that reveal glacial age life in the sub-arctic zone. Sometimes an almost complete body with flesh and skin intact has been found. During the advances of the ice, the unglaciated Alaskan tundra between the Brooks Range and the Alaska Range was a refuge for many large mammals. Mammoths and mastodons, giant ground sloths, horses, camels and bison grazed there in enormous herds but became extinct at the end of the Pleistocene. Some of the large carnivores such as the saber-toothed

CASE STUDY: How Did the Past Glacial Ages Affect the Distribution of Life Forms?

cat and a very large bear also disappeared. Certain other groups of animals all over the world became extinct – the Irish elk with antlers spanning 3.5 meters (11 feet), strange South American mammals, large marsupials in Australia, and the huge elephant bird of Madagascar – all creatures that were flourishing only a few tens of thousands of years ago. Strangely, these animals had already survived at least three similar climatic crises. Not all became extinct at once, although extinctions on any one continent occurred at about the same time.

It seems that climate alone could not have caused these extinctions. Recent theories point to humans – the efficient big-game exterminators. In each case, the largest mammals became extinct. The large carnivores might not have been exterminated by people directly but might have suffered because humans competed with them for game. Perhaps further research will show some tectonic or geographic cause for the extinctions and allow us to look back on early humans in a more favorable light.

Questions

1. **If the zone of accumulation is a region of snowfall, and the zone of ablation is a region of melting ice, what happens in between?**

   Snow converts to ice in a glacier through several stages (Figure 18.4, page 409). First, the snowflakes, which are crystalline water, compact under their own weight, melt under the pressure, and refreeze. They change to granular snow, then to *firn*, which consists of grains of ice. Firn, which has the texture of coarse sand, is usually white or grayish white, and the spaces between the grains are filled with trapped air. Eventually under continued pressure, the pore spaces disappear, and the firn changes to blue glacier ice made up of interlocking ice crystals. Blue ice is old ice, and the wondrous ice caves of old glaciers, such as the Rhone Glacier in Switzerland, gleam with an eerie blue light.

2. **How could tell if an area had been glaciated? List some of the features that would give you a clue.**

   The unique landforms created by valley glaciers are clues to "reading the landscape." Steep walls, as seen in Yosemite Valley; hanging valleys, many with waterfalls; glacial horns; broad U-shaped valleys; rounded cirques, some with a small lake within the amphitheater; and moraine ridges of unsorted till all bear witness to a previous glaciation (Figure 18.16, page 415).

3. **What would you look for to determine the southern limits of the North American continental glacier?**

   To determine the southern limits of the North American continental glacier, you would look for the looping terminal moraines marking the end of the last major glaciation (Figure 18.36, page 428). The moraine ridges are usually not more than 33 meters (100 feet) high and will be littered with *glacial erratics*, rocks that were not derived from the local bed rock but have been transported by the moving ice. Glacial erratics are found in Central Park in New York; the ice sheet was about 90 meters (300 feet) thick over Manhattan. In fact, Long Island is a terminal moraine.

Some erratics are called *indicators* because they have a distinctive composition and may be traced back to their source, sometimes hundreds of kilometers away. Some erratic boulders found in England were delivered in glacial ice from Norway. (The glacier went across the North Sea). One of the most challenging examples of glacial transport involves diamonds, some as large as small pebbles, found as far south as southern Indiana and having a presumed source north of the Great Lakes. Diamonds are not really *indicators*, however, because their source is unknown or at least a well-kept secret.

4. **How did the continental glaciers affect the landscape far beyond the actual limits of the ice?**

Continental glaciers affected the landscape beyond the limits of the ice sheet by the deposition of *outwash* material, the well-sorted sediments deposited by meltwater streams from the ice. Some of the best farmland on Earth is located on glacial outwash plains. The fine-grained dust and clay particles in some instances were carried great distances by the wind and deposited as *loess*, also a source of fertile soils.

Continental glaciers also change the landscape by creating many lakes. Finland is renowned for its lakes, 55,000 of which have been mapped. Canada and the United States also have networks of lakes and waterways. *Pluvial lakes* formed in the western United States far beyond the limits of the ice sheets. Some were probably fed by meltwaters from the mountain glaciers, but many were formed during the very wet climate that followed the glacial age. The Great Salt Lake in Utah is a small remnant of a much larger body of fresh water called Lake Bonneville (Figure 18.36, page 428). Many lakes occupied the down-faulted basins of the Basin and Range province, a region known for its present aridity.

Sea level was lowered during maximum advances of the ice, exposing the continental shelves of the world as dry land. Streams flowing across the shelves cut valleys that today exist as submarine canyons. Animals, such as the woolly mammoth and perhaps Pleistocene humans, roamed those cold windy lands during the final glacial period.

5. **How would you distinguish between glacial till and landslide colluvium in an ancient deposit? This is a tricky problem that still worries geologists.**

It is not always easy to distinguish between glacial till (Figure 18.25, page 418) and landslide deposits, especially in very ancient formations. In both cases, the deposit consists of angular unsorted rock material ranging in size from fine silt and clay-sized particles to large boulders. (Review Lesson 16 on mass wasting.) If some of the larger fragments are striated, polished, grooved, or faceted, the material may be glacial debris. One difficulty is that in very old deposits, weathering may have removed the striations or other marks on the surface. Sedimentary processes may have cemented the whole mass into hard *tillite*. The solution to the problem may be in discovering other glacial phenomena in the area.

6. How does a valley glacier move?

A valley glacier moves by sliding on meltwaters at the base of the glacier, by deformation of the ice crystals in the *zone of plastic flow*, and by movement of the *upper rigid zone* as it is carried by the ice below (Figure 18.8, page 411).

7. Why are some glaciers so dirty? What are the long streaks on the surface of valley glaciers called?

Some glacier surfaces transport rock debris derived from *frost shattering* and *frost wedging* of the cliffs above the glacier. The particles that fall near the sides of the glacier are carried downslope by the moving ice and form the *lateral moraines*. When a tributary glacier enters the main trunk, the lateral moraines will join and form a *medial moraine*. Very large glaciers may have several medial moraines, some of which are very convoluted because of surging or unusual conditions of flow. Near the *terminus*, rock debris may completely cover the ice, and if the glacier is slow-moving or stagnant, vegetation or even trees will grow in the rocky soil. Dust and ash and other wind-blown material also fall on the ice surface, all contributing to the "dirty" appearance. Core samples taken in glaciers and ice sheets reveal cycles of deposition of volcanic ash, dust, or even meteorites, possibly from the moon or Mars. (See Box 18.2, page 413.)

8. What is meant by crustal rebound?

During maximum buildup of continental glaciers, the land beneath the ice may be depressed several hundred meters by the weight of the ice. In the interior of Greenland, for example, seismic methods indicate the ice is over 3,000 meters (10,000 feet) thick. The immense weight has so depressed the rock floor, which now has a saucer-like shape, that the center lies below sea level! If the ice were to melt, the crust would slowly rebound or rise to some previous level. In Scandinavia, the uplift has been about 100 meters (330 feet) since the disappearance of the continental glacier that was centered there. Earth's crust is still rebounding 10,000 years after the ice disappeared. The shores of Hudson Bay are slowly rising, but some of the deeper depressions are still occupied by lakes or by the sea. The Great Lakes and the Baltic Sea are other outstanding examples of crustal rebound.

9. Why is the northern hemisphere ice sheet on Greenland instead of surrounding the North Pole in the way the Antarctic sheet surrounds the South Pole?

Ice sheets form on land surfaces. While the South Pole is located in the middle of the continent of Antarctica, the North Pole is in the middle of the Arctic Sea. The only land mass in the Northern Hemisphere with the proper conditions to form a continental glacier is Greenland.

Activities

1. Using Figure 18.34, page 426, which shows the extent of glaciation in North America, and an atlas or map of North America, determine which major cities would be under ice if the world were to experience another glacial episode similar to the last one, which ended about 10,000 years ago. Was your region glaciated? Notice the ice-free corridor in Alaska. This was probably part of the migration route between Asia and the Americas.

2. Using Figure 18.36, page 428, trace the boundaries of the Pleistocene ice sheet in North America. Study the location of the *pluvial lakes* in the southwestern United States. ("Pluvial" refers to rain; the lakes are evidence of a much wetter climate that followed the retreat of the continental ice sheets.) Locate Death Valley in eastern California, and note the elongate lakes that filled the down-faulted valleys.

3. A new industry has developed in Alaska based on the uses of glacier ice. Have you any ideas about how this ice could be used?

   Pure, clear, hard, glacial ice has been chopped out of floating ice in Alaskan bays, packaged in 22-kilo (50-pound) bags and exported to Japan. It makes a distinctive crackling noise as it chills a drink. Also, glacial ice has been used in the making of a beer that was judged the "Best Beer in America" in Denver in 1988.

Self-Test

1. Most of the glacial ice today occurs

   a. on major mountain chains all over Earth.
   b. around the North Pole.
   c. as icebergs off Greenland.
   d. on the Antarctic continent.

2. Most of the body of a valley glacier consists of

   a. snowflakes.
   b. recrystallized ice.
   c. firn or granule.
   d. meltwater.

3. If the *greenhouse effect* causes a warming trend, which of the following is most likely to occur?

   a. Sea level will fall because of increased evaporation.
   b. Glaciers will advance because of increased precipitation.
   c. Glaciers will recede because of increased ablation.
   d. Another glacial age will be started.

4. The *fiords* of Norway and British Columbia originated as

   a. glacially carved U-shaped valleys, now submerged.
   b. lakes formed behind high morainal dams.
   c. stream valleys drowned by rising sea level.
   d. tarns eroded from less resistant bed rock.

5. Glaciers move by all of the following except

   a. slippage along a bed rock base.
   b. active sliding within the rigid zone.
   c. plastic flow.
   d. internal deformation of the ice crystals.

6. In Antarctica, the most persistent form of *ablation* is

   a. melting of the surface of the ice sheet.
   b. loss by meltwaters at the base of the ice.
   c. calving of icebergs.
   d. recession of the rigid zone.

7. Continental ice sheets usually do NOT form

   a. cirques.
   b. eskers.
   c. drumlins.
   d. till.

8. Outwash beyond the edge of the continental ice sheets can be identified by

   a. thin deposits of ground moraine.
   b. thick deposits of till.
   c. large boulders that are scratched and polished.
   d. layered and sorted alluvial deposits.

9. The famous Matterhorn in Switzerland is a result of

   a. deposition by European ice sheets.
   b. headward erosion by several cirques.
   c. ice over-riding and shaping a deposit of till.
   d. crustal rebound after the end of the glacial age.

10. Pleistocene migrations of humans and animals to the Americas took place during

    a. summer, when the weather was better.
    b. the warm interglacial periods when sea level was high.
    c. the warm interglacial periods when sea level was low.
    d. maximum extent of the ice when sea level was low.

11. The bones and teeth of extinct mammoths dredged up from Atlantic continental shelves are used as evidence that

    a. sea level was much lower during the glacial age and the shelf was dry land.
    b. some mammoths migrated to new lands by swimming.
    c. mammoths lived near the edge of the ice, and their bones washed out to sea.
    d. mammoths became frozen in the ice and were left behind when the ice retreated.

12. In mountains at low latitudes, such as the Andes, the *zone of accumulation* will be

    a. lacking because of the location near the Equator.
    b. near sea level.
    c. the total length of the glacier.
    d. at very high altitudes.

13. The system of pluvial lakes developed in western U.S. states was a result of

    a. outwash from continental ice sheets.
    b. ponding of waters by moraines.
    c. a very wet and rainy period at the end of the glacial age.
    d. gouging out of the land by moving ice sheets.

14. One of the most likely of current theories about causes of glacial cycles is related to

    a. astronomic factors such as changes in Earth's orbit and tilt of the axis.
    b. increases in carbon monoxide in the atmosphere.
    c. changes in the position of the continents in the last 10,000 years.
    d. changes in the thickness of Antarctic ice.

15. Periods of widespread glaciation before the Pleistocene

    a. apparently did not occur.
    b. may have occurred, but there is no clear-cut evidence.
    c. did occur as seen in tillite deposits and ancient glacial striations.
    d. may have occurred but on lands now under the sea.

24

WAVES, BEACHES, AND COASTS

This lesson will help you appreciate the importance of waves in affecting coastal landforms.

GOAL

After reading the textbook assignments, completing the exercises in this study guide, and viewing the lesson's video portion, you will be able to:

LEARNING OBJECTIVES

1. Describe the movement of water in waves and the relationship of waves to the build-up of surf.

2. Describe the ways in which waves erode.

3. Describe the process of wave refraction and explain its relationship to the origin of longshore currents.

4. Sketch and label a cross section of a typical coastal beach.

5. Contrast the profile of summer and winter beaches, and indicate why the changes occur.

6. Indicate how sand is moved in longshore drift, both on the beach face and within the surf zone, and describe the coastal features that can result.

7. Recognize how longshore drift can be altered by jetties, groins, and breakwater.

8. Understand why coastlines are retreating in many areas of the world.

9. Explain the importance of river sediment and submarine canyons to beach sand supply.

10. Give examples of features found on erosional coasts, depositional coasts, drowned coasts, uplifted coasts, and organic coasts.

11. Suggest ways in which people are attempting to slow the effects of wave erosion, and list the unintended consequences of their efforts.

12. Describe the origin and character of barrier islands and estuaries, and give examples of each.

13. Recognize the relationship between uplifted or depressed coasts and plate tectonics.

INTRODUCING THE LESSON

The shore is a laboratory for the marine geologist. It is the most accessible part of the ocean, and on it the dramatic interactions among sea, land, and air can be observed and analyzed. Waves, tides, and currents from the sea; rivers and glaciers from the land; daily weather changes; and seasonal winter-summer cycles all play their part in the ever-changing coastal scene. During glacial age cycles lasting thousands of years, the continental shelves were repeatedly exposed as dry land and submerged again as the sea level rose. The longest cycles of all, plate movements, elevate some coasts and depress others; open and close ocean basins; create new coasts when continents split apart; and eliminate continental margins when plates collide.

Of all the agents of erosion active in the shore zone, waves are by far the most significant. Grinding and abrading the land bit by bit, the endless parade of wind waves cuts cliffs, erodes sea caves and sea arches, and builds beaches and sand bars. The *sea cliffs* are retreating during the current period of coastal erosion, and the shoreline is creeping landward. We might well wonder why the continents weren't planed off to sea level millions of years ago.

In this lesson you will study the motion of the water particles in a passing wave and learn how waves are generated. You will see that beaches are an important part of the geology of the coast. The beach is a dynamic zone in which the sand is in constant motion, sensitive to variations in winds, waves, tides, currents, and human intervention. Shoreline erosion and the loss of sand on many beaches is a major problem that is poorly addressed in our rush to develop and use the coastal zone.

In this lesson, you will learn to appreciate this ribbon-thin zone lying between land and sea. It not only supports the richest variety of life in the marine environment but is dynamic, unpredictable, and fascinating. And it frustrates the uninformed who try to inhabit its domain.

LESSON ASSIGNMENT

Completing the following eight steps will help you master the lesson objectives and achieve the goal for this lesson:

Step 1: Read the TEXT ASSIGNMENT, Chapter 20, "Waves, Beaches, and Coasts."

Step 2: Study the KEY TERMS AND CONCEPTS as noted in the study guide.

Step 3: Watch the VIDEO using the VIEWING GUIDE in the study guide.

Step 4: Read the study guide's PUTTING IT ALL TOGETHER section, which will help you summarize and integrate all of the information in this lesson.

Step 5: Complete any assigned lab exercises.

Step 6: Complete any assigned ACTIVITIES found in the study guide.

Step 7: Review the material in this lesson and complete the SELF-TEST found in the study guide.

Step 8: Go back to the LEARNING OBJECTIVES and make sure you have learned each one.

1. Read the Introduction to Chapter 20, page 453.
2. Read Chapter 20, pages 453-469, making sure to include the "Summary" in your reading.
3. Read Box 20.1, pages 456-457, on rip currents and Box 20.2, pages 466-467, on the dangers of rising sea level.

TEXT ASSIGNMENT

The key terms listed under "Terms to Remember," page 469 in your text, are important and will supplement your knowledge of waves learned from the chapter on waves, beaches, and coasts. Use the glossary and the summary in the text to help you understand the definitions. In addition, here are some additional terms to help you in this lesson:

KEY TERMS AND CONCEPTS

Wind waves: Waves generated by wind energy imparted to the sea surface, often in storms far at sea. (See Figure 20.2, page 454, to learn the anatomy of a wave.) The wave form moves out of the storm area as a swell, eventually expending its energy on the shore. The height a wave ultimately reaches depends on the force of the wind, how long the wind has been blowing, and how far the wave has travelled under the wind (the *fetch*). A storm-driven wave measured in the North Pacific reached a height of 34 meters (112 feet), about the height of a 10-story building!

Deep water waves: Wind waves in the open ocean where the depth to the bottom of the sea floor is greater than one-half the wave length. (Figures 20.3 and 20.4, page 455, are important in helping you understand the motion of water particles in a passing wave. Notice that the wave energy affects the water only to a depth equal to one-half the wave length; below that depth, most wave motion has ceased.) Most open-ocean wind waves do not "feel bottom" because wave lengths rarely exceed 300 meters (1,000 feet). The largest swell ever reported had a wave length of 792 meters (2,598 feet). Since the average depth of the ocean is about 4,000 meters (13,200 feet), most of the sea floor is below the depth of wave action. Wave energy therefore is expended in eroding and transporting sediment in the beach and coastal zone and not in modifying features of the deep sea floor.

Beach: A zone that extends from the low-tide line inland to a sea cliff, a zone of stabilized sand dunes or permanent vegetation, or sometimes to the edge of a parking lot. This is the zone in which the sand is in motion, both parallel to the beach face and more-or-less perpendicular to the beach, as in the *swash* (wave run-up) and *backwash* (wave return). (Learn the parts of a beach from Figure 20.7, page 458, and note the seasonal cycles shown in Figure 20.8, page 458. How does the sand get back on the beach in time for the summer visitor?)

Longshore current: A flow of water parallel to the beach face within the surf zone. Waves that approach the shore at an angle push the water up the beach, also at an angle, but the water recedes perpendicular to the beach face and flows down the foreshore. (Study Figures 20.9 through 20.12, page 459, to become familiar with the origin and effects of the longshore current and sand drift.) This generates a current parallel to the shore. The larger the waves, the stronger the current. Beyond the edge of the breaking waves, there is very little movement of the water parallel to the shore. The longshore current should not be confused with coastal currents such as the Gulf Stream or the California Current. The longshore current and the longshore drift of sand along the beach face are driven by the waves and generally flow in a southerly direction on both the East and West Coasts of North America.

Erosional coasts: Coasts marked by steep rocky cliffs and headlands, in which wave erosion is the dominant process. The rate of erosion is variable but is a function of the strength of the rock exposed at the shore and the energy of the waves. Wave erosion has a narrow upper and lower limit depending on the tides and works horizontally as seen in flat wave-cut platforms. These landforms differ greatly from those made by down-cutting agents of erosion such as streams and glaciers.

Depositional coasts: Broad gently-sloping plains found on the East Coast of the United States and in the Gulf of Mexico that are shaped predominantly by deposition rather than erosion by waves. In some areas, longshore currents have built long low barrier islands of sand that are separated from the mainland by narrow lagoons. On other coasts where the offshore slope is gentle, the waves may "feel bottom" a long way out and will bring in material over a wide stretch to be deposited just outside the surf zone.

Barrier islands: These islands may have formed from a reworking of sand dunes deposited on the continental shelf during the glacial age interval of sea-level lowering (Figure 20.22, page 465). Some major cities such as Atlantic City, Galveston, and Miami Beach, as well as many resorts and luxury hotels, have been built on these islands, most of which are less than 6 meters (20 feet) above sea level and are exceptionally vulnerable to storm-wave erosion (Figure 20.23, page 465). Barrier islands are said to be found on about 33 percent of the world's coastlines.

Drowned or submergent coasts: Coasts marked by drowned river valleys, estuaries and fiords (Figure 20.24, page 466). During Pleistocene intervals of sea-level lowering, streams flowed across the dry continental shelves, cutting valleys and lowering their bed upstream. When sea level rose as the ice melted, these valleys became long arms of the sea extending inland. The mouths of the valleys are estuaries, where fresh water from the river mixes with seawater. Estuaries are among the most biologically productive habitats of the marine environment. Here in the quiet, nutrient-laden waters, the larval forms of many marine organisms, including valuable food fish, are able to thrive and grow before returning to the open sea to become adults. Unfortunately, many estuaries are severely degraded by construction, pollution, chemical wastes, and oil spills, as well as other additives that have adversely affected this priceless habitat.

Uplifted or emergent coasts: Coasts elevated primarily by tectonic movements as evidenced by old beaches and wave-cut platforms that are now uplifted marine terraces (Figure 20.25, page 468). Crustal rebound following melting of the ice caps has also elevated shorelines. Elevated beach deposits on the southwest shore of Hudson Bay form remarkable patterns that demonstrate that the land has been rising since the disappearance of the ice sheets.

NOTE: As a result of the glacial age sea-level lowering and subsequent rise, many shorelines show features of both submergence and emergence. Also, both submergent and emergent coasts may be the result of tectonic activity, and extensive erosion may take place in both cases. Any system of classification of the coastal zone is valid only in a general way but is helpful in identifying the landforms and in unraveling the geologic history of the coast.

Once you have read the text material for the lesson and learned the terms, you are ready to view the lesson's video portion. Review the questions that appear in this section to help you watch for important points and to prepare you for what you will see in the video.

Before viewing, answer the following questions:

1. Describe how the water particles move in a passing wave.

VIEWING GUIDE

2. Describe the causes of the longshore current and indicate where it occurs.

After viewing, answer the following questions:

1. Based on the video animation, describe the motion of water particles in a passing wind wave. Learn the anatomy of a wave using the terms: wave base, wave length, wave height, and the significance of "depth=½ wave length."

2. How do the properties of waves change as they approach the shore? What happens to a wave to make it "break" in shallow water?

3. Compare a tsunami with a wind wave considering their origins, heights, lengths, and velocities in the open sea. Where was the place of origin of the destructive tsunami that devastated Hilo, Hawaii in 1946?

4. Why do waves coming toward shore at an angle bend or refract to become more nearly parallel to the coast? What is the effect on irregular coastlines over long periods of time? Where on an irregular coastline will the wave energy be concentrated (Note the animation)? (This is a difficult question, so you may need to watch the video more than once to answer it.)

5. What is the origin of the longshore current? Note in the video the movement of water shown by the yellow dye. Why is the beach called "a river of sand"? Why is knowledge of the longshore current important to maintenance of coastal installations? (This is another difficult question.)

6. How do dams upriver affect the "health" of beaches downstream? How do sea walls generally affect beaches? What might be a solution to the problem of loss of sand from beaches and increased erosion caused by emplacement of sea walls?

7. How do tides influence beach processes? What causes tides?

8. The disaster in Bangladesh in 1970 was unfortunately repeated in May 1991, under similar circumstances. What were the events that led to the flooding of the Ganges River?

9. Sea level is slowly rising. What changes are taking place worldwide that would account for this? What changes will take place in the coastal zone?

PUTTING IT ALL TOGETHER

You have read the text assignment and seen the video portion of the lesson. This section, made up of a SUMMARY, a CASE STUDY, the QUESTIONS, suggestions for ACTIVITIES, and a SELF-TEST, will help you pull the information together.

The coast is a dynamic environment in which most of the geologic processes of erosion and deposition are the result of wind waves expending their energy on the shore. As the depth of effective erosion is only one-half of the wave length, the deep sea floor is unaffected by even the greatest hurricanes and wind storms that roil the sea surface.

The water particles in a passing wave in deep water move in circular orbits, with the particles at the crest of the wave moving forward and the water in the trough moving backward. Thus, the wave form moves forward, transmitting the energy through the water, but the water itself has very little forward motion, except in the final stage of wave movement onto the shore.

In shallow water, the waves are influenced by friction with the bottom: the wave length shortens but the wave height builds. The faster-moving crest of the wave will eventually topple forward until the wave breaks in the surf zone.

Waves are refracted or bent as they approach an irregular coast, and most of their erosive energy is concentrated on headlands. Bays between the headlands are usually the sites of deposition of material eroded from the points of land. Wave action thus cuts back the headlands and fills in the bays, eventually straightening the coast. (Relate Figure 20.16 on page 462 to Figure 20.17 on page 463 to understand how wave action straightens an irregular coastline). What is the fate of the *sea stacks* and *sea arch*? Study the concept of wave refraction or bending of the wave front and relate it to straightening the coast. Wave refraction also produces a longshore current, one of the most important coastal processes that moves water and sediments parallel to the beach face. Remember: the longshore current is driven by the wind waves coming in at an angle to the shore and flows only within the surf zone.

The sand on most beaches is supplied by streams, although a small portion may be added by erosion of sea cliffs in the shore zone. Most continental beaches are composed of quartz sand with smaller amounts of feldspar and other land-derived minerals. Volcanic islands lack quartz, and the beach may consist of black basalt cinders or other volcanic debris. Around tropical islands that have carbonate reefs offshore, the white beach sand may be composed of particles of coral and other materials of organic origin.

While waves are the most important geologic agent in the beach zone, the tides increase the range over which the waves can act and generate strong erosive currents in narrow inlets and passages. If storm waves strike the coast at the time of the highest tides, the waves will rise exceptionally high on the beach, and man-made structures can suffer extensive damage. Seasonal cycles also affect the beaches. Large winter waves remove the sand and expose the rocks below, while small summer waves return the sand to the beach from offshore sand bars.

The loss of sand from the beaches is a major problem especially in the low-lying Southeastern states and the entire Gulf Coast. Rising sea level seems to be contributing to world-wide coastal erosion. In some areas, the land is actually sinking. The increasing use of low sandy

Summary

433

islands and shores for extensive development has resulted in emplacement of sea walls, jetties, breakwaters, and other structures. In most cases, they have caused more harm to an already threatened environment than natural processes would have had on the shores. Replenishment of sand to depleted beaches has been a costly but generally unsuccessful effort because most of the sand is carried away in a few months to a few years.

The coasts that are classified as erosional will have sea cliffs, arches, caves, and a wave-cut platform. Depositional coasts, which generally form on tectonically passive continental margins, are broad and gently sloping and have extensive deposits of sand. Barrier islands are characteristic of depositional coasts. Submergent coasts have drowned valleys, fiords, and estuaries where sea and river meet. The west coast of the United States has many uplifted marine terraces, which should dispel the myth that California is falling into the sea (Figure 20.25, page 468).

The Sea is Nibbling at My Door!

Coastal erosion seems to be a world-wide phenomenon that is threatening not only communities in low-lying areas but some of the major population centers located on sea coasts. Unfortunately, some of the efforts to control or protect establishments against the ravages of the sea are short-term solutions that only seem to make matters worse.

What are some of the causes of these changes in coastal environment? (Review Box 20.2, pages 466-467, on the dangers of high sea level.) There doesn't seem to be one simple answer. Sea level has been rising since the end of the glacial age, and the shoreline has retreated landward during thousands of years. Recent measurements indicate a rise of about 30 centimeters (13 inches) per century along the East and Gulf coasts of North America. This could cause the shoreline to retreat up the gently sloping coastal plain by as much as 150 meters (500 feet) each century. If the predictions of greenhouse warming are correct and more polar ice melts, sea level will probably continue to rise. But the future behavior of sea level is uncertain because our current knowledge still has some major gaps in it.

The problem of loss of sand from the beaches is crucial along many coasts, not only for the loss of recreation areas but because the sandy beach acts as a buffer against the erosive force of the waves. Beaches will reach an equilibrium between the amount of sand available and the variations in wave energy. Beach sand, during a summer and winter beach regime, will be returned to the shore face if the sand weren't funnelled into deep water. The damming of rivers that prevents sediments from reaching the shore, however, deprives the beach of sand. The moving longshore current will continue to move whatever sand is there, and eventually the beach will be "starved."

Most geologists agree that man-made structures have probably produced as many coastal erosion problems as rising sea level. Seawalls, groins, jetties, and offshore breakwaters all interfere with the movement of sand in the longshore current. At Cape May, New Jersey, for example, the culprit is a jetty that has cut off the natural supply of sand to beaches down the coast. In seeking solutions to the problem, people must consider the movement of sand over a great area under a variety of oceanographic conditions, some of which operate within a short period of time. (Figure 20.14, page 460 shows situations found near many shore installations. Can you determine why sand erodes on the right side of each jetty or groin in the diagram? And why is there a build-up of sand behind the breakwater?)

In many areas, the preferred solution to beach erosion is replenishment by trucking in new sand or dredging offshore deposits. This is a costly program that apparently doesn't really stabilize a beach since the new sand usually lasts from a few months to a few years. Some replenished beaches disappear after one heavy winter storm.

Coasts dominated by rocky cliffs are also experiencing erosion (Figures 20.18 and 20.19, page 463). The rates of retreat depend on the strength of the rocks exposed on the shore and the wave energy at any particular beach. At Cape Cod National Seashore in Massachusetts, the cliffs are composed of unconsolidated glacial deposits and are retreating about 1 meter (3 feet) per year. From the time the Romans invaded Britain about 2,000 years ago to the present, the soft chalky "White Cliffs of Dover" have retreated about 10 kilometers (6 miles) (Figure 20.15, page 461). These cliffs along the English Channel have retreated as much as 12 to 27 meters (39 to 89 feet) overnight during a single storm! While the sand on some beaches will return, sea cliffs, once eroded and removed, are gone forever. The problem now is how to keep the waves from nibbling away our beaches and homes while ensuring that our solutions don't make matters worse.

Santa Barbara is located on a fairly straight stretch of east-west tending coast with no natural harbors. Before 1930, coastal shipping was limited to using a wooden wharf that provided no shelter during stormy weather. Between 1880 and 1927, the city three times approached the federal government for advice and help in constructing a breakwater and harbor. Federal officials three times rebuffed the city, in part because the site was regarded as unsuitable for a harbor. In the late 1920s local pri- vate interests raised enough money to construct an L-shaped breakwater, parallel to shore, but with a gap on the western leg near the beach. It was expected that the open gap would allow the longshore current to move beach sand through the harbor to the beaches downcurrent to the east.

What actually happened was that the longshore current carried sand into the harbor and dropped it, widening the beach and shoaling the harbor. No one realized that waves generate the current and that breakwaters are designed to eliminate waves. Consequently, the current was slowed down enough that sand was not carried out of the harbor.

CASE STUDY:
Santa Barbara
Erosion and
Deposition
Within the Harbor

To combat this problem, the breakwater was extended to the shore, closing the gap on the west. This alleviated harbor shoaling for a while. However, in about three years, the longshore current had built up the beach west of the breakwater out to the bend in the L-shape and sand began to move along the outer leg and into the harbor entrance. While this was going on, the longshore current east of the harbor was causing severe beach erosion, having lost its normal load of sand to the breakwater built at the harbor. At one point, about 16 kilometers (10 miles) east of Santa Barbara, 55 meters (180 feet) of beach width including several buildings, was removed in a single winter storm.

The erosion and sand deposition in the harbor entrance forced the city to begin a regular dredging program. At first, they used a hopper dredge, a sort of scow with doors in the bottom. The fill dredged was towed to a site east of the city and the load dumped in about 5.5 meters (18 feet of water. Unfortunately this sand was dumped in water beyond the reach of the longshore current. The original dredge pile has remained almost undisturbed for more than 40 years, and the beaches to the east of the city continued to erode until the next dredging when sand was deposited directly on the beach face.

These problems led the U.S. Army Corps of Engineers to make a detailed study of sand transport at Santa Barbara in which they found that the average *daily* load of sand brought into the harbor by the longshore current was about 21 cubic meters (770 cubic yards or about 100 dump trucks full). During stormy periods in winter, the rate rose to as much as 141 cubic meters (4,600 cubic yards or about 600 trucks full), but in summer fell to as little as about 8 cubic meters (250 cubic yards or 35 trucks full).

Subsequent studies also showed that virtually all of this sand is delivered to the beaches by short streams draining the Santa Ynez Range as far west as Point Conception, about 80 kilometers (50 miles) from Santa Barbara. No sand appears to be getting around this headland from northern and central California beaches.

This "Case Study" was written by Dr. Robert Norris, Professor Emeritus, University of California – Santa Barbara. He is also one of the Academic Advisors for *Earth Revealed*.

Questions

1. **If water particles in a wind wave move in circular orbits, why do they break in shallow water?**

   Water particles in deep-water waves move in circular orbits; the water at the crest moves forward, and the water in the trough moves backward. As the wave form moves into shallower water, the orbits of the water particles flatten into ovals (Figures 20.3 and 20.4, page 455). The wave velocity decreases, and the wave lengths shorten or "bunch up." The energy now confined to a smaller area causes the wave height to increase. When the wave gets to a depth of about 1⅓ of the wave height, the top of the wave still moves forward while the base lags behind. The rounded wave form changes to a peaked top that rushes forward, and the wave "breaks." Since there is not enough water in the backwash to fill out the orbits, the wave curls over. The breaking waves make up the surf zone.

2. How do waves actually erode and shape a coastline?

Waves erode a coastline by different methods:

a. Waves erode mostly by abrasion just as streams do. During storms, when waves are highest and most capable of carrying such abrasive agents as sand, gravel, and even cobbles, their erosive power is enormous.

b. If the rocks at the shore are carbonates or other materials subject to solution, the waves may dissolve as well as abrade the cliffs.

c. Water will also enter cracks in the rocks. By compressing water behind a cliff face, the waves will cause blocks of rock to be shoved out. This is called hydraulic action.

d. On a cliffed shoreline, the base of the cliff will be undercut by the waves, and landslides will occur. In such cases, the landslides also can be considered the eroding agent of the upper parts of the cliff. (Review in Lesson 16 the effects of waves on unstable shorelines.)

3. Aren't the tides just as effective as waves in shaping a coastline? What causes tides? Are tides also waves?

Tides are not in themselves an effective agent of coastal erosion. They have other effects, however. By raising the water level, twice a day on most shores, they allow the waves to erode high up on the backshore and attack sea cliffs. They also expose the lower foreshore to wind and rain during periods of low tide. Tides are actually considered to be a wave, with the high tide as the crest and the low tide as the trough. Tides have a long wave length: from crest to crest may be a thousand kilometers or more. So tides move the whole water column from surface to the deepest sea floor. The average depth of the sea floor is about 4.2 kilometers (2.5 miles). However, the tidal range is so small at depth that there is little or no erosional effect on sea-floor rocks or sediments, except in narrow channels like the Golden Gate.

Tides, of course, are a cosmic event. The gravitational pull of the moon and the sun, combined with the rotation of Earth on its axis, are the principal forces that affect the movement of Earth's waters. Actually, there are perhaps 65 factors that will determine the tides in any one place – everything from bottom topography to depth of the water, from shape of the bays along the coast to tilt of Earth's axis. The range from high to low tide will vary from about 33 centimeters (1 foot) in Hawaii to over 132 meters (40 feet) in the Bay of Fundy in Nova Scotia.

4. Why do the longshore current and sand drift move generally from north to south on <u>both</u> coasts of North America? And what finally happens to that river of sand?

The longshore current and drift of sand generally move south on both coasts of North America because the major wave-generating areas are in the North Atlantic and the North Pacific, where the winds are fiercest and the storms most frequent. Remember – it is the breaking waves that drive the longshore current.

The sand moves south until it reaches a submarine canyon and is shunted down the canyon into deep water. This sand does not return to the beach during the summer and cannot replenish the beach. Sand carried by the longshore current is also deposited in harbors, behind breakwaters, near jetties or groins, or in sand spits and baymouth bars.

5. What features characterize an erosional coast? How will they differ from landforms of a depositional coast?

An erosional coast is generally characterized as rocky, steep, cliffed and irregular with small *pocket* beaches, sea caves and arches, and offshore isolated sea stacks. As cliffs retreat under the force of the waves, a flat wave-cut platform is formed, consisting of a horizontal bench of rock. Wave erosion acts like a horizontal saw with the lower limit at *wave base*, the lowest depth of effective erosion by the incoming breakers (Figure 20.19, page 463).

The predominantly depositional coasts of the Atlantic and Gulf Coast states are generally broad and gently sloping and have thick sand deposits. Some 282 barrier islands parallel the shore from Long Island, New York, around Florida and the Gulf of Mexico to the southern end of Texas. A similar chain of sandy barriers stretches along the low coasts of the Netherlands and Denmark around the southeastern margin of the North Sea. Quiet lagoons lie behind the islands, but strong tidal currents carry sediments both into the lagoons and out to sea (Figure 20.22, page 465).

6. What would you see from an airplane flying over a drowned or depressed coastline?

If you were flying over a drowned or depressed coast, you would see an irregular shoreline with deep indentations. Some coasts are the result of sea-level lowering and erosion by streams or valley glaciers of the exposed continental shelf. In each case, the rising sea level of the past 10,000 years has inundated the low places and created a coastline with long fingers of the sea extending inland. The fiord coasts of Scandinavia, New Zealand, British Columbia, Southeast Alaska, Puget Sound, and southern South America are all a result of glacial erosion. The estuaries and bays of the East Coast, such as Chesapeake Bay and the Gulf of Maine, and the southern coast of Brazil are drowned river valleys. Some coasts are subsiding, such as the delta coast of the Netherlands, which is sinking at a rate of nearly 10 centimeters (about 4 inches) per century. A complex of dikes and coastal defenses has been constructed to prevent the sea from entering the low land. The Mississippi River Delta is also subsiding, losing about 106 square kilometers (50 square miles) a year of Louisiana to erosion and encroaching sea. Other factors affecting the subsistence of the delta might include the lack of sedimentation caused by flood control, compaction of sediment, and subsidence caused by extraction of ground water and oil and gas.

7. **Beach sands come in a variety of colors, minerals, sizes, and shapes. What might be some determining factors? What are usually the most abundant minerals found on continental beaches? And what can you expect to find on the beach of your south sea island paradise?**

Beach sands seen under a microscope are a myriad of colors and shapes that rival some fine ancient mosaics. Continental beaches are usually covered with clear glassy quartz sand derived from granite or older sandstones exposed inland and brought to the coast by streams. Quartz is hard and tough, a survivor of the long journey to the sea. Some colorful beach sands also contain feldspar, hornblende, red garnet, pink quartz, black magnetite, green epidote, and other minerals exposed in the rocks inland or near the coast. The particle size may vary from fine sand to gravels, pebbles, even boulders, depending on the energy of the waves. Large waves on exposed headlands carry off the smaller or lighter fragments and leave the large rocks behind. *Shingle beaches* are covered with flat pebbles usually derived from a local source of slate or hard shale. (These are excellent "skipping stones.")

Black sand beaches in Hawaii are composed of volcanic cinders or shattered lava mixed with bits of black obsidian or abraded basalt pebbles or cobbles depending on the wave energy. The black sand beaches of New Zealand are composed of the iron mineral magnetite, and some are rich enough to be mined as iron ore. The red sand beaches on Maui in Hawaii are made of porous or vesicular particles of scoria that are sometimes ejected from a coastal cinder cone. The beautiful green sand beaches on the big island of Hawaii contain grains of green glassy olivine. (Review the different types of volcanic rocks in Lesson 14 on volcanism.) The white sand beaches of Tahiti and the other islands and atolls of French Polynesia form where the waves break off bits of the carbonate fringing or barrier reefs and deposit them on the shore. White sand (which is fascinating to observe under a low-power microscope) consists primarily of particles of coral and calcareous algae but also includes purple and green sea urchin spines, tiny delicate snail shells, clams, microscopic forams, and abraded hard parts of various inhabitants of the reef. (Take a hand lens or magnifying glass with you on your next trip to a tropical island.)

Activities

1. If you live near a coast, look at the beach with your new geologic point of view. Notice from what direction the wind waves are coming. This will be a clue to the direction of flow of a longshore current. Look at the landforms. They should tell you if you are on an erosional or depositional coast. Try timing the breaking waves. Those that formed locally will probably break every 4 to 6 seconds. The waves that were generated by a storm in the open sea will take about 10 to 12 seconds to break. Your new knowledge will give you an added appreciation and enjoyment of the beach and shore.

    If waves are starting to come in about 20 minutes apart, what should you do? (RUN FOR HIGH GROUND – it may be a tsunami! Review Lesson 8 on earthquakes for a discussion of tsunamis.)

2. If you live near a river, use a map to try to find the mouth of your river. Does it empty into another river, and where does the major river empty? What would happen to the beach or the delta at the mouth if all the streams in your drainage basin had dams on them?

3. Suppose you have always wanted to build your dream house on the beach. Use some of the photographs in your text, such as Figure 20.1, page 454, Figures 20.18 and 20.19, page 463, and Figure 20.23, page 465, to decide what to look for and what to avoid. Draw a cross section of a beach, and sketch in where you will put your house.

Self-Test

1. The motion of water particles in a passing wind wave is felt

    a. from the surface to the deep sea floor.
    b. to a depth equal to the wave height.
    c. to a depth equal to 33 percent the height of the wave.
    d. to a depth equal to 50 percent of the wave length.

2. Waves erode sea cliffs by all of the following methods except

    a. abrasion by sand and rocks carried by the waves.
    b. solution of granite and quartzite rocks in cliffs.
    c. hydraulic action compressing water in cracks.
    d. dissolving away of limestone rocks in sea cliffs.

3. The height of a wave in the open sea from crest to trough is equal to

    a. the wave length.
    b. the diameter of the orbit of the water particles.
    c. 33 percent the wave length from crest to crest.
    d. 50 percent of the depth of the water.

4. Waves will break when the

    a. wave lengths increase in deep water.
    b. depth of the water is greater than one-half the wave length.
    c. depth of the water is less than 4/3 the wave height.
    d. wave form becomes rounded on top.

5. The longshore current is driven primarily by

    a. force of the wind.
    b. the moving river of sand on the beach face.
    c. the distance to the wave-generating area.
    d. the wind waves striking the shore at an angle.

6. The height a wave will ultimately reach in the open sea depends on all of the following except the

    a. force of the wind.
    b. duration of the storm.
    c. direction the wind is blowing from.
    d. fetch.

7. In North America, the longshore drift generally deposits sand on the

    a. north side of a jetty.
    b. east side of a jetty on the east coast.
    c. west side of a jetty on the west coast.
    d. south side of a jetty.

8. The source of most of the sand on the beach is

    a. the bottom of sea canyons brought by winter waves.
    b. erosion of sea cliffs by waves.
    c. river sediment brought down to the ocean.
    d. sediments left by melting glaciers.

9. On an irregular coast, most of the wave energy will be concentrated on the

    a. sandy beach.
    b. baymouth bar.
    c. barrier islands.
    d. rocky headlands.

10. Depositional coasts are usually characterized by

    a. broad gently sloping plains.
    b. wave-cut platforms.
    c. uplifted marine terraces.
    d. sea stacks and sea arches.

11. Narrow berms and steep beach faces are characteristic of

    a. drowned coasts.
    b. winter beaches.
    c. wave refraction.
    d. long, low waves.

12. Of the following, the best line of evidence that the west coast of the United States is not falling into the sea is the presence of

    a. sea cliffs along the coasts.
    b. thick beach sands demonstrating coast stability.
    c. drowned valleys and estuaries.
    d. uplifted marine terraces.

13. Deep inlets with steep walls are the result of all of the following except

    a. rising sea level.
    b. tectonic uplift.
    c. glacial erosion of a coastal valley.
    d. melting of ice following the glacial age.

14. Estuaries are important and worth saving because

    a. they are the nurseries for many marine organisms.
    b. they are suitable and safe sites for factories and chemical plants.
    c. good water circulation prevents pollution.
    d. they are not affected by problems of rising sea level.

15. Of the following, the most likely place to find barrier islands is

    a. the California-Oregon coast.
    b. British Columbia.
    c. the southern Atlantic coast.
    d. the Maine coast.

MODULE VI

LIVING WITH EARTH

25

LIVING WITH EARTH, PART I

This lesson will help you understand human responses to destructive natural phenomena such as earthquakes and landslides.

After reading the textbook assignments, completing the exercises in this study guide, and viewing the lesson's video portion, you will be able to:

1. Describe ways to minimize loss of life and property caused by volcanism, earthquakes, mass wasting, and flooding.

2. Describe the present status of earthquake prediction.

Volcanoes, earthquakes, floods, and landslides have always been important to human history. The annual flooding of the Nile was an integral part of the cultural and religious foundation of early Egypt. In fact, almost every mythology tells of a great flood. The Romans incorporated *Vulcan*, god of volcanoes, into their Olympian pantheon. The Hawaiians revered and respected *Pele*, the goddess that dwelt in the active volcano Kilauea. In recent decades, with better understanding of the planet, scientists have looked for means to predict, prepare for, and lessen destruction from these natural events.

Not long ago (geologically) – only about 50,000 years – modern humans *Homo sapiens* (Man the Wise) joined the multitude of living things on this planet. Carrying their environment with them in the form of culture, they invaded every habitat on Earth. As populations burgeoned and technologies advanced rapidly, cities grew, sometimes in areas unsuited to habitation. Perhaps humans have not been on the planet long enough to understand that what are called *catastrophes* or *disasters* are part of long natural cycles that will take place regardless of the effect on human lives or structures.

In previous lessons, you learned about the interaction between Earth's internal heat engine and external forces powered by solar energy that are constantly reshaping the face of the planet. This lesson reviews the major geologic hazards, their relation to moving tectonic plates, and the effects of some of the more catastrophic events. The status of forecasting and prediction are discussed, as well as steps to minimize loss of life and property.

LESSON ASSIGNMENT

Completing the following seven steps will help you master the lesson objectives and achieve the goal for this lesson. Note that the steps are different for this lesson from the steps you have followed in most of the previous 24 lessons.

Step 1: This lesson is divided into four major segments – volcanoes, earthquakes, mass wasting, and flooding. Each segment includes a separate text assignment. So for each of the four segments, read the TEXT ASSIGNMENT before going to the next step.

Step 2: Review any KEY TERMS AND CONCEPTS listed in the text assignment sections.

Step 3: Read each of the four "Summaries" in the study guide's PUTTING IT ALL TOGETHER section.

Step 4: The CASE STUDY in this lesson will be the video portion. Watch the video portion of the lesson using the viewing guide in the study guide *after* you have read the Summaries and the text assignments.

Step 5: Complete any assigned lab exercises.

Step 6: Review the material for the lesson and complete the study guide's SELF-TEST.

Step 7: Go back to the LEARNING OBJECTIVES and make sure you have learned each one.

PUTTING IT ALL TOGETHER

Here is a TEXT ASSIGNMENT and a SUMMARY for the four main geological phenomena discussed in this lesson. After the fourth summary, QUESTIONS, a CASE STUDY with the VIEWING GUIDE, and a SELF-TEST are included to help you integrate your learning of the material in this lesson.

VOLCANOES
Text Assignment

1. Review Chapter 11, "Volcanism and Extrusive Rocks," pages 239-265, and go over the "Terms to Remember," page 265 in your text, to refresh your memory.

2. Study Box 11.1, pages 240-243, on North America's Eruption of the Century – Mount St. Helens.

3. Review Lesson 13 in this study guide.

In discussing geologic hazards, we tend to forget that these great natural phenomena are part of the dynamic, evolving planet and in some cases provide certain benefits. Volcanoes, over geologic time, have produced much if not all of the waters of the surface of the planet, as well as all of Earth's atmosphere, later modified by biologic activity (See Lesson 2). They have built new island surfaces such as the Hawaiian Islands, Iceland, and the Galapagos Islands – all of which are islands available for settlement by many life forms. Volcanic arcs have added new material to the edges of continents. Chains of volcanic islands have served as stepping stones for migrations of life to continents and to other islands. Fertile volcanic soils have produced coffee beans in South America, grapes and olives in the Mediterranean region, and pineapples and sugar cane in Hawaii.

Where are Active Volcanoes Most Likely to Occur?

Volcanoes are not scattered at random over Earth. They most frequently occur in the Pacific Rim of Fire and in the Mediterranean Belt. The Hawaiian volcanoes are located over a *hot spot* in the middle of the Pacific Ocean. All volcanic activity is related to the moving plates of the lithosphere, especially the boundaries where plates collide, separate, or grind past one another.

Volcanoes as Natural Disasters

Volcanoes can erupt explosively at Earth's surface, taking a heavy toll in lives and property. Mont Pelée on the island of Martinique erupted in 1902 and emitted a glowing cloud of dense incandescent gases, hot ash, and pumice that swept like a hurricane across part of the island; 30,000 people died in a matter of minutes. Krakatoa, in the Indian Ocean between Java and Sumatra, erupted in 1883, one of the most violent eruptions in history. If Krakatoa had erupted in a densely populated area, millions of lives would have been lost. As it was, the reverberations unleashed a *tsunami* or seismic sea wave that reached about 36 meters (120 feet) as it crashed against the coasts of Java and Sumatra and flooded over the low-lying islands of southeast Asia. Over 36,000 people were swept to their deaths. Krakatoa is not dormant by any means. A new cone arose above the sea and apparently is still growing. Tambora, a *composite volcano* in Indonesia, erupted in 1815 with an explosion that may have exceeded the size and power of Krakatoa. As a result of the massive ash and block fall, 10,000 people perished and about 80,000 starved from crop loss and famine. World climate was affected; 1816 was the year of *no summer* – snowfalls in June, July, and August caused crop failures in New England and Northern Europe where famine was widespread as the result of this one eruption half an Earth away.

Forecasting Volcanic Eruptions

Forecasting the time, place, and character of future volcanic eruptions is one of the major goals in volcanology. The statistical record of a volcano, as well as the day-to-day signs such as local earthquakes, surface deformation, temperature changes at hot springs, changes in ratios of certain gases at steam vents, even small changes in Earth's magnetic and electrical fields near volcanoes are promising in their usefulness in forecasting eruptions.

Each volcano or volcanic region has its own unique record of eruptive activity that provides a basis for evaluating the probable hazards of future eruptions. This geologic record tells what kinds of eruptions have occurred in the past (quiet or explosive), what areas have been affected, and how often the eruptions have taken place. For example, small eruptions occur much more frequently than large ones. On this basis we can say that a small eruption of one of the Cascade volcanoes will probably occur about once every 100 years, one of large volume about once every 1,000 to 5,000 years, and an eruption of very large volume about once every 10,000 years. (Refer to Figures 11.5 and 11.37, pages 246 and 263, showing the volcanoes of the Cascade Range.)

Extremely large volcanic eruptions also occur at some volcanoes and throw out material hundreds of times more voluminous than the largest known historical eruption. Such eruptions are quite uncommon, occurring once every few hundred thousand years. Nevertheless, they are potentially extremely hazardous. During the last two million years, such eruptions have occurred in Yellowstone National Park, Wyoming; at Long Valley, California; and in the Jemez Mountains of New Mexico. These eruptions deposited thick blankets of ash over very large portions of the western United States. If one were to occur today, its effects on the entire country would be devastating. Unfortunately, (or fortunately, some would say), only a small fraction of the world's volcanoes have been active during historic times or have been studied long enough to establish timetables for eruptive patterns.

In addition to historical and geologic records, the seismograph is one of the most effective instruments used to monitor a volcano. In Hawaii, several small earthquakes per day can almost always be recorded by sensitive seismographs on an active volcano, regardless of whether it is erupting. Prior to an eruption, seismic signals indicate the movement of magma into a chamber a few kilometers below the surface. (Of the two types of body waves, S and P, the S waves are most revealing, because they cannot pass through a liquid.) The site of the magma chamber is confirmed by a local variation in Earth's gravitational field. (The density of magma is less than the density of solid crustal rock.) Another important indicator is the swelling of the terrain or the rate of swelling, as was seen on the north flank of Mount St. Helens.

A new automated observatory is being planned to monitor the gradual birth of the next Hawaiian island. It will be complete with submerged microphones, seismometers, thermal and chemical sensors, and video cameras . The observatory will be in place on Loihi seamount, a rugged, steep-sided submarine volcano about 35 kilometers (22 miles) southeast of the island of Hawaii. Loihi rises about 3,394 meters (11,200 feet) above the surrounding sea floor, but its cratered summit is still about 1,000 meters (3,300 feet) beneath the surface of the Pacific. Since about 75 percent of the world's volcanoes are on the sea floor, long-term monitoring of this active vent will provide valuable information about the growth of submarine volcanoes. The data will help confirm plate tectonic theories about the origin of the Hawaiian Islands as well as provide information for forecasting future eruptions at this site.

Mount St. Helens: Forecasting Saved Thousands of Lives

The eruption of Mount St. Helens in the Cascade Range is a good example of the existing state of eruption forecasting. In general, the forecasts were accurate and saved thousands of lives. The exact timing and magnitude of the eruption, however, was unexpected. The impending eruption was marked by a swarm of earthquakes, some of magnitude 4 or greater, that began on March 20, 1980, and by a series of small steam and ash eruptions. In mid-April, a more ominous sign was noted. There was major deformation occurring across the north flank of the volcano. The area bulged out about 100 meters (330 feet) and was continuing to push out at a steady rate of about 1.5 meters (5 feet) *per day* during late April and early May. Because evidence from prehistoric eruptions suggested danger, the forest and the mountain resort region on the north flank of the volcano were evacuated.

On May 18 at 8:32 A.M., a magnitude 5.1 earthquake was accompanied by a tremendous avalanche that ripped apart the entire north flank of the cone. This was the largest avalanche ever witnessed. The gaseous magma chamber of the volcano was exposed as the northern flank fell away. It exploded at once. A dense hurricane of gas, hot ash, and debris, including up-rooted trees, spread out from the volcano across the landscape to the north, reaching as far away as 24 kilometers (15 miles). Huge flows of ash and mud swept down the surrounding river valleys, burying everything in their path. About 65 people were killed, including a volcanologist who was making deformation and gas measurements from an observation post 10 kilometers (6 miles) from the summit. (See Box 11.1, pages 240-243.)

On reviewing all the data prior to the sudden gigantic eruption – seismic information, deformation measurements, volcanic gases, and visual observations – there was no evidence that the activity was about to take a tragic turn. The sudden change from small explosions to the giant blast was not foreseen, since the scale of the 1980 eruption was unprecedented in the volcano's 30,000 years of geologic history.

What Can be Done to Reduce Losses from Volcanism?

The activities of the United States Geological Survey (USGS) during the Mount St. Helens eruption were admirable and can be credited with saving many lives. The knowledge of past eruptive activity defined the potential hazard, including the scale, location, extent, and severity of a future eruption, as well as the *hazard zones*, the geographic limits of destruction. The seismic and geologic monitoring systems were accurate in providing a warning of the impending eruption. The emergency evacuation was generally successful. The people who perished were, for the most part, in the warning area: those whose jobs included a calculated risk, like volcanologist Dave Johnston, two researchers, and several loggers; homeowners who wouldn't leave their property; and volcano watchers who had skirted the barriers and gone into the warning area on logging roads. If the timing could have been more precise, perhaps these lives would not have been lost either.

Protection from volcanic hazards can be effective in reducing losses for some events but not for others. Providing dust masks and goggles can protect individuals from respiratory damage and eye irritation; changing oil and air filters can reduce damage to vehicles from ash falls. Heavy ash falls on flat roofs may cause them to collapse. Risk assessment and intelligent land-use planning allow a community to make decisions that are consistent with their goals of public safety. Pre-disaster planning and preparedness are important for relief and rehabilitation, especially in areas known to be at risk. The best we can do at present is to consider the probabilities of an eruption, use all the modern techniques of forecasting, concentrate on planning and preparedness, and spread the knowledge of the capricious and sometimes deadly nature of volcanoes.

Review the "Case Study" on Yellowstone in Lesson 14. Consider the effects if an eruption that size happened today in that area. The ash blanketed a region from Canada to Texas and east almost to the Mississippi River.

EARTHQUAKES
Text Assignment

1. Review Chapter 7, "Earthquakes," pages 147-172, including the "Terms to Remember," page 172, and the chapter "Summary."

2. Re-examine Figure 7.20 A and B and Figure 7.21, pages 163 and 164, showing the distribution of earthquakes.

3. Inspect Figure 7.11, page 156, to see if you live in an area of high seismic risk.

4. Study Box 7.3, pages 170-171, "Waiting for the Big One," and examine the photograph facing page 147.

5. Review Lesson 9 in this study guide.

Summary

Humans have long sought to explain such cataclysms as earthquakes by putting the blame on angry gods or mythical gigantic beasts. The lurching of a huge tortoise that balanced Earth upon its back was believed by ancient Hindus and many other Eastern peoples to cause earthquakes. Others believed in a monstrous elephant, while the ancient Japanese pictured a giant carp holding up the world and shaking it with every flick of its tail.

Some people thought the theory of plate tectonics was equally bizarre! The scientific evidence of plate motions is now so persuasive that modern Earth scientists feel they have a credible model that correctly explains the working of this planet, including causes and locations of earthquakes. Pinpoint-predictions of the next quake, however, are still not easy.

Where do Earthquakes Occur?

Earthquakes, like volcanoes, are most frequent in the Pacific Rim of Fire. They also occur in the Mediterranean-Himalayan belt and along the crests of the mid-ocean mountains. These are the boundaries of the great lithospheric plates, especially where plates are converging and subducting (such as around the margins of the Pacific basin), where continental masses are colliding, or along the San Andreas fault, where the Pacific plate is sliding northwestward against the margin of the North American plate. The earthquakes in the ocean basins mark the zones of plate divergence and creation of new sea floor. In fact, one of the great attractions of plate tectonics is that the location of plate boundaries is precisely defined by the distribution of earthquakes.

Earthquakes as Natural Disasters

An earthquake – the sudden motion or trembling of Earth caused by an abrupt release of slowly accumulating strain in rocks along a fault zone – is one of the deadliest of natural hazards. Each year, several million earthquakes occur throughout the world, varying in size from minor tremors to a few great earthquakes that cause extensive damage, injuries, and loss of life.

The degree of damage caused by ground shaking at a given site depends on (1) the severity and duration of the shaking, (2) the lateral and vertical distance to the earthquake source, (3) the type and thickness of the geologic materials that underlie the site (such as solid rock or unconsolidated deposits), and (4) the type of building construction at the site. Seismic waves emitted from the point of focus are the body waves, termed P and S waves. Of these, S waves can be extremely damaging to buildings because they cause the structure to vibrate strongly from side to side in a snapping motion. Surface waves, which arrive last, cause the ground and the buildings to vibrate in a complex manner (they undulate like a ship in a storm at sea!) and can be the most destructive. Surface waves may move great distances, and buildings located many miles from an epicenter can be seriously damaged. This was seen in the Loma Prieta earthquake near Santa Cruz, California, in October 1989, where buildings, bridges, freeways, and other structures in San Francisco were strongly affected by earthquake waves generated at least 96 kilometers (60 miles) away. It was also seen on January 31, 1991, in Afghanistan, when a 6.8 earthquake killed over 700 people and was felt as far away as Delhi, India, about 1,000 kilometers (600 miles) from the epicenter.

During an earthquake, tremendous damage and loss of life is caused by the collapse of houses, churches, and commercial buildings. Serious damage also occurs to hospitals, railroads, dams, bridges, canals, police and fire stations, schools, storm drains, and water wells, in addition to water, gas, and sewer lines. The secondary effects of earthquakes are often equally or even more destructive than the earthquakes themselves and include fires, floods, landslides, sometimes epidemics, and, in coastal areas, tsunamis. It was fire following the 1906 earthquake that destroyed most (90 percent) of San Francisco. At Yungay, Peru, in 1970, it was an avalanche and landslides started by an earthquake that destroyed the town, where 17,000 people perished.

What Can be Done to Minimize the Loss of Life and Property from Earthquakes?

Over the long run, the incorporation of earthquake resistance into building design and construction is the first line of defense. Building codes in the United States and other developed countries have been devised for all major structures and impose rigorous guidelines based primarily on experience and engineering analysis for schools, hospitals, and dams. Special studies are required for buildings to be constructed near known active faults. But unfortunately, attention to detail, such as placement of reinforcing steel, proper welding of joints, and strengthening of foundations – details crucial to the success of the design – are sometime sacrificed in the effort to speed construction or reduce costs.

Large modern skyscrapers are generally designed to provide emergency access and strength and a certain flexibility within the steel skeletons that allow them to withstand considerable shaking. Among the more vulnerable buildings are those with roofs extending over a broad span, such as supermarkets, bowling alleys, auditoriums, arenas, and large churches. If the roof of the structure is arched, it exerts an outward thrust against the bearing walls, which may be pushed down and out during violent shaking until they collapse.

Earthquakes in underdeveloped countries can be deadly because of poor construction practices. Extensive use of thick brittle masonry such as adobe results in walls and ceilings that are very heavy; lacking strong internal reinforcement, they will collapse when shaking starts. An inexpensive but effective method of reinforcement uses plant reeds mixed into the mortar; they add strength and hold the material together. Overcoming the habits of building the old way, unfortunately, is the greatest difficulty.

Single-family wood-frame houses seem to be inherently earthquake resistant. Strong internal frames where the walls, ceilings, and foundations are firmly attached to each other tend to resist damage if all the parts vibrate together. Chimneys, however, are frequently a weak point since they are rarely reinforced and are not well attached to the rest of the house. They are often the first portions of a house to suffer damage.

The selection of building sites is a crucial factor, as was so clearly demonstrated during the October 1989 Loma Prieta earthquake. (Refer to Figures 1.1, 1.2 and 1.3, pages 4 and 5.) In general, buildings that will be most severely damaged during an earthquake are those that stand on filled, unconsolidated, or water-soaked ground. The least damaged structures are those that stand on solid rock. This was seen in the Marina District in San Francisco, 96 kilometers (60 miles) from the epicenter of the Loma Prieta quake. This section of the city was built on rubble, uncompacted mud, silt, and sand used to fill a portion of San Francisco Bay. The soft surface deposits in some places underwent *liquefaction* (a temporary loss of strength when silts and sands behave as viscous fluids rather than as solids) caused by shaking, and the ground was unable to support a load. The buildings settled, tilted, or shifted from their foundations, and many fine private residences were severely damaged or destroyed. Underground utilities such as gas and water lines were severed by the movement of the liquified substrate. It is ironic that *sand boils* – upwellings of sand and debris – in the Marina district erupted pieces of buried charred redwood, tar paper, and building rubble from the 1906 earthquake that had been used as fill, a bitter reminder of the city's earthquake history. San Francisco's modern tall buildings that stood on firm ground came through the quake in good condition. Unfortunately, many highways, schools, hospitals, and public buildings have been constructed adjacent to faults or over other pockets of unconsolidated materials. The collapsed section of a freeway in Oakland, where 41 people were killed and many others injured, failed partly because of its design and construction and partly because it was built on San Francisco Bay fill. (See the photograph facing page 147. Read pages 4-6 in Chapter 1 on the Loma Prieta quake.) Evidently, some of the lessons taught by the 1906 shock never were learned or had been forgotten or ignored.

The Present Status of Earthquake Prediction

Earthquake prediction is a rapidly emerging scientific field offering great promise for loss reduction. There is a certain urgency to the subject because of the predicted *big one* that threatens the great urban centers of California – San Francisco and Los Angeles. Nevertheless, accurate predictions of the magnitude, time, and location of future earthquakes may still be years away.

Some of the interesting methods of prediction that have appeared in the popular press include the restless behavior of animals, which according to Chinese scientists may take place prior to earthquake activity. This has not been verified, although some of the Chinese predictions have been extremely successful, while others have failed. (They have also used other earthquake precursors to predict occurrences.) A popular notion concerns *earthquake weather*, but the study of disastrous shocks in the last few hundred years has revealed no correlation between weather and earthquakes. Other predictions have been based on the relative position of the moon, planets, and sun in relation to Earth or on the effects of the tides. No positive correlation can be seen, however, nor have the forecasts made by people outside the field, such as the prediction in 1990 of a "big one" on the New Madrid fault, proved accurate.

Turning to some of the more scientific methods of forecasting shocks, we note that historic records are of great value in predicting volcanic eruptions; records are also essential in determining on which faults earthquakes are likely to occur. There are also many features on the surface of Earth that indicate recent active faulting, such as *fault scarps* (small steps or cliffs produced by vertical offset along a fault), troughs, ridges, diverted streams, and displacement of young rock units.

Areas in the United States where young surface faults are known to exist have been mapped. The remaining portions of the country are thought less likely to experience earthquakes as frequently. In addition, most of the faults east of the Rockies are deeply buried, making them difficult to study. Estimates can be made of the maximum surface length and displacement that a specific fault can produce; this is useful in predicting how a fault will behave.

Most other scientific methods of forecasting involve monitoring the accumulation of long-term strain and the slight changes that occur next to a fault before rocks break and move. Before a quake, small cracks may open, causing small tremors or *micro-earthquakes*. Properties of the rock may change, especially the magnetism or electrical resistivity or the local speed of seismic waves. Small cracks change the porosity of the rock, and the water level in some wells may rise or fall. Emissions of the radioactive gas *radon* moving through opened pathways in rock may increase in certain wells and are now considered a possible indication of an impending quake. A series of vigorous *foreshocks* in 1975 alerted authorities in China to an impending earthquake. Millions of people were safely evacuated before the 7.3 magnitude earthquake struck about 5 hours later. The Chinese forecasters were less successful when they failed to predict a 7.6 earthquake in 1976 that produced no foreshocks at all. The quake caused a loss of life officially estimated at 240,000 to 500,000.

Some earthquakes have been predicted on the basis of *gaps* in seismic activity. These are places along active faults where no earthquakes have occurred for a significant period and therefore are considered most likely to move. This method has predicted some smaller earthquakes on the San Andreas fault in California as well as the probability of the 1989 Loma Prieta shock.

Historical records show that earthquakes in some places occur at fairly regular intervals, allowing seismologists to make long-range predictions. A monitoring program has been established on the San Andreas fault at Parkfield in central California (shown in the video in Lesson 9) where earthquakes of magnitude 6 occur on an average of every 21 or 22 years at the same epicenter and rupture area. (The frequency of the quakes has varied from 9 to 36 years apart). An array of instruments including *creep meters* (buried wires stretched across the fault), *strain meters*, water levels in wells, seismometers, and laser beam distance-measuring devices, is now in place. The hope is to learn exactly what geophysical changes and precursors occur before the next quake, predicted for sometime between 1983 and 1993. Parkfield may be running a little late from its most expected time of occurrence (about January 1988 ± 5 years). It may be that the 6.7 Coalinga earthquake on May 2, 1983, about 40 kilometers (24 miles) northeast of Parkfield affected the timing of the next Parkfield event.

MASS WASTING
Text
Assignment

1. Review Chapter 13, "Mass Wasting," pages 285-301. Go over the "Terms to Remember," page 301, and the "Summary."

2. Study Figures 1.7 A and B, pages 12-13, and the photograph facing page 285, showing disastrous mudflows.

3. Re-examine Box 13.1, pages 286-287, "Disaster in the Andes," noting the photographs.

4. Review Lesson 16 in the study guide.

Summary

While there are about 400 to 500 active volcanoes on Earth, and although thousands of earthquakes are felt each year, (Figure 7.10, page 155), mass wasting occurs much more frequently than either of these other two disasters and is a significant hazard in many of the regions of the world. Within the United States, the most susceptible areas are mountain and shore areas of the western coastal states, the Appalachian Mountains, the Rocky Mountains, and along major rivers such as the Ohio and the lower Mississippi. Large parts of Alaska are also severely affected. In addition, disastrous landslides take place in conjunction with other hazards such as earthquakes, volcanoes, and floods.

Because landslides are so widespread and vary from subtle forms such as *creep* to the dramatic collapse of whole mountainsides (seen at Armero, Colombia, Figure 1.7, pages 12-13, or Yungay, Peru, Box 13.1, pages 286-287, and shown in the video for Lesson 16), an accurate estimate of the cost of damages from this geologic hazard is very difficult to make. Direct costs relate to losses or damage to property or various kinds of installations. Indirect costs are varied and may be substantially larger than direct costs. These include loss of productivity of agricultural or forested lands, loss of industrial productivity (because landslides affect transportation and communication systems), and loss of property values in known landslide areas. The sum of direct and indirect costs has been estimated at more than one billion dollars per year in the United States alone. The Portuguese Bend landslide in California, for example (Lesson 16), cost more than ten million dollars between 1956 and 1959. The loss of life in the United States has not been as calamitous as in some other countries since most of the catastrophic slope failures have taken place in unpopulated areas. Nevertheless, it was estimated in 1976 that total loss of life in the United States from landslides averages more than 25 per year. And, as noted in Lesson 16, on a world-wide basis, more lives are lost from mass wasting events each year than from either volcanoes or earthquakes.

Causes of Mass Wasting

There are many kinds of mass wasting, varying in velocity of movement and in type of material. There are likewise many causes of mass movement. The undercutting of cliffs by stream and river erosion, glaciers, waves, and longshore or tidal currents are among the most widespread natural causes of mass wasting. Landslides on oversteepened slopes resulting from construction of canals, quarries, mountain roads, reservoirs, dams, and highways are common. High relief in mountainous areas, especially where bedding planes or foliation in metamorphic rock is parallel to the slopes, forms unstable conditions. Accumulations of volcanic debris, ash, or cinders; thick debris over bed rock; and stockpiles of ore or rock waste are all subject to landslides. The nature of the substrate is also a determining factor – whether it is unconsolidated material such as soils, weak shales, or extensively fractured, or solid rock. Finally, rain or snow adds weight, infiltrates permeable material, and introduces pressure (pore pressure) between rock particles, gradually forcing them apart and weakening the entire mass.

Identifying Landslide Areas

As with earthquakes and volcanoes, historical records are invaluable in identifying unstable areas. Geologists study suspected land surfaces for evidence of past movements and try to forestall development in these terrains. The town of Yungay in Peru, for example, was built on known landslide material, but unfortunately, the geologic signs were ignored. The mud and ash flow from the Mount St. Helens explosion devastated the land for miles around but was largely predicted by geologists studying the region. Scarps, earthflows, mudflows, tilted telephone poles, and old landslide scars are all useful in identifying areas of previous mass movement. The presence of *hummocky topography* and slump in Portuguese Bend, California, was clear evidence that the housing development there was built on a moving hillside.

What Can be Done to Reduce Losses from Landslides?

In some cases, the causes are natural – erosion, fractured rock, steep slopes, and accumulations of ice and snow – and can be triggered by an earthquake. In these cases, it is impossible to remove all the physical causes of a slide. Efforts are made to reduce the losses on a small scale intermittently or continually as they occur, and of course, not to build in the slide area.

Unfortunately, some of the most damaging landslides are closely related to human activities. One of the greatest dam disasters was not the result of dam failure but of mass movement of surrounding hillsides into the reservoir behind the dam. On the night of October 9, 1963, a sudden torrent of water, mud, and rocks plunged down a narrow gorge, shot out across the wide bed of the Piave River in the Italian Alps and up the mountain slope on the opposite side. Several towns were demolished and 2,600 people perished. When the flood was over, it was assumed the Vaiont Dam that impounded water in the steep-sided valley had failed, but upon inspection the dam proved to be intact. The reservoir behind the dam had not been even half full, so what caused the water to rise up over the dam and surge down the canyon? Geologists knew that one shoulder of the dam was supported by Monte Toc, nicknamed *la montagna che cammina* – "the mountain that walks." The engineers who designed the Vaiont dam apparently did not take into account three crucial factors: the weakness of the cracked and deformed layers of limestone and shale that walled the reservoir; an obvious geologic history of landslides; and evidence of a small rockslide three years before.

On that night in October, the mountain did more than walk: about 250 million cubic meters (330 million cubic yards) of mountainside slid into the lake behind the dam. A wave rode 260 meters (850 feet) up the valley wall opposite the slide. Another wave topped the dam, which was 265 meters (870 feet) high, and dropped into the deep gorge below. There, constricted by the narrowness of the valley, the water, carrying tons of mud and rocks, raced on a destructive path through the towns below. It was all over in seven minutes. The landslide was not preventable; from the history of the region, however, the dam should never have been placed in a geologically unstable site.

Regulating land use after careful geologic and engineering studies and *before* construction is the best way to minimize losses from landslides. Such regulation would include mandatory run-off and drainage controls, improved grading procedures, terracing cliffs on highway cuts, planting proper vegetation, building effective retaining walls, and preventing water from entering unstable areas whether from irrigation, sprinkling of home gardens, or use of home water disposal systems. Pumping out water from a sliding mass has proven effective because it increases the frictional resistance to downslope movement. Landslides in many areas can be prevented. Through the use of geological and engineering principles, the costs of landslide repairs in the seven years prior to 1976 were reduced by as much as 90 percent in parts of New York state.

FLOODING
Text
Assignment

1. Review Chapter 16, "Streams and Landscapes," pages 349-381.

2. Re-read the "Summary" and study the "Terms to Remember," page 380 in the text, to refresh your memory.

3. Study "Flooding," pages 352-353, and note the causes of the Big Thompson River flood in north-central Colorado.

4. Review Lessons 19 and 20 in this study guide.

Summary

Floods have been and continue to be one of the most destructive natural hazards in the United States as well as other places such as on the Indian subcontinent and in Europe. A flood is any abnormally high stream flow that tops the natural or artificial banks of a stream. Flooding is a natural characteristic of rivers; *flood plains* on either side of a stream are normally dry-land areas that act as a natural reservoir and temporary channel for flood waters. Flood plains are important to agriculture, and because of the flat land and easy access to food and water resources, have become the sites of major cities. Unfortunately, people have built in flood-prone areas faster than flood-protection works have been constructed. The increased losses in recent years to life and property are less related to greater floods than to increased encroachment on flood plains.

About 4.8 million kilometers (3 million miles) of streams exist in the continental United States, and about 6 percent of the land area is prone to flooding. More than 20,800 communities are in flood-prone areas, and about 6,100 of these communities have populations larger than 2,500. (These statistics are from the U.S. Water Resources Council, 1977. With the population growth in this country, these community and population figures have increased substantially.) The annual cost of floods, not to mention loss of life and property, is expected to be greater than four billion dollars by the year 2000, even with application of flood-plain management measures.

What Causes River Floods?

River floods are caused by heavy precipitation over large areas or by the melting of the winter's accumulation of snow and ice, or both. They take place in river systems whose tributaries drain large geographic areas. The flooding may range from a few hours to many days. The flow is influenced by the amount and distribution of the rainfall or snow cover, as well as the amount of soil moisture, vegetation, and the amount of ground covered by urbanization.

Where Do Floods Occur?

Floods take place somewhere in the United States in all seasons. Seasonal winter floods caused by rainfall take place in the east progressing northward from the Gulf Coast states in January to the Ohio River valley in March. Winter floods from storms originate on the western-facing slopes of mountain ranges in California, Oregon, and Washington.

Spring floods are common in the Great Lakes area, the Missouri and the Ohio River basins, the lower Mississippi River, the eastern slopes of Washington and Oregon, and the mountains in California and Arizona. Summer floods, which are usually local in extent, are likely to take place in any part of the United States (except on the west coast) and may occur as flash floods in arid regions and in the Rocky Mountains. On the basis of present knowledge, however, the exact size, time, and location of floods cannot be predicted much in advance, although the time of arrival of the crest of a flood at a particular site downstream can be estimated.

What are "Flash Floods"?

Flash floods are local floods of great volume and short duration and generally result from *cloudbursts* associated with severe thunderstorms in localized areas. The run-off results in high waters that carry a heavy load of sediment and rock debris. This flow can destroy roads, bridges, homes, buildings, and other community development in its path. The discharge can reach a maximum quickly and diminish almost as rapidly. Flash floods can take place in almost any area of the country. They are particularly common in the mountainous areas and desert regions of the West, where the slopes are steep, surface run-off rates are high, streams flow in narrow canyons, and thunderstorms are not infrequent.

The 1976 flood on the Big Thompson River in north-central Colorado resulted from thunderstorms and heavy rains in the Rocky Mountains. As much as 51 centimeters (20 inches) of rain fell in a short time on about 156 square kilometers (60 square miles), far more than could soak into the ground or be carried off by the river. The volume of water in the Big Thompson River swelled to four times its previously recorded maximum. The disastrous flood lasted only a few hours and ended quickly but left 139 people dead and did more than 35 million dollars in damages (Figures 16.4 and 16.5, page 353).

What Can be Done to Reduce Losses from Floods?

Efforts have been made by federal and state governments to reduce losses from floods. Flood-prone areas have been identified on more than 13,000 topographic maps. Also, detailed surveys have been made on sites of future urban development to evaluate the risk of flooding.

Flood control engineering, especially in the lower Mississippi valley, has been practiced for decades in hope of reducing the hazards of overbank flooding. The Mississippi is controlled by a vast levee system that has been continuously improved and now includes over 4,000 kilometers

(2,500 miles) of levees in places up to 10 meters (33 feet) high. Most levees are broad earth embankments, although walls of reinforced concrete are used where a city lies close to the river channel. During a major flood, the levee may be topped by the high fast-moving water. The overflow can rapidly erode the bank and cut a deep channel in the levee through which the flood waters quickly inundate the low flood plain.

Another engineering program to reduce the height of the river surface consists of creating channels in the river by artificial meander cutoffs (Figure 16.9, page 355). This shortens the course of the river and steepens the gradient, which lowers the height of the water at flood crest. There is no way to prevent other meanders from forming, however, and the process must be repeated at intervals to maintain security from floods.

Some of the projects to control floods have included reforestation of watershed slopes and crop planting in contour belts and terraces to increase the capacity of the surface to absorb rainfall and to reduce the rate of run-off. Construction of many small dams, usually of compacted earth that store flood waters temporarily can thus distribute the discharge over a long period. This also reduces soil erosion and increases recharge of the ground water zone.

In some areas, providing a new channel through which any dangerous excess of water can be diverted can forestall a flood. Mesopotamia, including modern Iraq, has a long record of floods dating back to the biblical time of Noah. The melting of snowfields in the mountains to the north has been an almost annual menace to Baghdad. At the time of Baghdad's great flood of 1954, a channel was already being dug from the Tigris to a large depression 70 meters (230 feet) deep between that river and the Euphrates. In 1957, Baghdad would have suffered an even greater disaster had not the rising water been drawn aside before the danger level was reached.

Despite various programs, flooding still occurs, and predictions of flooding are not yet completely effective because of the vagaries of weather forecasting. It seems that many people in high flood areas are uninformed about the risks they face. They seem to be vastly overoptimistic that their property will not be flooded, or maybe they simply accept that floods will occur where they live and that they must put up with it.

Questions

1. Hawaiian volcanoes are some of the most active on Earth, yet we rarely hear about loss of life caused by volcanic activity there. Why?

Hawaiian volcanoes usually have *non-explosive eruptions* with minimum explosive pyroclastic activity. They emit fluid outpourings of basaltic lavas, which usually move at a few kilometers per hour, allowing time to evacuate the premises. Lava, like water, flows downslope controlled by the topography of the land and is thus somewhat predictable. Barriers have been erected that divert lava flows, and sprays of cold water have chilled some flows into immobility. Even bombing from aircraft has been used with some success to slow the progress of flowing lava. During 1989 and 1990, however, flank eruptions from Kilauea inundated some fields and villages. With explosive volcanoes, such as Mont Pelée, Krakatoa, even Mount St. Helens, which emit the more viscous silica-rich lavas with high gas content, there is little defense except immediate evacuation at the first indication of activity. Indirect damage caused by extensive ash and mudflows can also occur swiftly and without warning. Ash and cinders are loose material and, with the rain that frequently accompanies volcanic eruptions, become mud that can flow even on very gentle slopes. Sometimes the direction of mudflows can be predicted, but lives are lost because many people refuse to leave their homes and fields.

2. How reliable is short-term earthquake forecasting?

Earthquake forecasting is a technique seismologists are working hard to perfect, not only in California, where major shocks are anticipated along much of the San Andreas fault, but in China, the U.S.S.R., southeastern Europe, the Mideast, Latin America, and Africa, places that are equally or even more at risk. Long-term forecasts are improving, but the key to short-term earthquake prediction is poorly understood. Historical records are of great value in predicting which segments of a fault are most likely to move. Maps have been made showing areas of ground movement and liquefaction around San Francisco Bay which indicate where the greatest destruction is likely to take place. There are many other faults in California, however, some of which have moved in recent years with few if any precursor events. Although most of the young faults have been mapped, the probability of earthquakes on most of them is still being investigated. All the predictive techniques require repeated surveys which are time-consuming, expensive, and limited to a few areas of study at a time. So far, no one method alone can predict the time, place, and severity of the next shock. The best we can do is prepare for the inevitable by making our personal environment as safe as we can.

3. Can we control earthquakes rather than just trying to predict them?

Some earthquakes have been accidentally triggered, which suggests that they might be controlled. Many small to moderate earthquakes, with magnitudes up to about 5.5, occurred near Denver beginning in 1962 and were believed to be caused by pumping waste fluids into a disposal well. Until that time, there had been few earthquakes in the Denver area. A series of seismic stations were set up to monitor the well. Analysis of the data indicated that the movements resulted from reduced friction that the injected fluids caused along faults. Artificial earthquakes have also occurred near

Lake Mead, since the building of Hoover Dam. It was suggested that the weight of the water and perhaps lubrication had caused movements on old faults.

As a result of the Denver experience, it has been suggested that a series of fluid injection wells be used to dissipate the strain along California's San Andreas fault by setting off a number of small harmless earthquakes or creeping movements. But the San Andreas crosses many densely populated areas. There is no way to accurately predict the magnitude of an artificial earthquake and the damage that might follow. Since it is not possible at present to distinguish between natural earthquakes and those caused by fluid injection, the lawsuits that might result could blame all damages on the injection wells. It seems prudent to seek better methods of forecasting and, for the present, let nature take its course.

4. **Dam breaks occur occasionally and cause catastrophic flooding. Can these be prevented?**

Some of the worst flood disasters in modern history have been caused by dam failures. In some cases, the breaks were the result of exceptionally heavy rainfall, but in others the problems were caused by poor construction or lack of understanding of the geologic conditions at the dam site. (Refer to the Vaiont dam disaster earlier in this lesson.) Most dams in the U.S. were built in the 1930s, and over 3,000 dams (mostly small ones) were considered to be inadequately maintained and repaired and likely to fail under stress, according to a federal assessment made in the late 1970s.

The May 1889 flood in the vicinity of Johnstown, Pennsylvania, resulted from the failure of a large dam about 21 kilometers (13 miles) upstream from Johnstown. The day before the dam failed, streams were swollen as a result of 15 to 20 centimeters (6 to 8 inches) of storm rainfall. The water level behind the dam rose above normal, and leaks formed in the earth embankment. About 3:00 P.M. on May 31 the dam suddenly burst. The resulting flood wave was estimated at 9 to 12 meters (30 to 39 feet) high. Seven small towns and Johnstown were in the path of the flood and were almost totally destroyed. More than 3,000 lives were believed lost.

Almost 90 years later, the failure of the Teton Dam in Idaho on June 5, 1976, caused an enormous flood upstream from American Reservoir. Water spread over more than 466 square kilometers (180 square miles), and damages reportedly were about 400 million dollars. The Teton Reservoir behind the 40-meter (130-foot) earth dam was being filled for the first time during spring 1976. About 7:30 A.M. on June 5, two seepage leaks were observed; by 11:57 A.M., water breached the dam and cascaded into the canyon. Some of the downstream communities were devastated by a 5-meter-high (16-foot) wall of water; other were buried under 7 feet of water, large trees, and swirling debris. By midnight on June 7, the crest of the main flood reached American Falls Reservoir about 160 kilometers

(100 miles) downstream from Teton Dam. American Falls Reservoir stored the entire flow in the river. Despite widespread destruction, only eleven lives were lost. Flood warnings by the U.S. Water and Power Resources Service, local radio stations, and law enforcement agencies enabled nearly all of the people to vacate the flood plain; this action saved hundreds of lives.

The case study for this lesson is the video portion, which concerns the earthquake at Loma Prieta, California, on October 17, 1989. Review the questions in this section to help you watch for important points and to prepare you for what you will see in the video.

CASE STUDY & VIEWING GUIDE:
The Loma Prieta Earthquake

Before viewing, answer the following questions:

1. Where was the epicenter located in relation to San Francisco?

2. Why were buildings in certain areas in San Francisco severely damaged while others a few blocks away survived almost unscathed?

3. What is meant by liquefaction, and why did it occur in San Francisco?

After viewing, answer the following questions:

1. What was the magnitude of the Loma Prieta quake?

2. What features were seen in the ground near the town of Santa Cruz during and after the shock?

3. Was there a clear break on the San Andreas fault? How was movement on this fault different from previous breaks along the San Andreas?

4. Was Loma Prieta the "Big One"? Will this relieve the possibility of another major earthquake during the foreseeable future? Explain.

5. Were there any precursors to the Loma Prieta quake? What was Dr. Anthony Smith investigating, and what did he discover?

6. What are some suggestions to make earthquake-prone areas at least relatively safe?

1. In terms of plate tectonic activity, the line of Cascade volcanoes is located near

   a. a continent-continent collision zone.
   b. a hot spot.
   c. a subduction zone.
   d. the San Andreas fault.

Self-Test

2. The prediction of a major eruption at Mount St. Helens was based on all of the following except a

   a. flurry of thousands of small earthquakes.
   b. flood of lava erupting from the central vent.
   c. series of ash and steam eruptions.
   d. bulge developing on the north flank.

3. The most destructive events at Mount St. Helens were

   a. the earthquakes during the eruption.
   b. the fast moving lava flows.
   c. boulders ejected from the central vent.
   d. flows of gases, ash, and mud down the flanks.

4. The destructiveness of an earthquake in any one area depends primarily on all of the following except the

   a. time elapsed between major shocks.
   b. severity and duration of the shaking.
   c. distance to the epicenter.
   d. nature of the substrate or ground.

5. An earthquake monitoring program has been established at Parkfield, California. This area was chosen because

   a. the largest earthquakes recorded in California took place there.
   b. even though Parkfield is on the San Andreas fault, earthquakes have never been detected there.
   c. earthquakes seem to occur here on fairly average intervals.
   d. this is an area of high population density, and accurate forecasting is especially important.

6. Mass wasting, as a geologic hazard,

   a. is a relatively rare natural event.
   b. happens only because of careless human activities.
   c. occurs widely, in every state in the nation.
   d. is restricted to mountainous areas where slopes are steep.

7. In the long run, the most effective method of control of mass wasting is

   a. terracing of highway cuts.
   b. preventing river and wave erosion.
   c. preventing water from infiltrating into the ground.
   d. regulating land use before human activities take place.

8. Major flooding by rivers is a hazard that must be considered as

   a. primarily a result of urbanization and paving of the ground.
   b. a natural and recurrent event.
   c. an event that is localized and a hazard only in restricted areas.
   d. well controlled and soon to be removed from the list of geologic hazards.

9. The increased financial losses caused by river flooding are mostly a result of

   a. increased encroachment of human activities on flood plains.
   b. larger floods in recent years, based on historic records.
   c. poor engineering and increased numbers of dam breaks.
   d. global warming and increased precipitation, especially in the southwestern states.

10. From what was learned in the Loma Prieta earthquake, one of the most important methods of preventing damage from earthquakes in the future would be to avoid building

    a. skyscrapers where earthquakes take place.
    b. in areas noted for many days of earthquake weather.
    c. near a tectonic plate boundary.
    d. where the substrate is fill or unconsolidated sand and silt.

11. All of the following methods have been used or suggested as a means of predicting earthquakes. The least dependable seems to be

    a. analysis of accumulated strain along a fault.
    b. the increase of emitted radon gas from water wells.
    c. analysis of restless animal behavior.
    d. increase in low frequency sound waves before a quake.

12. According to geologists with the U.S. Geological Survey, the most important information to relay to the public after an earthquake is the

    a. magnitude of the earthquake.
    b. strong possibility of aftershocks.
    c. location of the epicenter.
    d. fault on which the earthquake occurred.

13. During the Loma Prieta earthquake, the greatest damage and loss of life was caused by

    a. opening of fissures and cracks in the ground.
    b. damage to water, gas, and sewer lines.
    c. collapse of buildings, bridges, and other structures.
    d. landslides.

14. Major volcanic eruptions occurred in Japan and the Philippines during the summer of 1991. Although the volcanoes had not erupted for hundreds of years. geologists did not feel the eruptions were completely unexpected because

    a. both volcanoes are located near active subduction zones.
    b. both volcanoes had formed large mountains.
    c. volcanic lavas were abundant around the cones.
    d. the exact date of eruption had been predicted two years before.

15. Of all the geologic hazards, the kind most related to human activities is

    a. volcanic eruptions.
    b. river floods.
    c. earthquakes.
    d. landslides.

26

LIVING WITH EARTH, PART II

This lesson will help you understand how civilization affects Earth.

After reading the textbook assignments, completing the exercises in this study guide, and viewing the lesson's video portion, you will be able to:

1. Discuss how the continual increase in world population is affecting Earth.

2. Discuss the origin and significance of the observed changes in the ozone layer.

3. Explain why the atmosphere is warming and how that will affect Earth.

4. Give examples of resources being used indiscriminately.

5. Discuss some possible solutions to the negative environmental effects of uninformed use and development of global resources.

6. Recognize the relatively short period of time in which humans have had an impact on Earth's resources.

7. Give examples of positive steps that people can take to renew the environment.

Before we close this last chapter of *Earth Revealed*, we should look again at this beautiful blue planet, ancient as compared to the time span of modern human populations and powerful as compared to the activities of humankind. Yet, in a few sensitive places, the growing human population is making an impact on such a planetary scale that it will ultimately have an adverse effect on the thin skin of life called the *biosphere*.

Two million years ago, when humans beings first appeared, we inherited a world rich in natural resources: lush rain forests, good soil for farming, plentiful mineral supplies. It came with a protective covering – the ozone layer – that prevented ultraviolet light from harming the life below. And it was stocked with an astonishing diversity of species.

As recently as 10,000 years ago there were only from 5 to 10 million human beings living in small scattered Neolithic villages, not enough to exercise much influence on the environment in which they lived and worked. That condition lasted for almost 10,000 years, for it is only in the last decades that populations have increased to such an extent that major changes have taken place. In 1987, the population passed the five billion mark! In the next 35 years, there is an expected increase to 8.5 billion, of which 95 percent will be in the less developed countries. Some experts anticipate a doubling of the five billion figure in about 40 years! What changes can be anticipated by then?

For example, consider the question of energy: the human economy runs on a power flow of about 10 trillion watts, trivial compared with the *80,000 trillion watts* that flow to Earth's surface from the sun per second. The fossil fuels we burn for energy, however, have given off enough carbon dioxide (CO_2) to raise the atmospheric concentration of that gas by about 25 percent so far. To maintain the heat balance, the 80,000 trillion watts flowing in from the sun must flow out again. Carbon dioxide, although a minor gas in the atmosphere, is blocking the outflow of heat from Earth, and the atmosphere is apparently warming. Thus, while our energy use may seem trivial, it may well upset a critical condition that alters the heat balance of the entire planet.

In this lesson, you will examine some of the critical challenges we face, such as increased population, indiscriminate use of non-renewable resources such as petroleum and certain minerals, the growing hole in the ozone layer, the *greenhouse effect*, and the extinction of species at an unprecedented rate. The problems are not being ignored, but although some are being addressed, there is no intensive, coordinated plan to grapple with these issues and their devastating effects. Action must be taken to preserve and sustain the environment to assure the quality of life in the future.

LESSON ASSIGNMENT

Completing the following eight steps will help you master the lesson objectives and achieve the goal for this lesson:

Step 1: Read the TEXT ASSIGNMENT, Chapter 21, "Geologic Resources."

Step 2: Study the KEY TERMS AND CONCEPTS as noted in the study guide.

Step 3: Watch the VIDEO using the VIEWING GUIDE in the study guide.

Step 4: Read the study guide's PUTTING IT ALL TOGETHER section, which will help you summarize and integrate all of the information in this lesson.

Step 5: Complete any assigned lab exercises.

Step 6: Complete any assigned ACTIVITIES found in the study guide.

Step 7: Review the material in this lesson and complete the SELF-TEST found in the study guide.

Step 8: Go back to the LEARNING OBJECTIVES and make sure you have learned each one.

1. Read Chapter 21, pages 471-497. Be sure you understand the environmental effects of extracting and using the various resources.

2. In Box 18.3, pages 424-425, under "Changes in the Atmosphere," read about the greenhouse effect.

3. Study Box 21.2, page 485, "Alternate Sources of Energy."

4. Review the chapter "Summary." Consider the "Questions for Review" and the "Questions for Thought."

TEXT ASSIGNMENT

The key terms listed under "Terms to Remember," page 496 in your text, will aid your understanding of the lesson's material. Make sure you look up the definitions of these terms in the glossary or in the chapter.

KEY TERMS AND CONCEPTS

Once you have read the text material for the lesson and studied the terms, you are ready to view the lesson's video portion. Review the questions that appear in this section to help you watch for important points and to prepare you for what you will see in the video.

VIEWING GUIDE

Before viewing, answer the following questions:

1. What are the source materials of petroleum?

2. Are the source materials of coal the same as those for petroleum? Explain.

3. Explain the causes of the greenhouse effect. What are the beneficial effects, and what might be some of the harmful effects?

4. What chemical is believed to be the cause of the *ozone hole*, and what are its sources? What is the importance of the ozone global shield?

After viewing, answer the following questions:

1. What is the geologic origin of petroleum?

2. Describe some innovative techniques, such as the horizontal drilling project shown in the video, being used to increase our supply of petroleum.

3. Why is horizontal drilling so difficult?

4. Discuss the positive and negative aspects of alternative energy sources.

5. What can you do, as an individual, to maintain or improve environmental conditions in your community?

PUTTING IT ALL TOGETHER

You have read the text assignment and seen the video portion of the lesson. This section, made up of a SUMMARY, a CASE STUDY, the QUESTIONS, suggestions for ACTIVITIES, and a SELF-TEST, will help you pull the information together.

Summary

In the "Introduction," you learned of some of the effects of over-population. In this section, you will examine some of the specific problems brought about by the increased use of certain materials, some of which had never been in Earth's environment before the last few decades. It is as a global species that we are transforming the planet. It is only as a global species – pooling our knowledge, sharing the resources, and taking concerted action – that we have any prospect of managing the changes and directing them toward sustaining life. This is one of the great challenges facing humanity as it approaches the 21st century.

The Challenge of an Increasing Population

The tremendous growth of population and its assault on the environment is so recent that it is difficult to appreciate how much damage is being done. Through long ages, many societies encouraged population growth. Increased numbers of people added to the strength of the family and to the kingdom in which they labored. Death rates, however, were so high that the population increased very slowly. With the advent of machines, the productivity per worker increased, while the demand for products increased even faster. With the improved standard of living, the death rate declined and the population grew at an unprecedented rate.

Negative environmental changes became apparent, however, with the increased use of resources. Disastrous flooding, as one example, is caused in many areas by deforestation. Overlogging is directly related to the demand from expanding populations for building material, firewood, additional farmland, and foreign capital. Heavy floods have recently led Thailand to ban all logging, and Malaysia is considering doing the same, even though both countries depend heavily on timber and its products as an important source of employment and foreign exchange.

While the increase in population and urbanization presents serious problems, especially for the less developed countries, many scientists believe that absolute growth cannot continue forever. There must be some natural limit, some ultimate constraint. In Nigeria, for example, the population will double in about 22 years. If this were to continue for

470

the next 140 years, its population would be equal to that of the whole world today! The birth rate has slowed in some industrialized nations, as well as in Asian nations that have family planning programs. But even with careful management, rapid industrial development pollutes air and water and depletes natural resources at an ever increasing rate. Each country, as part of the world community of nations, will have to solve its problems of development without destroying the environment, an eventuality that may limit both population and economic advance.

The Antarctic Ozone Hole

In 1985, atmospheric scientists of the British Antarctic Survey published a completely unexpected finding: the springtime amounts of ozone in the atmosphere over Antarctica had decreased by 40 percent between 1977 and 1984. There was, in essence, an ozone "hole" in the polar atmosphere. A sudden decline in the ozone layer was indeed a cause for much concern.

Ozone – a three-atom molecule of oxygen – constitutes less than one part per million of the gases in the atmosphere. Ozone is continually formed and destroyed in the upper atmosphere. Its importance lies in the fact that ozone absorbs most of the ultraviolet rays from the sun, preventing them from reaching Earth. Until the ozone layer developed – created from the oxygen produced by primitive plants in the ocean – Earth was literally uninhabitable. Ultraviolet light, an extremely energetic form of radiation, can break apart important biological molecules, including DNA. It can increase the incidence of skin cancer and cataracts. It appears to damage the human immune system, and this makes us more vulnerable to infections. It may reduce crop yields and harm aquatic ecosystems.

As a result of the British report, investigations of the ozone hole were started using measuring instruments carried by balloons, satellites, and airborne laboratories, while other measurements were made from the ground. Various pollutants were cited as possible factors in the decrease of ozone, but in 1974, it was determined that the increasing use of compounds known as *chlorofluorocarbons* (CFCs) (which consist of chlorine, fluorine, and carbon) were the culprits. Before CFCs were developed in the 1930s, refrigerator compressors contained ammonia or sulfur dioxide; both chemicals were toxic, and leaks killed or injured many people. CFCs saved lives, saved money, and provided such niceties of modern life as air-conditioned buildings, untainted food, aerosol sprays, cleaners for electronic parts, foam used in home insulation, and that mainstay of modern society, Styrofoam™ fast-food containers. CFCs have hundreds of uses because they are relatively nontoxic, nonflammable, and do not decompose easily.

Only later did atmospheric scientists learn that CFCs not only contribute to warming of the lower atmosphere but destroy ozone in the upper atmosphere through a series of chemical reactions. Because millions of tons of CFCs were being released into the environment, it was feared that the compounds would accumulate in the atmosphere to a level capable of seriously eroding the *global* ozone shield. In 1978, the United States banned the use of CFCs in aerosol products. Later, an agreement was signed by 23 nations to freeze consumption of CFCs at 1986 levels by mid-1990 and to halve usage by 1999.

Even if all CFC emissions stopped today, chemical reactions causing the destruction of stratospheric ozone would continue for at least a century. There has also been some evidence that a hole is now appearing over the Arctic, a major concern since that would indicate a worsening of the decline of the ozone protective shield.

Atmospheric Warming and the Greenhouse Effect

While the depletion of global ozone seems to be the work primarily of a single class of industrial products, CFCs, several different emissions into the atmosphere combine to raise the possibility of a rapid greenhouse warming of Earth. There is some evidence already that the world is warming. Shifting climatic zones, changing precipitation patterns, receding glaciers, and rising sea level, make up a good part of this evidence. If these changes accelerate over the next few years, it would cause severe disruption of natural and economic systems throughout the world.

The changes in climate may be slow, and by all accounts they are irreversible for any time span of interest to us or to our children. There is no way to quickly grow great forests, cool Earth, freeze more ice in glaciers, or lower sea level. We cannot return immediately to an atmosphere with lower concentrations of greenhouse gases. The best we can do is to reduce current emissions and try to maintain the current balance. Even if we succeed, a further warming can be expected just from the effect of the heat-trapping gases already present. It is interesting to note that Congress has approved a Coastal Zone Management Act that requires states to plan for sea-level rise caused by global warming.

One problem with controlling atmospheric warming is that records have been kept over a relatively short period. Some scientists (and countries) want to move slowly, saying that the warming may be part of a natural cycle and that putting restrictions on emissions of carbon dioxide will curb industry and economic growth. This is the point of view adopted by the United States at present. Others agree, however, that if we wait for absolute proof that the greenhouse effect is real, we may have waited too long to control it. There may be some as yet unknown critical threshold in carbon dioxide concentration which, if crossed, will cause a rapid, perhaps catastrophic climatic change.

Global conferences on the issues are under way, especially in the developing nations that are poised for a massive increase in fossil-fuel consumption. Positive international action, such as an effective ban on atmospheric tests of nuclear weapons, has occurred. In 1987, the Montreal Protocol moved the world far along toward the elimination of CFCs. The issues of global warming, however, will persist through the next century and may dominate scientific, political, and technological considerations well into the future.

Heating the Atmosphere

To understand the greenhouse effect, we must understand how it works. The sunlight that reaches Earth does not heat the atmosphere directly but passes through it in the form of short wavelength radiation that heats Earth's surface. You have no doubt touched your car roof on a pleasantly warm day and found it hot enough to fry eggs. The inside of your closed car may become like an oven while the outside air is no more than comfortable. The sun's rays penetrate the glass of your car and warm the objects inside. The heated objects radiate long infrared wavelengths (heat energy) that cannot pass out through the glass. The radiation and re-radiation continue to warm objects and raise the temperature of the air within your car. This is the principle on which a greenhouse operates, heating the air inside while the air outside can be quite cold.

It is understandable why the inside of a car heats up, but why doesn't Earth lose the daily heat back into space? We must realize that it is the atmosphere's natural greenhouse effect that makes Earth a habitable planet, by modifying the daily variations in temperature and preventing excess heat loss at night. The moon and some of the inner planets suffer wide temperature ranges because they lack an atmosphere. Venus, of course, with its heavy CO_2 atmosphere and proximity to the sun, has a tremendous greenhouse effect and a surface hot enough to melt lead. Certain gases in Earth's atmosphere, such as carbon dioxide, methane, CFCs, nitrous oxide, ozone, and trace gases accumulating through human activities, act like the glass in a greenhouse and absorb infrared radiation from Earth's heated surface. The warmed gases radiate heat back to Earth, further warming the surface. This effect has been seen over cities for years, where burning of fossil fuel and high carbon dioxide emissions have created *heat islands* that average as much as 4 degrees hotter than outlying areas.

The total amount of carbon dioxide in the atmosphere has increased by about 25 percent since the Industrial Revolution. The actual amount of carbon dioxide in the atmosphere is a little more than 0.03 percent by volume. In contrast to nitrogen and oxygen, which make up over 99 percent of the atmosphere, it is that small amount of CO_2 and other trace gases that play an important role in the greenhouse effect. Because the total amount of these gases is so small, their concentrations are easily changed. In recent years, the concentration of other gases such as methane, nitrous oxide, and CFCs has increased to such levels that their combined effect approaches that of CO_2. And it is this imbalance that causes problems.

Sources of Carbon Dioxide

The largest source of carbon dioxide in the atmosphere is the burning of fossil fuels (coal, oil, and gas), which releases about 5 billion metric tons

(5.6 billion tons) of carbon into the atmosphere annually. Industrial nations contribute about 75 percent of the emissions. Therefore, steps toward stabilizing the composition of the atmosphere must begin in the industrialized world. (See Box 18.3, pages 424-425.)

The burning of Kuwaiti and other oil wells during the Persian Gulf War in the spring of 1991 spewed vast amounts of carbon dioxide, hydrocarbons, and other possibly noxious gases into the air. Most scientists believe this catastrophe will surely affect the atmosphere world-wide. This may be the test of whether global temperature is indeed strongly regulated by these greenhouse gases.

The other known major source of carbon dioxide is deforestation, primarily by burning, and decay of the organic material. The rate of destruction has increased in the last decade. The rain forests – which naturally absorb carbon dioxide at the rate of 13 pounds per year per tree – are being systematically destroyed, at the alarming rate of 27 million acres a year. One solution would be reforestation in millions of acres that could be returned to its previous state of forested land. The forests must be maintained, however, and never harvested or cleared for other land use. Tree planting has started in some urban areas, and the effect there might be even greater than in rural areas. A program to send fruit trees to Ethiopia, Kenya, Tanzania, and Cameroon is in effect; new nurseries will provide the double benefit of additional food for thousands of people and CO_2 removal from the air.

Geologic Resources are Non-Renewable

Geologic resources, except for ground water, are non-renewable. They form so slowly over geologic time that they cannot be replenished at the same rate as they are being used. This means that eventually these resources will run out or will become too costly for most uses. Recycling, finding substitutes, or making new discoveries will extend the life of some resources, but it is inevitable that some will run out.

The demand for energy, for example, has been increasing, especially in the developing countries, as they seek to industrialize, raise standards of living, and accommodate population growth. The demand is occurring even though the effects and the threat to the environment are becoming well known.

Humans are expending in one year an amount of fossil fuel that took nature roughly a million years to produce. Oil and gas originate from marine organisms such as diatoms and other algae that accumulate slowly in sea-floor muds. Sedimentation buries the organic matter and prevents oxidation; the organic matter is later compacted and subjected to elevated temperatures. The original algae and other organisms are no longer recognizable after long periods of disintegration, decomposition, and compaction. Eventually, droplets of oil and gas form and are squeezed out of the mud to migrate into more porous and permeable sandstones. This takes place over millions of years; and while oil may be forming today in some isolated basins, the process is so slow that such oil cannot be considered a resource. (See Figure 21.2, page 474, show-ing traps for oil and gas.)

Is the world running out of oil? There are estimates of enough world reserves of already-drilled oil to last about 42 years at present rates of use. The resources still in the ground should last well over 100 years. (See Figure 21.5, page 478.) Estimated reserves in the United States at the current rate of production constitute only a 25-year supply of oil unless new fields are discovered, alternate sources of fuel are developed, or we import more oil. In the past, similar forecasts of world oil deple-tion have been made, only to be revised upward with the discovery of new oil. Experts now believe few, if any, great new fields are left to be discovered, especially in easily accessible regions. With deep drilling and exploration of the more difficult geographic regions, perhaps this prediction will also have to be adjusted upward.

By the year 2030, 10 billion people are likely to live on this planet. If nations consume critical natural resources such as copper, cobalt, molybdenum, nickel, and petroleum at current U.S. rates, and if new resources are not discovered or substitutes developed, the resources will probably last until no later than 2040. This is not meant to be a "doom and gloom" forecast for the future but rather an incentive to recycle, conserve, and switch to alternative materials.

Positive Steps

With so many problems facing the world today, what are some of the solutions that can be applied? Of central concern is whether we can control the exploding population. Most of the population growth will be in the newly developing nations. To achieve any control will require coming to terms with numerous political, economic, religious, and other cultural issues. Nevertheless, education and inexpensive and easily available methods for population control will have to be carried to people in every corner of Earth. Some countries already have consid-erably slowed their birth rate through various kinds of incentive plans.

It has been suggested that replanting forests, on a global scale, would slow down atmospheric warming. Growth of enough plants to remove adequate amounts of carbon dioxide from the air will take con-siderable time, however, but it should be started now to enhance any future actions taken to slow down greenhouse warming.

Energy sources besides oil, gas, and coal must be seriously investi-gated, not only for conservation of the fossil-fuel resources but also to reduce their effects on global warming, acid rain, and other concerns. Wind power is used in places with some success. Tidal energy has been used as an alternate energy source, though on a rather small scale because places where facilities could be installed successfully are rela-tively uncommon. Geothermal energy is being produced in a few areas, and solar energy has some productive installations. Hydroelectric facil-ities are used world-wide and are important in certain areas for the generation of electricity. Perhaps ironically, increased use of nuclear power plants may be the most effective way to reduce carbon dioxide emissions from coal- and oil-fired power plants. Actually, what is needed is a global environmental organization with plenty of money that would be immune to political winds of change. A tax on certain uses of common world resources has been suggested as a means to this end.

On the other side of the ledger of resource use is the mounting (or rather "mountainous") problem of waste disposal. Using the predicted population figure of 10 billion people on Earth by 2030, and at the current rate of production in the United States, about 360 billion metric tons (400 billion tons) of solid waste will be produced every year, enough to bury greater Los Angeles 100 meters (330 feet) deep!

Many industries today are involved in recycling in such a way that the natural resource progresses to a finished product which, after use, enters the cycle again as a raw material. For example, a chunk of steel could potentially show up one year in a tin can, the next year in an automobile and 10 years later in the skeleton of a building. Thus, there are developments that could be cause for optimism in the concept of manufacturing recycling. Many communities, and even countries, have initiated garbage-sorting programs to reduce the amount of unrecycled waste. An effective system for collecting and segregating various consumer wastes can dramatically improve the efficiency of the industrial ecosystem without causing any environmental problems or hardships for the consumer.

We must realize that environmental concerns are a comparatively recent issue. NATO, the World Bank, and multinational corporations all have fairly short histories. The great international institutions in today's world are concerned with money, with trade, and with national defense. When concern for the environment becomes equally pressing, perhaps comparable institutions will be developed.

This is a beautiful planet, and we hope that through *Earth Revealed* you have come to appreciate the wonderful way in which the natural systems work. Since this is the only place in the universe that we know supports life, it is well worth every effort to keep the planet able to sustain life. Human ingenuity has led us from crude stone tools to space travel, communication systems, benefits, and comforts that could not have been imagined even a century ago. Undoubtedly, human ingenuity will again prevail and lead us to solve the problems that are affecting the quality of life on our planet home. It may take time, effort, and money, but isn't it worth it?

CASE STUDY:
Renewable
Energy Sources

Renewable energy sources have several advantages over fossil fuels. They are enormous and inexhaustible, produce little or no pollution or hazardous waste, and pose few risks to public safety. The United States already relies on renewable energy sources for about 7.5 percent of its energy needs. The majority of this energy comes from hydroelectric power.

Solar technologies have considerable potential since they can provide both electricity and heat. Home heating systems, solar water heaters, solar-thermal electric systems, and photovoltaic cells that generate electricity directly have all met with some commercial success in the 1980s. Most solar technologies are much less expensive today than just a few years ago, and they are likely to become even more competitive in the next few years.

Wind power – one of the oldest and most environmentally compatible energy sources – has become one of the most successful alternative energy systems. Between 1981 and 1988, over 15,000 wind turbines were installed in California, where most of the world's wind-power development has taken place. By 1987, these turbines generated about as much electricity as San Francisco consumes in a year. *Wind farms* are a wonderful sight and offer the possibility that at least 20 percent of the current United States electricity demand could be met by the whirling blades. Installations could be made on the Great Plains, along coasts, and in other windy areas. The persistent low-latitude trade winds have also been tapped as a source of wind energy in such places as the Hawaiian Islands.

Although there is considerable public concern about nuclear power and the waste generated, there is no doubt that nuclear power has the capability of meeting much of the present demand now satisfied by burning fossil fuels. To classify nuclear power as renewable assumes that we will build breeder reactors that produce more nuclear fuel than they burn and reprocess spent fuel from present reactors. If we were to follow this logic, we could stop uranium mining for 200 years while we used up supplies already mined.

Safety considerations have thus far prevented development of a national program for isolation of nuclear waste, but again, although the waste is dangerous and must be very carefully isolated, technical problems are well within our capability to solve. Politically, the problem is much more difficult.

Nuclear power is furnishing 70 percent of the energy needs in France today and 66 percent in Belgium. In the United States, about 20 percent of our electric power comes from nuclear power plants. Environmentally, the biggest plus for nuclear power is that it releases *no* greenhouse gases into the atmosphere and *less* radioactivity than coal-fired plants. It could well play a major role in our worldwide effort to reduce emissions of greenhouse gases. (Thanks to Dr. Robert Norris, professor emeritus, University of California – Santa Barbara, for contributing these paragraphs on nuclear power.)

Another important energy source is *biomass* use, which now accounts for 4 to 5 percent of annual United States energy consumption but could rise to about 15 to 30 percent. Already biomass is a major source of energy in some countries, such as Brazil. The raw material is agricultural, forestry, and municipal waste, plus plants and trees grown on *energy plantations*. Ethanol, derived from corn and other grains, is added to about 8 percent of gasoline and sold in a mixture called *gasohol*.

Biomass-derived gases such as methane and a mixture of carbon monoxide and hydrogen can be readily burned in existing natural gas-fired steam boilers and furnaces. Perhaps by the turn of the century, biomass fuels will become competitive with petroleum-based fuels. But as you are probably anticipating, there are certain problems with biomass energy sources. The biomass must be replaced by new growth so that the carbon dioxide released by combustion is reabsorbed and does not contribute to global warming. There are some other environmental risks, such as air pollution, toxic residues and ash, and depletion of soil nutrients. These issues will be studied and appropriate standards will be developed if biomass is to become a truly viable alternative energy source.

Renewable energy sources have other advantages. They are entirely a domestic resource and would be immune to foreign disruptions such as embargoes and wars. They would also tend to control the price inflation that will occur with the depletion of fossil-fuel reserves. In terms of global development, renewable resources with improvements in energy efficiency are estimated to slow the pace of global warming by 50 percent over the next 100 years – a very worthy goal.

Questions

1. How is the increase in human population affecting the rest of the biological world?

Even paleontologists and geologists – accustomed to the vast extinctions and climatic changes of the geologic record – are alarmed by the exceptional rate of human-caused extinctions on land. The human species came into being at the time of greatest biological diversity in the history of Earth. Today as human populations expand and alter the natural environment, they are reducing biological diversity to what will be its lowest level since the mass extinction at the end of the Mesozoic, 65 million years ago, when the dinosaurs disappeared. More than half of the species on Earth live in moist tropical forests. Already, the rain forest has been reduced to approximately 55 percent of its original cover and is being further reduced at a rate in excess of 100,000 square kilometers (36,000 square miles) a year, more than the area of Switzerland and the Netherlands combined. The global loss that results from deforestation could be as much as 4,000 to 6,000 species a year.

The loss of diverse species is unfortunately wholly irreversible. The value of the plants and animals of Earth remains largely unstudied. The living world is a potential source for immense untapped wealth in the form of food, medicine, and other important substances that may disappear before we have time to learn about their wonders.

2. How does coal form?

Coal is our third major energy resource (after oil and gas). It is used principally for generating electricity. One major advantage to coal is that it can be crushed, mixed with water to form a *slurry*, and transported through pipelines to be burned as a liquid fuel. Coal forms on land from plant growth in swamps, bogs, and lowlands, usually in tropical or temperate climates. In the coal beds are found plant fossils, roots, twigs, and trunks of trees – indicating that the coal forms where it is found (differing from oil, which migrates from its source materials to a reservoir rock).

Burial by sediment slowly compresses the plant material, which undergoes partial decay in the swampy environment. Increased and continuous compression drives out the water and various gases. The coal changes from a low rank called *peat*, to *lignite* or brown coal, to soft black coal, called *bituminous*, to *anthracite* or hard coal (Figures 21.8 A, B, and 21.9, page 482). Anthracite has limited distribution but is very valuable because it burns hot and smoke-free with very little ash (Figure 21.11, page 483). (Hard coal is actually a metamorphic rock associated with regional compression and folding. See Lesson 18 on metamorphic rocks.)

3. Are any environmental effects caused by extracting and using coal?

In your study of ground water contamination (Lesson 21), you learned that mine drainage from coal and metal mines can affect both surface and underground water. Mine drainage contains sulfuric acid formed by the oxidation of sulfur in the mineral pyrite (iron sulfide) when exposed to the air during mining. When coal is burned, it not only adds to the CO_2 in the atmosphere but also adds sulfur gases that contribute to acid rain and a host of minor contaminants, including some radioactive materials.

Mining of shallow coal beds by stripping off the surface overburden (Figure 21.10, page 482) is particularly destructive to the landscape. The process leaves piles of soil and rock waste that pose problems of erosion, surface and ground water quality degradation, and loss of the land for productive activities. Since some of the dumps contain sulfide minerals, they present a major problem for re-vegetation efforts.

New techniques have helped restore the overburden and topsoil so the mined areas can be used for other purposes. A strip mine area in Australia has been reshaped, covered with topsoil, and successfully restored to pasture. The area has attracted a large number of kangaroos and wallabies, and the ponds created for water recycling are supporting bird life, including ducks, ibis, and swans. Reforestation in the area also has been used in surface rehabilitation. With research and effort, the effects of strip mining can be mitigated in many places.

4. If global warming does occur, what realistically would be some of the results?

It is impossible to list all of the results, but some are fairly certain to occur. The waters in the heated ocean would expand; the ice caps of Greenland and Antarctica would diminish in size; glaciers around the world would recede. This would cause a rise in sea level, which is estimated to be about 1 to 2 meters (4 to 7 feet) by the year 2100 (Box 20.2, pages 466-467). Ocean water might encroach into fresh water systems, which would jeopardize drinking and agricultural water. Flooding would occur around the world. The dikes of low-lying Holland would have to be raised, and coastal cities, especially those built on or near deltas around the world would have to take special measures to prevent being inundated. Many of the beautiful low tropical islands would be swallowed up by the rising sea. Beach erosion would become a major danger, and areas like Miami Beach, with its luxury hotels, would be uninhabitable.

As severe changes in weather patterns occur, agricultural production will become increasingly uncertain. Will tropical and temperate conditions extend northward, possibly increasing the range of agricultural products? Will plants flourish with all the extra CO_2 in the air? On the other hand, what will happen to the present warm equatorial belt on Earth? The changes in agriculture will cause economic strains on many parts of the world.

There are other effects of climatic warming on the living world that must also be addressed. While habitat loss is most destructive to tropical life, warming will deeply affect life of the cold temperate and polar regions. A poleward shift of climate reasonably estimated at the rate of 100 kilometers (60 miles) per century would leave wildlife preserves and entire species behind, as many plants and animals could not migrate fast enough. Plants, especially, do not migrate readily. Will hundreds of thousands of species be displaced, or will they stay behind and adapt to the changing climate? These are serious questions to be asked when we consider global warming.

Activities

We have been discussing serious global problems caused by the effects of civilization on Earth. What can we, as individuals, do to help? The following list provides some simple activities which may sound trivial, but if done by great numbers of people would go far to mitigate some of the adverse effects of human civilization.

1. Stop junk mail. If you stopped all the junk mail you received this year, it would be the equivalent of saving one-and-a-half trees. If 100,000 people stopped their junk mail, it would be equal to 150,000 trees; one million people equals 1.5 million trees. Recycle your junk mail, and stop your name from being sold to large mailing list companies.

2. Plant a tree.

3. Snip six-pack rings. They are deadly in the marine environment since they are invisible underwater. Birds get caught in them and strangle or drown. Young seals get them around their necks, and as the animals grow, the rings slowly suffocate them. Cut them up before you put them in the trash, and become a committee of one to pick them up on the beach, in parks, or wherever you see them.

4. Save energy. Turn your water heater down to 130° F instead of the usual setting of 140° F. Insulate your water heater. Keep your house and office a little warmer in the summer so your air conditioner doesn't have to run so much. Microwave ovens use only a third to a half as much energy as conventional ovens. Use full loads in your dish and clothes washer. Don't use hot water: warm water, with a cold rinse cleans just as efficiently on clothes.

5. Cut down on CFCs and save the ozone layer. Avoid using polystyrene foam or Styrofoam™ made with CFCs as it is completely non-biodegradable. This includes form-fitting packing for electronic parts, coolers, egg cartons, disposable picnic goods, coffee cups, etc. If you don't know whether the product contains CFCs, ask. Eventually retailers will pass on your concerns to manufacturers. Don't buy aerosol cans containing CFCs. Lots of products come with non-aerosol vacuum pumps; they don't need gases and are just as easy to use.

6. Help save endangered species. Don't buy ivory jewelry or other elephant products. At the rate elephants are being slaughtered, they may become extinct in Africa by the year 2000! Don't buy tortoise-shell; reptile skins for shoes, purses, or belts; leopard or other wild cat skins; or other products from endangered species. Don't buy tuna that doesn't state it is dolphin safe. Try albacore or bonita – equally tasty and not caught in the *purse seines* that slaughter the dolphins.

7. Recycle. Get into the habit of sorting newspapers, glass, aluminum cans, tin cans (which are 99 percent steel) plastic soda bottles, telephone books. In the U.S. and Canada call the Environmental Defense Fund Hotline: 1-800-CALL-EDF. They can tell you about recycling programs in your area.

8. Change your transportation habits. Use public transportation wherever possible. Join a car pool.

9. Write your legislators about your environmental concerns.

10. Remember – every little bit counts. It's worth a try. Don't give up. Pass the word.

Self-Test

1. In 1987, the world population reached a critical point. The population is now closest to

   a. 5 million.
   b. 500 million.
   c. 5 billion.
   d. 50 billion.

2. The greatest population increases in the future are likely to be in

   a. the United States.
   b. the most industrialized countries.
   c. central Europe.
   d. the undeveloped nations.

3. The global ozone shield is an important defense against

   a. carbon dioxide emissions.
   b. infrared back radiation.
   c. ultraviolet radiation from the sun.
   d. excessive warming of the atmosphere.

4. Various substances have been suggested as affecting the ozone layer. As far as we know, the most important is

   a. carbon dioxide.
   b. chlorofluorocarbons.
   c. methane.
   d. nitrous oxide.

5. Important sources of the substances in Question 4 that are contributing to the ozone hole are

   a. use of Styrofoam containers and aerosol products.
   b. burning of petroleum products.
   c. burning and decay of forest products.
   d. mining activities.

6. The natural greenhouse effect

   a. modifies the temperature range on the planet.
   b. contributes to acid rain.
   c. was of minimal importance before the spread of humans on Earth.
   d. screened out harmful rays and permitted life to start.

7. The increase in what is known as the *greenhouse effect* correlates well with the increase in

   a. oxygen caused by activities of marine plants.
   b. reforestation in tropical countries.
   c. desertification in undeveloped countries.
   d. burning of fossil fuels.

8. As a result of the present greenhouse effect, all of the following seem to be occurring except

    a. atmospheric temperature is increasing.
    b. sea level has dropped by 1 to 2 meters (3 to 6 feet).
    c. glaciers are melting back.
    d. climatic zones are moving.

9. The causes of the present greenhouse effect include

    a. changes in sea level.
    b. increase in the amount of carbon dioxide in the atmosphere.
    c. increase in ultraviolet on Earth's surface.
    d. increased direct radiation from the sun.

10. Oil and natural gas originate from

    a. plants in swamps, bogs, and ancient forests.
    b. remains of dinosaurs and other giants of the land.
    c. deep sea deposits of volcanic origin.
    d. marine deposits of algae and planktonic organisms.

11. Based on the known oil reserves and the rate of world consumption, experts believe that there is adequate petroleum for about

    a. 7 years.
    b. 18 years.
    c. 42 years.
    d. 100 years.

12. The U.S. has about 6 percent of the world population, but it uses what percent of the world's oil production each year?

    a. Less than 6 percent
    b. About 30 percent
    c. Almost 66 percent
    d. Over 78 percent

13. Coal is a major energy resource. It is used principally for

    a. generating electricity.
    b. home heating.
    c. manufacturing of plastics.
    d. making of coke used in soft drinks.

14. Strip pit mining is one method of recovering coal. It

    a. does not require burning and is thus environmentally safe.
    b. requires deep mining and is rarely used today.
    c. does not have adverse affects on ground water supplies.
    d. disturbs the surface, removes topsoil, and may add sulfur products to the ground water.

15. The great attractions of wind, solar, and tidal power include all of the following except that they

    a. are renewable.
    b. are not dependent upon the politics of other nations.
    c. contaminate the land but are manageable.
    d. do not consume fossil fuels.

Appendix A

LABORATORY EXERCISES

Procedures and Guidelines for the Laboratory Component

To the Student:

Why does this course have a lab component, and what do we expect you to gain from doing the lab exercises?

As we developed this telecourse in geology, one suggestion that kept coming from advisors and geology instructors across the United States was that in order to get the most from this geology course, *Earth Revealed* must have a laboratory component. Geology is a very visual science, and you will gain much of the knowledge you need from the videos and from the textbook. But geology is also a science that requires hands-on involvement. To develop a true understanding of *geology*, you must begin to develop some of the basic skills of a geologist – mineral and rock identification, map interpretation, and interpretation of geologic formations. Understanding the work of geologists is perhaps more important to the citizens of our planet now than it ever has been. Geologists are studying Earth, hoping to find solutions to the problems of the environment that concern all of us. Work being done by geologists now will have a vital impact on the future of this planet.

You will be performing laboratory exercises at home using a laboratory manual, a set of required laboratory materials (either purchased or checked out from your school), and the set of additional instructions and guidelines on the following pages. Most of you will be off campus when you complete the lab activities and therefore unable to receive the close personal instruction which is often required to develop geology skills. We hope to help with this by providing you a procedure and some guidelines to approach the laboratory component of this course. The guidelines will supplement the laboratory manual and provide assistance in completing the laboratory exercises.

Recommended Procedure for Completing the Laboratory Exercises

This procedure was developed with the help of professional educators as well as students who actually performed the exercises. If you follow it carefully, you should master the information and skills necessary to complete the lab exercises successfully.

1. Check for any required materials that may be listed at the beginning of the AGI lab manual exercise. If you don't have all the materials, contact the telecourse instructor.

2. Locate the questions for each exercise in your AGI lab manual, and read them carefully. While you read the exercise, refer to the set of additional instructions for each lab exercise we have provided in this Appendix.

3. Reread the textbook material that covers the same topic as the exercise. Again, keep in mind the questions you are going to have to answer so that you can focus your reading.

4. Be sure to follow both the AGI lab manual and the additional instructions as you work through the exercise.

5. When you have finished the questions in the AGI lab manual, check to see if the lab guidelines we have given you include any additional questions to enhance your understanding of the exercise you are doing.

Required Laboratory Text

Laboratory Manual in Physical Geology, Second Ed., AGI/NAGT Merrill Publishing Company

Order of Laboratory Exercise

Lab Exercise No.	AGI Exercise No.	Exercise Title	AGI Page No.
1	5	Topographic Maps	62
2	15	Structural Geology	154
3	18	Introduction to Plate Tectonics	179
4	17	Earthquakes	174
5	14	Relative Ages of Rocks	147
6	1	Mineral Properties and Identification	1
7	2	Igneous Rocks and Processes	22
8	3	Sedimentary Rocks and Processes	33
9	4	Metamorphic Rocks and Processes	52
10	7	Ground Water	82
11	10	Work of the Wind	118
12	12	Alpine Glaciation	130
13	9	Shorelines and Coastal Processes	105

LABORATORY EXERCISE 1:

Topographic Maps

Use Lab Exercise Five: "Topographic Maps," in your AGI lab manual for this lesson.

We have a few suggestions that should help you as you begin this lab activity.

1. Be sure to follow the recommended procedure for completing the laboratory exercises on page 486 of this appendix.

2. Look through the information presented below before attempting to complete any questions. There may be some practice questions or additional examples for you to look at. Remember that if you can do the practice examples, you will perform better on the required questions.

3. Don't try to answer the questions by writing in your book; there isn't enough room. Use a separate piece of paper.

4. When you need to mark on a diagram in your laboratory manual, first make a copy and work out an answer on your copy. Then complete the question in your manual as neatly as possible.

In this section, you will find a list of numbers corresponding to the questions in your AGI lab manual. After each number you will find information on how to proceed with the question. If you have enough information to proceed without difficulty, you will see "No further information needed." If you need additional information, the necessary information will be provided for you. If the question is one that would be difficult to complete without assistance from an instructor, you will see "Omit this question."

1. Remember that mean sea level has an elevation of zero.

2. Plotting contour lines is a skill that will develop with practice. Read AGI Figure 5.8 to learn the rules for contour lines. Then before you complete the contour lines on AGI Figure 5.13, we would like you to work on Figures 1, 2, and 3 at the end of this section of lab notes. Figures 1 and 2 will help you get started; use a pencil and remember to draw smooth curving lines that do not cross. Figure 3 is similar to the exercise in your lab manual in that you are not given the actual points where the contour lines are to be drawn. Remember

that for Figure 3 the required contour interval is 10 feet – your lines should be at 10, 20, 30, 40, and 50 feet. Now complete AGI Figure 5.13.

3. *This question will probably require the most time to complete.* You will be plotting topographic profiles in many of the lab exercises during this course. Since you will get better with practice, complete the following before attempting the profile on AGI Figure 5.14: Go to AGI Figure 5.11 and use your own paper to develop the profile for section line A-A'. Read the instructions at the bottom of page 73 in your AGI manual to help you. Your profile should match the profile on AGI Figure 5.11. Before continuing, be sure to practice on this profile until you understand the procedure for plotting profile sections. *Now* complete the profile on AGI Figure 5.14. Project the contour line locations straight down to the profile sheet.

4. Refer back to AGI Figure 5.1.

5. No further information needed.

6. No further information needed.

7. No further information needed.

8. The *distance* is the difference between the answers for questions 4 and 5.

9. No further information needed.

10. No further information needed.

11. No further information needed.

12. No further information needed.

13. The AGI lab manual text does not explain the symbol for magnetic north located at the bottom of your map.

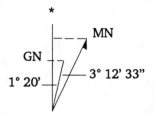

    * – Indicates the location of true north as located by the star, Polaris.
GN – Geographic North.
MN – Magnetic North.

The angles between true, geographic and magnetic north are given in terms of degrees (°), minutes (') and seconds ("). Example: 3° 1 2' 3 3" indicates an angle of 3 degrees, 12 minutes and 33 seconds.

14. No further information needed.

15. No further information needed.

16. Review the section on *scale* on page 66 of your lab manual before attempting this question.

Questions 17, 18, and 19 require basic math skills and the use of a calculator.

17. Remember, depending on the scale of the map, the map ratio scale will show you that one unit on the map will equal a specific number of units on the ground. These units can be either inches or centimeters for this lab exercise. To make conversions, follow these guidelines:

To convert inches on the map to ground distances:

1 inch on the map x map scale = inches on the ground.

Example: 3.2 inches on the map is how many inches on the ground?

(3.2) x (24,000) = 76,800 inches on the ground.

To convert this to feet: 76,800 ÷ 12 = 6,400 feet.

To convert this to miles: 6,400 ÷ 5,280 = 1.21 miles.

Therefore, 3.2 inches on the map equals 1.21 miles on the ground.

To convert centimeters on the map to ground distances:

1 centimeter on the map x map scale = centimeters on the ground

Example: 24 centimeters on the map is how many centimeters on the ground?

(24) x (24,000) = 576,000 centimeters on the ground.

To convert this to meters: 576,000 ÷ 100 = 5,760 meters.

To convert this to kilometers: 5,760 ÷ 1,000 = 5.76 kilometers.

Therefore, 24 centimeters on the map equals 5.76 kilometers on the ground.

18. No further information needed.

After completing question 18, use the map scale at the bottom of the map to check the results. Measure 3 inches on the map scale. Does it equal approximately 6,000 feet, 1.13 miles, and 1.88 kilometers?

19. Use the values from question 17 to calculate these answers.

20. No further information needed.

21. No further information needed.

22. No further information needed.

23. The names of the adjacent maps are given along the edges of the map and are in parentheses. In AGI Figure 5.4, the adjacent map to the east is San Bernardino.

24. No further information needed.

25. No further information needed.

26. No further information needed.

> For questions 27-29, review the discussion of *relief* on page 70 in your lab manual. You will have to hunt to find some of the elevations, but they are there.

Before continuing to questions 30-42, study the example below to be sure you can come up with the same location descriptions as the ones given in the example.

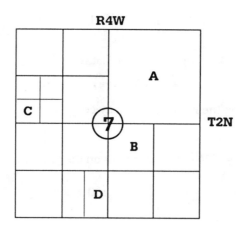

Location of A: NE¼, Sec 7, T2N, R4W

Location of B: NW¼, SE¼, Sec 7, T2N, R4W

Location of C: SW¼, SW¼, NW¼, Sec 7, T2N, R4W

Location of D: E½, SE¼, SW¼, Sec 7, T2N, R4W

> For questions 33 and 34: If using the map scale to measure the map distances does not seem sufficiently accurate, make conversions as you did in questions 17, 18, and 19.

33. Use the scale presented on the map (AGI Figure 5.4) and measure the required distance.

34. Since this is not a straight line, use a piece of string to measure along the creek and then measure the string using the scale on the map.

35. No further information needed.

36. No further information needed.

37. No further information needed.

38. Review the discussion of *relief* on page 70 in your lab manual.

39. No further information needed.

40. No further information needed.

41. No further information needed.

Here is another question that is not in the AGI manual. It should help your learning and application of the information.

Review the section on compass bearings on page 63 and in Figure 5.2 on page 64 in your AGI manual.

42. Using AGI Figure 5.4 on page 67, determine the bearing of a line which begins at Grapeland and extends to the Day Canyon Guard Station.

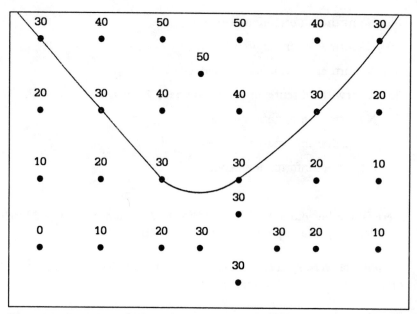

Figure 1   CONTOUR INTERVAL = 10 ft.

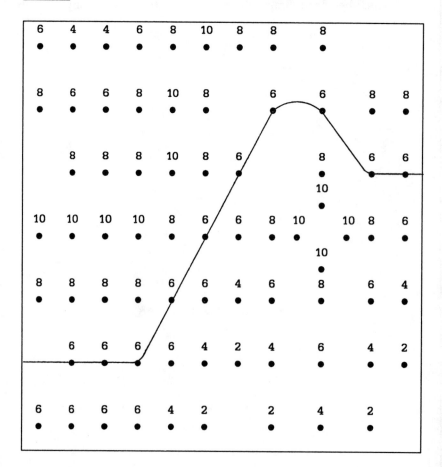

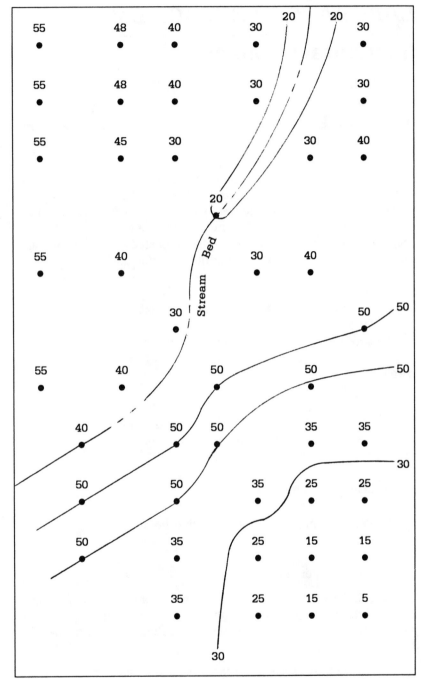

Figure 3   CONTOUR INTERVAL must be equal to 10 ft.

LABORATORY EXERCISE 2:

Structural Geology

Use Lab Exercise Fifteen: "Structural Geology," in your AGI lab manual for this lab activity.

> An understanding of structural geology is vital to an understanding of subsequent laboratory exercises. In terms of time requirements, this exercise is one of the longest you will perform. Activities and questions could easily take five hours. Be sure to allow ample time for this exercise. It does not have to be completed all at once; perhaps one hour a day for five days might fit your schedule better.

ADDITIONAL INSTRUCTIONS

Here is a listing of information not in the AGI lab manual that will help you perform the activities in this exercise.

Be sure to read Chapter 15, on geologic structures in your course textbook; this chapter provides excellent background material for this exercise.

PART 1: STRUCTURE MODELS

The six structure models are in the back of the lab manual. Carefully remove them and follow the instructions on page 159 in the AGI lab manual to form the models properly.

Model 1 General Comments

The general comments are not numbered because they do not apply to a specific question, but to the whole activity.

A protractor is required for this activity.

On page 159 in the AGI lab manual, the second formation which is discussed is not the gray formation directly to the east (right) of the yellow formation. It is the thin vertical blue formation with the black dash lines.

Be sure to review AGI Figure 15.1 for strike and dip information and AGI Figure 15.10 for map symbols.

Model 1 Question

1. No further information needed.

Model 2 General Comments

See AGI Figure 15.1 for an explanation of how to measure the dip angle and how to write the angle on a strike and dip symbol. Be sure to measure the dip angle at locations I, II, III, and IV and to write these angles at their proper locations on the model.

Model 2 Questions

2. No further information needed.

3. No further information needed.

4. No further information needed.

Model 3 General Comments

Again, you will need to measure dip angle at locations I, II, III, and IV and write these angles at their proper locations.

Model 3 Questions

5. No further information needed.

6. No further information needed.

7. No further information needed.

8. No further information needed.

9. No further information needed.

Model 4 General Comments

For Models 1, 2, and 3, the strike direction at each point was the same and we assumed that the dip at each location was the same as the end cross-sections. For Models 4, 5, and 6, the strike and dip at each location will be different, so be careful. The dip angle can be determined only if you can see or assume the shape of the cross-section. (See number 12 , AGI Figure 15.9.) Therefore, you need to indicate only the dip direction and should not try to measure the dip angle.

Model 4 Questions

10. Compare the north and south ends of models 2 and 4 and describe the differences on the strike and dip at points I, II, III, and IV.

11. The unconformity is not at the base of H as stated in the AGI lab manual. Answer the question for the structure labeled I at the southeast corner of the model.

12. No further information needed.

13. View the model from the south and observe the horizontal bedding of the unconformity. Place the proper symbol for this orientation on top of the model at the location labeled I.

Model 5 General Comments

The strike and dip symbol at each location (I-V) will be in different orientations.

Model 5 Questions

14. No further information needed.

15. No further information needed.

Model 6 General Comments

Before answering questions 16 and 17, look in your text on page 136 for the fault which has both strike-slip and dip-slip components.

Model 6 Questions

16. Answer for the type of vertical movement.

17. Answer for the type of lateral movement.

18. Omit this question.

19. Omit this question.

PART 2: GEOLOGIC MAPS:

> Note on geologic maps: As you look at the geologic maps in AGI Figures 15.11 and 15.12, it would be easy to be overwhelmed by their "busyness." Don't panic. There is a vast amount of information contained in these maps, and if you will proceed slowly, you can extract that knowledge. The colors you see on the maps may not match the colors in the explanation key exactly. You will notice that the geologic units in the key have letter symbols on them. These also appear on the map, and you should rely on them more than the colors when locating a formation. We know this exercise is long and difficult, but if you keep at it, you should complete it successfully.

Geologic map, page 163 in the lab manual: For this map, the direction of north is toward Hillers Butte and the direction of south is along the bottom of the map toward Coconino Plateau. This is the opposite of what is stated in the legend for AGI Figure 15.11 on page 162.

20. Look for the symbols for the units, and remember that the colors do not match exactly. Review "The Geologic Time Scale" presented on page 216 in the AGI lab manual. For this question, locate the oldest formation on AGI Figure 15.11 and then locate the same formation on the map. This formation occurs at two locations, but only one is marked with the proper symbol.

21. No further information needed.

22. You must use color to find this formation. Look in the southwest portion of the map.

23. No further information needed.

24. No further information needed.

25. *Cliff-forming units* are units that are resistant to erosion, and they typically form cliffs or near-vertical changes in elevations. To help you answer this question, look at the spacing of the contour lines at Bradley Point, Sumner Point, and Johnson Point.

26. Omit this question.

27. No further information needed.

28. No further information needed.

29. No further information needed.

> For questions 30 and 31: Remember that north on the map on page 163 of the lab manual is toward Hillers Butte.

30. No further information needed.

31. No further information needed.

32. There is a very small displacement in the pipe fault near the letter "P" in pipe. Since there is no displacement south of this point, when was this fault active?

33. Refer to "The Geologic Time Scale" on page 216 in your lab manual. Note the location of the Silurian Period and then try to find it on AGI Figure 15.11. What are the possible reasons you cannot find any Silurian Period formations in the area?

Geologic map on page 165: Be sure to read the information after question 41 in your AGI lab manual to gain an understanding of what the red lines represent.

34. No further information needed.

35. No further information needed.

36. No further information needed.

37. No further information needed.

38. You must use the information from AGI Figure 15.12 (the geologic map on page 165) and the relationship of the contour lines to complete this question.

39. No further information needed.

40. No further information needed.

41. No further information needed.

42. No further information needed.

43. No further information needed.

44. No further information needed.

45. No further information needed.

Stop after question 45 unless you are requested to work on the "Additional Questions" by your instructor.

LABORATORY EXERCISE 3:

Introduction to Plate Tectonics

Use Lab Exercise Eighteen: "Introduction to Plate Tectonics," in your AGI lab manual for this lab activity.

> In this exercise you will enhance your understanding of plate tectonics. You will use information obtained from previous exercises, particularly AGI Exercise Fifteen on structural geology. You may wish to review the information on faulting that you have learned before.

QUESTION-BY-QUESTION INSTRUCTIONS

In this section, you will find a list of numbers which correspond to the questions in your AGI lab manual. After each number you will find information on how to proceed with the question. If you have enough information to proceed without difficulty, you will see "No further information needed." If you need additional information, the necessary information will be provided for you. If the question is one that would be difficult to complete without assistance from an instructor, you will see "Omit on this question."

When reviewing AGI Figure 18.5, be sure to look for the names of each plate and not just at the colors. There are more than seven colors on this figure, but there are only seven major plates shown.

For questions 1 and 2: "Complete" means fill in the blanks.

1. For additional information, review the figures presented on pages 136 through 138 in your textbook.

2. No further information needed.

> For questions 3, 4, and 5: Review Figure 7.20 on page 163 of your textbook.

3. No further information needed.

4. No further information needed.

5. No further information needed.

6. For a check of your answer: How do your arrows compare with the arrows on Figure 7.20 on page 163 of your textbook?

7. Make sure you review your lab manual and textbook reading assignments.

8. No further information needed.

9. Measure the distance on AGI Figure 18.7 using the scale at the bottom of the figure. This will only be an approximate distance. If you need help with the conversion of kilometers to centimeters, review the guidelines for question 17, AGI Lab Exercise Five, Topographic Maps. Then study the conversion chart opposite page one of your AGI lab manual to see how 1 kilometer = 3,280 feet. To calculate the rate of movement, divide the distance measured by 25 million (25,000,000) years.

10. No further information needed.

11. How many 5 meter movements are required to equal the distance you measured in question 9? Use the total number of 5 meter movements and the 25 million years to find how often this type of movement must occur.

12. Review the procedure you used for question 9. The distance you should measure is from the center of the island of Hawaii to the center of the island of Kauai. Try to measure a distance which passes through the centers of the remaining islands. You will, therefore, be measuring a curving line rather than a straight line. Refer to the additional instructions for AGI Laboratory Exercise Five for guidelines for measuring distances that are not straight.

LABORATORY EXERCISE 4:

Earthquakes

Use Lab Exercise Seventeen: "Earthquakes," in your AGI lab manual for this lab activity.

Before you begin working on the questions for this exercise, you will need to work through the Dallas example on page 177 of your AGI lab manual. Work this example in detail, and you should not have any problems completing this exercise. Be sure to draw your lines on Figure 17.1 for pratice. Also, review the information presented on pages 151 and 152 in your textbook.

In this section, you will find a list of numbers which correspond to the questions in your AGI lab manual. After each number you will find information on how to proceed with the question. If you have enough information to proceed without difficulty, you will see "No further information needed." If you need additional information, the necessary information will be provided for you. If the question is one that would be difficult to complete without assistance from an instructor, you will see "Omit this question."

1. Remember that you are estimating your times to the nearest tenth of a minute.

2. No further information needed.

3. a. Using the latitude and longitude information given, you are required to locate the three seismic stations on AGI Figure 17.4. Take your time and remember the approximate locations of North Carolina, Alaska, and Hawaii.
   b. You will be using a drafting compass to draw your circles. Your three circles may not meet exactly at a single point.

4. No further information needed.

5. For the Dallas example, the L-waves would begin to arrive approximately 9.8 minutes after the origin time of the earthquake or at approximately 8:14.5. You can determine this value by projecting the vertical line on AGI Figure 17.1 upward until it intersects with the L-wave curve. At this point, project a line horizontally to the time axis. This will give you the number of minutes after the origin time of the earthquake that the L-waves began to arrive at the seismic station.

LABORATORY EXERCISE 5:

Relative Ages of Rocks

Use Lab Exercise Fourteen: "Relative Ages of Rocks," in your AGI lab manual for this lab activity.

> Be sure to allow yourself enough time to read the captions and to study the figures for this exercise. They are full of information and are an excellent way to develop your understanding of relative ages of rocks and of geologic events. Be sure to review Chapter 8 on time and geology in your text.

QUESTION-BY-QUESTION INSTRUCTIONS

In this section, you will find a list of numbers which correspond to the questions in your AGI lab manual. After each number you will find information on how to proceed with the question. If you have enough information to proceed without difficulty, you will see "No further information needed." If you need additional information, the necessary information will be provided for you. If the question is one that would be difficult to complete without assistance from an instructor, you will see "Omit this question."

1. No further information needed.

2. This question asks you to summarize the 13 items presented on pages 148 through 150 of your lab manual. These items refer to the relative ages of sedimentary, igneous, and metamorphosed rocks. Be sure to include a description of how faults can be used in relative dating and summarize the principles presented on page 147 of the lab manual.

3. Omit this question.

4. AGI Figure 14.15 has two "C" units of different colors. Relabel the lower right, lighter colored unit with "O." This is an excellent exercise. Be sure to identify the zones of contact metamorphism. Mark these zones on the figures, and write a short description for each zone. You should also describe any unconformities that are present.

> For questions 5 and 6: These questions require you to describe how the pages of a book and the layers of a rock body are similar. This is meant to be a thought question; think about how you would use each one to gain information.

5. No further information needed.

6. No further information needed.

7. No further information needed.

8. This question asks if there must always be surface erosion for all unconformities, or if there might be another reason for the missing strata.

LABORATORY EXERCISE 6:

Mineral Properties and

Identification

Use Exercise One: "Mineral Properties and Identification," in your AGI lab manual for this lab activity.

Before you begin this exercise, you should read the section "The Physical Properties of Minerals" on pages 204-210 of your textbook. Be sure to pay close attention to the discussions on luster, hardness, and streak as you will need to use this information when you begin to identify mineral samples. There are many new terms contained in your laboratory manual reading for this exercise and you may need to make yourself a list of the words which are in **bold** type.

BEFORE YOU GET STARTED

We have a few suggestions that should help you as you begin this lab activity.

1. Study the section on "Hardness of Some Common Objects" on page 3 in your lab manual.

2. Read all the captions beneath the figures in this exercise in your lab manual.

3. AGI Figures 1.22 and 1.23 are good flow charts. Study them to help you in your identification of minerals.

4. AGI Figures 1.25, 1.26, and 1.27 contain a lot of valuable information. Be sure you know how the listings differ from each other.

5. For your safety, you will not be using hydrochloric acid as listed in the materials on page 1 of your lab manual.

QUESTION-BY-QUESTION INSTRUCTIONS

In this section, you will find a list of numbers which correspond to the questions in your AGI lab manual. After each number you will find information on how to proceed with the question. If you have enough information to proceed without difficulty, you will see "No further information needed." If you need additional information, the necessary information will be provided for you. If the question is one that would be difficult to complete without assistance from an instructor, you will see "Omit this question."

1. No further information needed.

2. No further information needed.

3. Rather than naming two additional properties, make a table of mineral properties for both halite and calcite.

4. No further information needed.

5. No further information needed.

6. No further information needed.

7. To identify your mineral samples, follow the outline presented on page 10 of your laboratory manual. Be sure you record the identification letter of each sample. Since you will be performing this process on your own, here are a few instructions to help you identify your samples correctly.

    a. Follow the outline on page 10, but remember: *your most important resources are AGI Figures 1.25, 1.26, and 1.27.*

       The main identification items are:

       | Luster | Streak |
       | Hardness | Cleavage |

       The remaining identification items discussed in your reading will be used in cases where you need more information to identify the samples properly.

    b. **Luster:** As directed in procedure A on page 10 in your lab manual, you should first separate the samples into two main groups: metallic luster and nonmetallic luster. Remember, if you are unsure about a sample's luster, then it probably has a nonmetallic luster. Once you have these two groups separated, you will need to divide the nonmetallic samples into groups of light- and dark-colored nonmetallic samples. Be sure to review the pictures of the mineral samples contained in your lab manual and pay close attention to the luster descriptions.

       Do not try to identify your samples from the pictures Many beginning students try this, but they often end up identifying samples incorrectly.

    c. **Hardness:** To get correct results, you must do this test firmly. Press the edge of one sample against the smooth surface of another; then slowly and firmly pull the edge across the smooth surface. Use care and be sure that you identify which sample is doing the scratching. Sometimes a softer material will rub off on a harder one and will appear to have scratched the harder one. Rub your finger across the scratch; if it remains, then there truly is a scratch. Determine the approximate hardness of your samples by trying to scratch them with the items supplied in your lab kit. Remember to study the hardness of common objects on page 3 of your lab manual.

d. **Streak:** As with the hardness test, remember to do this test firmly. Pull the sample across the streak plate (the small white plate in your lab kit) and observe the color of the powdered mineral. If the mineral is harder than the streak plate (about 6 hardness), then there will be no streak and the mineral sample will scratch the plate. As you will note from AGI Figures 1.25 through 1.27, the streak test is used mainly for minerals having metallic or submetallic luster.

e. **Cleavage:** This is a difficult property for beginning students to understand. Review the pictures of the mineral samples in your lab manual and note the comments on cleavage and fracture. Also use AGI Figure 1.24 as a guide on the type of cleavage a sample may display. If a sample does not have a defined cleavage pattern, then you must consider the property of fracture. Ask yourself "Does this sample show cleavage?" If it does, then determine the type of cleavage. If it does not, then look for the type of fracture the sample may display. Be sure to review AGI Figure 1.12 for a discussion of basal cleavage.

You should be able to identify your samples using the identification items discussed above. If you need additional information for your identification, look at the remaining properties discussed in your text and laboratory manual reading.

8. No further information needed.

9. and 10. Draw your conclusions for these questions from AGI Figures 1.25 through 1.27 and the information on luster presented above.

11. No further information needed.

12. No further information needed.

13. No further information needed.

Stop after question 13.

LABORATORY EXERCISE 7:

Igneous Rocks and Processes

Use Lab Exercise Two: "Igneous Rocks and Processes," in your AGI lab manual for this lab activity.

Follow the recommended procedure for completing the laboratory exercises on page 486 of this Appendix. This the first of three laboratory exercises which require you to identify rock samples. Follow the instructions carefully, and use your hand lens to look at the crystals closely.

BEFORE YOU GET STARTED

We have a few suggestions that should help you as you begin this lab activity.

1. Carefully read the text in your AGI lab manual. Make a list of the new terms included there (printed in **bold**) and learn their meanings.

2. Study the figures containing photographs and read the captions closely. The information in the captions is very important. The smaller photomicrographs presented along with the photographs of the rocks are enlargements of the actual crystalline structure of the rocks. These photomicrographs are taken using a microscope and a special light called polarized light. Note the difference in the crystal sizes and be aware that the color of the crystals may be distorted due to the polarized light.

3. Use AGI Figure 2.1 when you begin identifying your rock samples. This figure contains most of the information you will need to make your identifications. Make sure you know all the terms and percentages given.

QUESTION-BY-QUESTION INSTRUCTIONS

In this section, you will find a list of numbers which correspond to the questions in your AGI lab manual. After each number you will find information on how to proceed with the question. If you have enough information to proceed without difficulty, you will see "No further information needed." If you need additional information, the necessary information will be provided for you. If the question is one that would be difficult to complete without assistance from an instructor, you will see "Omit this question."

1 and 2. Refer to the top of AGI Figure 2.1. You are looking for a range of values rather than a specific number. Try to match the composition of the rocks to the range on the color index. For example, andesite is an intermediate rock with a color index ranging from 15 to 40 percent.

3. No further information needed.

4. Refer to the section on porphyritic textures in your textbook.

5. AGI Figure 2.2 is Bowen's reaction series and illustrates the sequence in which minerals crystallize as high temperature magma or lava cools to lower temperatures. The magma types are along the right edge and the temperatures are along the left. Refer to AGI Figure 2.1 for the types of rock which can develop from each type of magma.

   a. AGI Figure 2.3 indicates the location of the asthenosphere. You should assume that the temperature of the magma is above the 1,200° C in AGI Figure 2.2. Look at the composition of the rock in AGI Figure 2.4 and try to describe the steps in its formation as the very hot magma begins to cool.

   b. For this part of the question, substitute the word composition for the word chemistry. Use the color index on AGI Figure 2.1 to find the light colored composition. Using this composition, refer to AGI Figure 2.2 and estimate the lowest melting temperature for this material.

   c. Compare the specific composition determined from AGI Figure 2.2 to the composition of granite from AGI Figure 2.1.

6. Follow the outline on page 23 of your lab manual for identifying your samples of igneous rocks. The captions and photographs are a good resource for identifying unknown samples.

Remember: Do not simply try to match your rock samples to the pictures. This does not work, because color alone is not reliable for identification purposes.

> Stop after question 6 unless your instructor requests you to go on.

LABORATORY EXERCISE 8:

Sedimentary Rocks and Processes

Use Lab Exercise Three: "Sedimentary Rocks and Processes," in your AGI lab manual for this lab activity.

Sedimentary rocks are the most abundant type of rock on or near the surface of the earth. This exercise is very important in developing your knowledge of sedimentary rocks and sedimentary processes. Make a list of the terms printed in **bold** type in your AGI laboratory manual and be sure you know their meanings. Your textbook and study guide have good background information on sedimentary rocks and processes. Although much of this information is duplicated in your lab manual, reviewing the figures in the text may be helpful.

We have a few suggestions that should help you as you begin this activity.

1. Carefully read the laboratory text. Pay close attention to the sections on sedimentary structures and sedimentary environments.

2. Study the figures and photographs in detail and read the captions carefully.

3. Use the outline at the top of page 42 in your lab manual when you begin identifying your rock samples.

BEFORE YOU GET STARTED

In this section, you will find a list of numbers which correspond to the questions in your AGI lab manual. After each number you will find information on how to proceed with the question. If you have enough information to proceed without difficulty, you will see "No further information needed." If you need additional information, the necessary information will be provided for you. If the question is one that would be difficult to complete without assistance from an instructor, you will see "Omit this question."

QUESTION-BY-QUESTION INSTRUCTIONS

12. To get a good introduction to sedimentary environments and sedimentary structures, complete this question first. You may need to refer to AGI Figures 3.1 through 3.17. Spend enough time to complete this question, and it will make the remaining questions much easier to do.

1. Omit this question. Instead of question 1, complete the following:

   a. Describe the movement of the grains being transported in AGI Figures 3.4 and 3.5. Do this by imagining a rounded particle on top of each figure and then asking yourself how this particle will move according to the direction of water movement. Describe the difference in cross-bedding for each figure based on the movement of the particles.

   b. Look at the end section at the right of each figure. Describe how the two sections differ and how they were formed.

2. Read the captions carefully.

3. a. No further information required.

   b. Briefly describe the items which might be found in the surrounding rocks to support your answer to part a.

4. Refer to AGI Figures 3.5 and 3.17 for locations of cross-stratification.

5. Review the discussion on flute casts beginning on page 34 of your lab manual.

6. No further information needed.

7. Do you see any ripple marks? What does this tell you about the depth or movement of the water?

8. Try to imagine the environment in which the three-toed dinosaur lived. Some suggestions for the answer are ripple marks, tool marks, or cracking.

9. What do you think happened to the smaller dinosaur?

10. Omit this question.

11. Follow the outline on the top of page 42 in your lab manual. AGI Figures 3.18 and 3.19 contain the names of many sedimentary rocks which you do not have or will not be able to identify. For AGI Figure 3.18, identify the detrital samples in terms of conglomerate, breccia, sandstone, or mudstone. For AGI Figure 3.19, identify the rocks composed mainly of calcium carbonate in terms of the major groups of skeletal, oolitic, or chemical limestone.

> Stop after question 12 unless your instructor requests you to go on.

LABORATORY EXERCISE 9:

Metamorphic Rocks and Processes

Use Lab Exercise Four: "Metamorphic Rocks and Processes," in your AGI lab manual for this lab activity.

As with the three previous laboratory exercises, your laboratory manual contains many new terms and excellent photographs. Read the laboratory text information in detail and study the captions beneath each figure.

BEFORE YOU GET STARTED

We have a few items to call to your attention that should help you as you begin this lab activity.

1. AGI Figure 4.1 on page 53 contains a cross section of a mountain range and a granite intrusion. This figure is meant to show two separate examples of metamorphic processes – the metamorphic process associated with mountain building and plate tectonics and the metamorphic processes associated with the granite intrusion.

2. On page 56 of your lab manual, the text material on gneissic texture erroneously indicates that AGI Figure 4.7 is an example of gneiss. The sample is actually a chlorite schist, and it is correctly identified in the caption under the figure.

QUESTION-BY-QUESTION INSTRUCTIONS

In this section, you will find a list of numbers which correspond to the questions in your AGI lab manual. After each number you will find information on how to proceed with the question. If you have enough information to proceed without difficulty, you will see "No further information needed." If you need additional information, the necessary information will be provided for you. If the question is one that would be difficult to complete without assistance from an instructor, you will see "Omit this question."

1. a. First determine whether the rock samples are foliated or nonfoliated. AGI Figure 4.15 refers to foliated rocks, and AGI Figure 4.16 refers to nonfoliated rocks.

   b. You cannot answer this question for samples which have a fine-grained groundmass. For those samples with sufficiently large crystalline groundmass, identify the minerals in the groundmass using AGI Figures 4.2 and 4.17.

c. You may not have any samples which contain porphyroblasts.

d. Use AGI Figures 4.2, 4.15, and 4.16.

e. For foliated rocks, refer to the captions for AGI Figures 4.3 through 4.8.

f. Use the figures indicated.

2. AGI Figure 4.1 is meant to represent more than one metamorphic process. Refer to the text material in your laboratory manual for explanations of metamorphic processes other than contact metamorphism.

3. Omit this question.

4. Omit this question.

5. Refer to AGI Figure 4.16.

LABORATORY EXERCISE 10:

Ground Water

Use Lab Exercise Seven: "Ground Water," in your AGI lab manual for this lab activity.

> Ground water is affected by how we pump it out of the ground, how and what we build on the ground surface, and by the types of subsurface materials in which it flows. As with the previous exercises, reading the laboratory text in detail and following the instructions carefully is the key to doing a good job on the lab exercise.

In this section, you will find a list of numbers which correspond to the questions in your AGI lab manual. After each number you will find information on how to proceed with the question. If you have enough information to proceed without difficulty, you will see "No further information needed." If you need additional information, the necessary information will be provided for you. If the question is one that would be difficult to complete without assistance from an instructor, you will see "Omit this question."

1. The blue numbers in the lakes are elevations in feet.

2. Refer to AGI Lab Exercise Five for a review of this technique. Remember to draw smooth curving lines. A 5-foot contour interval requires a line for every five foot change in elevation; intervals are at 55 feet, then 50 feet, then 45 feet, and so on.

3. Draw three or four arrows on each contour line.

4. Count the contour lines in the sinkholes carefully. You may need to use your hand lens.

5. Draw a straight line on AGI Figure 7.3 from the north of the Blue Sinks to Sulphur Springs. Use the map distance scale to measure the distance in miles. You will need the following conversions:

    1 mile = 5,280 feet
    = 1,600 meters
    = 1.6 kilometers

6. How has building a parking lot or more houses affected rainfall and ground water? What is the effect of additional pumping to supply water to more houses?

7. Think about where the sinkholes have already occurred.

8. The contour lines in AGI Figure 7.10 are brown rather than black, as stated in the caption.

9. No further information needed.

10. No further information needed.

11. No further information needed.

12. No further information needed.

13. No further information needed.

14. No further information needed.

15. No further information needed.

16. Knowing the definition of consolidated rocks will help you answer this question.

17. No further information needed.

18. No further information needed.

19. No further information needed.

20. No further information needed.

21. No further information needed.

Stop after question 21 unless your instructor requests you to go on.

LABORATORY EXERCISE 11:

Work of the Wind

Use Lab Exercise Ten: "Work of the Wind," in your AGI lab manual for this lab activity.

This is a good time to evaluate how you are doing on your lab work. If you are having difficulties, refer back to the recommended procedure for completing the laboratory exercises on page 486 of this Appendix. Then read the laboratory exercise carefully, and pay close attention to the figures in the lab manual.

We have a few suggestions that should help you as you begin this lab activity.

1. You will not need a pocket stereoscope for this exercise, contrary to the information on page 118 of your laboratory manual.

2. Pay close attention to AGI Figures 10.1 and 10.2.

3. Several questions in this exercise do not have specific answers which can be found by reading the laboratory manual. Answer these questions by considering the type of materials involved and then by asking yourself what is happening in the question: Is water flowing? Is the wind blowing?

BEFORE YOU GET STARTED

In this section, you will find a list of numbers which correspond to the questions in your AGI lab manual. After each number you will find information on how to proceed with the question. If you have enough information to proceed without difficulty, you will see "No further information needed." If you need additional information, the necessary information will be provided for you. If the question is one that would be difficult to complete without assistance from an instructor, you will see "Omit this question."

QUESTION-BY-QUESTION INSTRUCTIONS

1. Ask yourself what is missing: major streams, surface water movement, etc.?

2. For a review on how to determine *relief*, refer to AGI Laboratory Exercise Five: Topographic Maps.

3. Remember that closely spaced contour lines indicate steep slopes. Look at AGI Figure 10.1. Which side of this dune has closely spaced contour lines?

4. This is a straightforward question, so keep your answer simple.

5. This is another straightforward question, so keep your answer simple.

6. and 7. Refer to pages 443 and 444 in your textbook, and pay close attention to Figure 19.17 on page 444.

8. This is another straightforward question, so keep your answer simple.

9. This is another straightforward question, so keep your answer simple.

10. Compare AGI Figure 7.1 on page 83 of your laboratory manual to what you observe in the topographic map in AGI Figure 10.3.

For questions 11-14: Refer to the topographic map symbols on page 69 of your laboratory manual.

11. No further information needed.

12. No further information needed.

13. No further information needed.

14. Try to imagine the flow of water inside the dashed blue lines below the area in question. How would the levees, channels, and blowing sand interact?

15. Be sure to check the contour interval of AGI Figure 10.4.

16. Refer to AGI Figure 10.2.

17. Refer to the assigned reading in Chapter 13 in your textbook and the text material in your laboratory manual. Give a general description rather than searching for a specific answer.

18. Refer to AGI Figure 10.2.

Stop after question 18 unless your instructor requests you to go on.

LABORATORY EXERCISE 12:

Alpine Glaciation

Use Lab Exercise Twelve: "Alpine Glaciation," in your AGI lab manual for this lab activity.

> You will find that your lab manual and your textbook use two different terms to describe the same area of a glacier. When your laboratory manual refers to *zone of ablation*, it is talking about the feature called *zone of wastage* in the textbook. Refer to Chapter 18 in your textbook and specifically to Figure 18.6 for clarification.

QUESTION-BY-QUESTION INSTRUCTIONS

In this section, you will find a list of numbers which correspond to the questions in your AGI lab manual. After each number you will find information on how to proceed with the question. If you have enough information to proceed without difficulty, you will see "No further information needed." If you need additional information, the necessary information will be provided for you. If the question is one that would be difficult to complete without assistance from an instructor, you will see "Omit this question."

1. Refer to the section on glacier erosion on page 414-417 of your textbook.

2. Remember to look only for A through H. AGI Figure 12.4 is dark at the top so the formations there may be hard to observe.

3. Make a list of the glacier landforms and characteristics you observe. Identify the moraines. Ask yourself if the landforms you observed could be present if the area were completely covered with glaciers.

4. There may be more than one location possible.

5. No further information needed.

> For questions 6-9: Refer to AGI Figure 12.2.

6. No further information needed.

7. No further information needed.

8. The area in question is just north of the large X.

9. Again, the area in question is just north of the large X.

10. Review AGI Lab Exercise Five: Topographic Maps, for a refresher on how to draw a topographic profile.

11. Read the section on glacial valleys on page 414 of your textbook.

12. Refer to the text material on pages 93 and 94 of your laboratory manual.

Stop after question 12 unless your instructor requests you to go on.

LABORATORY EXERCISE 13:

Shorelines and Coastal Processes

Use Lab Exercise Nine: "Shorelines and Coastal Processes," in your AGI lab manual for this activity.

Let's take some time to go to the beach. You have completed all the rest of the laboratory exercises and are just about to finish this telecourse in geology. You have worked very hard to get to this point, and you deserve a break at the beach. You'll start your trip to the beach by reading the text material in your lab manual carefully. Then review Chapter 20 on waves, beaches and coasts in your textbook and the accompanying lesson in the study guide. Knowing the textbook chapter thoroughly is critical to your ability to complete this final exercise. We'll see you at the beach!

We have a few suggestions that should help you as you begin this lab activity.

1. You will not need a pocket stereoscope for the activity, contrary to the instruction on page 105 of your lab manual.

2. The text on page 105 in your lab manual contains a number of new terms. It will help you to learn the terms more easily if you relate them to the figures on page 106 in the lab manual and to the material in Chapter 20 of your textbook.

3. The terms berm crest, washover fan, and wave-cut platform in the list on page 107 in your lab manual are not shown in AGI Figures 9.1 and 9.2. These terms are covered in Chapter 20 of your textbook.

4. Some of the questions in this lab are vague, and you should not take too much time trying to understand them. Keep your answers simple and to the point.

In this section, you will find a list of numbers which correspond to the questions in your AGI lab manual. After each number you will find information on how to proceed with the question. If you have enough information to proceed without difficulty, you will see "No further information needed." If you need additional information, the necessary information will be provided for you. If the question is one that would be difficult to complete without assistance from an instructor, you will see "Omit this question."

For questions 1-4: Refer to AGI Figure 9.1 and be sure to use Chapter 14 in your textbook to help with your explanations.

1. No further information needed.

2. No further information needed.

3. No further information needed.

4. No further information needed.

5. You may not be able to answer this question using only the information in your lab manual and textbook. Here's some additional observations to help you understand the question: Greenhill is a drumlin, and Greenhill Beach is a baymouth bar. Greenhill Beach is thicker to the west because the debris from Greenhill is being moved west.

6. This is one of the questions referred to in the introduction to this activity. It may be too vague for you to give a complete answer. Try for a simple answer that seems reasonable to you.

7. Let's break this question down into the following parts:

    a. What type of feature is Revere Beach and the Point of Pines?
    b. What type of feature is the area contained behind Revere Beach?
    c. Explain why the Pines River appears to have changed its course to flow northward.

8. This is a general question; include the following points in your answer:

    a. How are the two areas similar with respect to development, type of materials, and type of features present?
    b. How are the two areas different with respect to type and size of the features and the elevations of the near shore features?

11. Refer to page 459 in Chapter 20 in your textbook.

12. Omit this question.

13. Refer to the information on drowned river mouths on page 465, in your textbook.

14. Look on page 69 of your lab manual for topographic map symbols.

15. Use the reading in your textbook and study guide to help you with your explanation.

16. Use the reading in your textbook and study guide to help you with your explanation.

17. Refer to AGI Figure 9.2 on page 106 of your lab manual.

18. Omit the question in your manual and answer the following instead:

    a. Describe the development of the features between Carlsbad and South Oceanside. Be as detailed as possible, including descriptions of the sand movement and the development of the tidal flat. Use the correct terminology for each feature.

    b. Look on the map north of South Oceanside, close to the edge of the map. Notice the feature which extends into the ocean and how the waves curl around this feature. Refer to your textbook reading to name this feature and describe a possible purpose for it.

19. No further information needed.

For questions 20-22: Refer to AGI Figure 9.2 and to your textbook reading.

20. No further information needed.

21. No further information needed.

22. No further information needed.

23. Refer to AGI Figure 9.2.

24. No further information needed.

25. No further information needed.

26. No further information needed.

27. No further information needed.

28. No further information needed.

Stop after question 28 unless your instructor requests you to go on.

Appendix B

ANSWER KEY

1. c	9. c	**LESSON 1 –**
2. a	10. b	**Down to Earth**
3. d	11. d	
4. b	12. a	
5. c	13. c	
6. a	14. b	
7. d	15. d	
8. c		
1. b	9. a	**LESSON 2 –**
2. d	10. b	**The Restless**
3. a	11. d	**Planet**
4. c	12. c	
5. b	13. b	
6. b	14. c	
7. c	15. a	
8. d		
1. c	9. d	**LESSON 3 –**
2. b	10. c	**Earth's Interior**
3. a	11. d	
4. c	12. a	
5. b	13. b	
6. a	14. d	
7. c	15. c	
8. a		

LESSON 4 – **The Sea Floor**	1. d 2. b 3. c 4. c 5. a 6. b 7. a 8. c	9. c 10. b 11. d 12. a 13. c 14. b 15. d
LESSON 5 – **The Birth of a** **Theory**	1. c 2. b 3. c 4. d 5. c 6. d 7. c 8. a	9. d 10. c 11. a 12. b 13. a 14. c 15. d
LESSON 6 – **Plate Dynamics**	1. c 2. a 3. d 4. c 5. a 6. c 7. c 8. d	9. b 10. b 11. d 12. a 13. d 14. c 15. a
LESSON 7 – **Mountain** **Building**	1. d 2. b 3. c 4. d 5. a 6. b 7. b 8. c	9. a 10. c 11. b 12. c 13. a 14. b 15. b
LESSON 8 – **Earth Structures**	1. c 2. d 3. b 4. a 5. c 6. a 7. d 8. a	9. b 10. a 11. c 12. b 13. d 14. c 15. b

1. d	9. a	**LESSON 9 –**
2. b	10. b	**Earthquakes**
3. c	11. d	
4. a	12. c	
5. d	13. d	
6. b	14. a	
7. c	15. c	
8. c		

1. c	9. b	**LESSON 10 –**
2. d	10. c	**Geologic Time**
3. a	11. d	
4. c	12. a	
5. d	13. c	
6. a	14. c	
7. b	15. d	
8. d		

1. c	11. d	**LESSON 11 –**
2. d	12. b	**Evolution**
3. a	13. b	**through Time**
4. c	14. d	
5. b	15. c	
6. c	16. a	
7. d	17. d	
8. b	18. c	
9. d	19. c	
10. a	20. b	

1. c	9. c	**LESSON 12 –**
2. b	10. b	**Minerals: The**
3. c	11. d	**Materials of Earth**
4. d	12. a	
5. a	13. d	
6. c	14. c	
7. b	15. b	
8. d		

1. c	9. c	**LESSON 13 –**
2. b	10. d	**Volcanism**
3. a	11. b	
4. d	12. a	
5. c	13. c	
6. d	14. b	
7. a	15. d	
8. b		

LESSON 14 – Intrusive Igneous Rocks	1. b 2. c 3. d 4. c 5. b 6. b 7. d 8. b	9. d 10. c 11. a 12. d 13. b 14. c 15. b
LESSON 15 – Weathering and Soils	1. c 2. d 3. b 4. c 5. d 6. b 7. b 8. a	9. d 10. c 11. c 12. b 13. c 14. d 15. c
LESSON 16 – Mass Wasting	1. d 2. b 3. c 4. a 5. d 6. c 7. a 8. b	9. c 10. d 11. c 12. a 13. c 14. b 15. c
LESSON 17 – Sedimentary Rocks: The Key to Past Environments	1. d 2. b 3. c 4. a 5. d 6. d 7. b 8. d	9. a 10. b 11. b 12. c 13. d 14. b 15. c
LESSON 18 – Metamorphic Rocks	1. c 2. b 3. d 4. b 5. c 6. a 7. d 8. c	9. d 10. a 11. b 12. c 13. d 14. b 15. c

1. d	9. d	**LESSON 19 –**
2. c	10. c	**Running Water I:**
3. b	11. b	**Rivers, Erosion,**
4. d	12. d	**and Deposition**
5. a	13. c	
6. c	14. a	
7. b	15. b	
8. a		

1. a	9. c	**LESSON 20 –**
2. a	10. d	**Running Water II:**
3. b	11. b	**Landform**
4. d	12. d	**Evolution**
5. c	13. b	
6. a	14. a	
7. d	15. d	
8. b		

1. c	9. d	**LESSON 21 –**
2. d	10. c	**Ground Water**
3. a	11. b	
4. c	12. c	
5. d	13. d	
6. b	14. a	
7. c	15. b	
8. a		

1. c	9. c	**LESSON 22 –**
2. b	10. a	**Wind, Dust, and**
3. a	11. d	**Deserts**
4. a	12. d	
5. c	13. b	
6. a	14. a	
7. b	15. c	
8. b		

1. d	9. b	**LESSON 23 –**
2. b	10. d	**Glaciers**
3. c	11. a	
4. a	12. d	
5. b	13. c	
6. c	14. a	
7. a	15. c	
8. d		

**LESSON 24 –
Waves, Beaches,
and Coasts**

1. d
2. b
3. b
4. c
5. d
6. c
7. a
8. c

9. d
10. a
11. b
12. d
13. b
14. a
15. c

**LESSON 25 –
Living with
Earth, Part I**

1. c
2. b
3. d
4. a
5. c
6. c
7. d
8. b

9. a
10. d
11. c
12. b
13. c
14. a
15. d

**LESSON 26 –
Living with
Earth, Part II**

1. c
2. d
3. c
4. b
5. a
6. a
7. d
8. b

9. b
10. d
11. c
12. b
13. a
14. d
15. c